Concise
Encyclopedia of
Geography

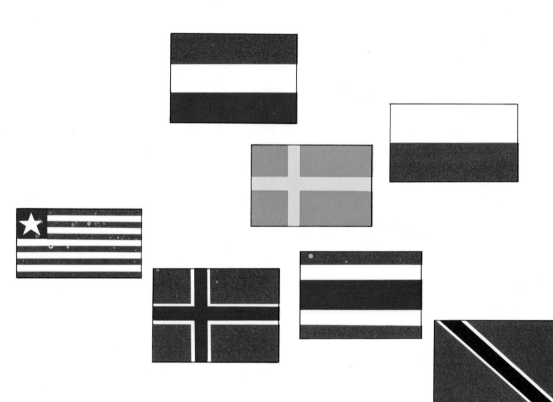

Consultant Editor
J.E.B. Wrighton B.A.

Contributors
Keith L. Lye B.A., F.R.G.S.
T. Rowland-Entwistle F.R.G.S.
A.S. Butterfield
Diane James
Lionel Grigson B.A.
Marion Bloch Ph.D.

SBN 361 03261 7

Copyright © 1976 Purnell and Sons limited
Published 1976 by Purnell Books,
Berkshire House, Queen Street,
Maidenhead, Berkshire.
Copyright © Relief Globe by Rand Mc Nally & Company

Printed in Italy by Vallardi Industrie Grafiche.

Reprinted 1978

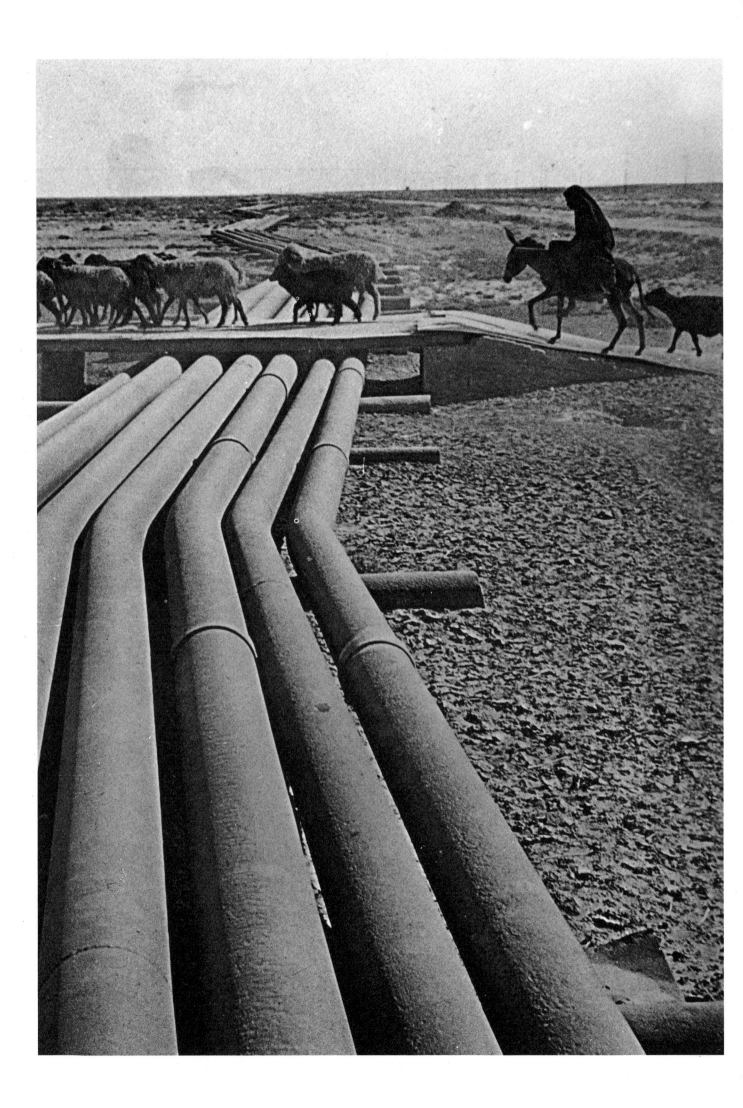

Concise Encyclopedia of Geography

C.J. Tunney M.A.

Purnell

CONTENTS

Inland Waters. Above left: The desolate shore of Lake Titicaca, high in the Andes. Above right: Old houses line a canal in Ghent, in Belgium.

THE VARIED WORLD

A traveller from Outer Space studying our planet might be forgiven if he called it by his word for 'Ocean' or 'Water'. The large volume of water on Earth makes our planet unique in the Solar System. Some 70.8 per cent. of the Earth's surface is covered by five great, interconnected oceans—the Pacific, the Atlantic, the Indian, the Antarctic, and the Arctic. The oceans are important to life on Earth because their currents act like a thermostat, cooling some warm areas and warming some cold areas.

Land covers only 29.2 per cent. of the Earth's surface. A space traveller would be impressed by the variety of habitats on our planet. He would be struck by the fact that, while some plant life and some animal life are adapted to live almost everywhere, the distribution of Man is very uneven. Most of the world's people live on about 10 per cent. of the Earth's land surface.

The continent of Antarctica is a vast ice sheet which is so inhospitable to Man that people can survive there only if they import everything they need. In the Northern Hemisphere, around the polar region, there are vast tracts of almost uninhabited treeless plains, called *tundra*. To the south of the tundra lie cold, coniferous forest regions.

Other parts of the Earth are too hot and too dry for much human settlement. The world's largest desert, the Sahara, is populated by a small number of nomads. Its only farmers cluster around a few scattered oases. Tropical forests and mountain regions are also environments hostile to Man.

However, in regions where it is not too hot, too cold, or too dry, Man has created great civilizations. His ingenuity has made it possible for farming and manufacturing industries to support the needs of the world's enormous and fast-growing population.

The Eternal Hills. Above: A prosperous farm on its hillside in France. Below: A lonely mountain road in Peru, flanked by inhospitable snow-clad peaks.

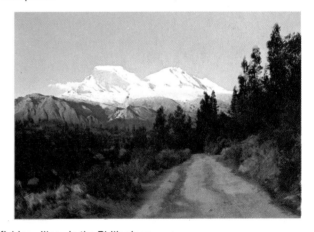

Borders of the Sea. A fishing village in the Philippines, built on stilts. Men have established settlements wherever they can manage to live.

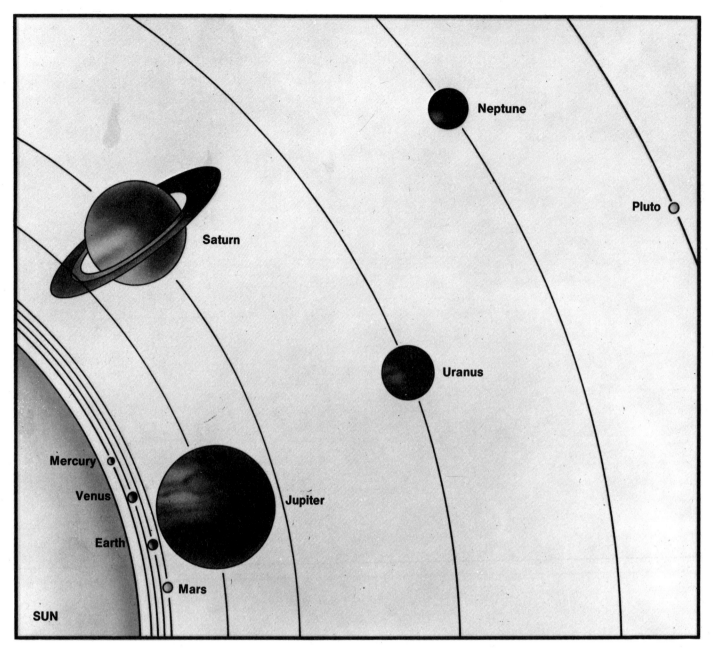

THE EARTH IN SPACE

The nine planets of the Solar System are bound to the Sun by the force of gravity. Because of gravity, all the planets orbit around the Sun. The time taken for a planet to orbit once around the Sun is called the planet's 'year'. The orbits are elliptical in shape and lie roughly in the plane of the Sun's equator. Each of the planets also spins on its own axis and the time it takes to spin once is called the planet's 'day'. All the axes of the planets are tilted from the vertical.

In the Solar System, the Earth is the third planet from the Sun and it is the fifth largest of the planets. However, by comparison with the Sun, the Earth is very small. The Sun's mass is 330,000 times as great as that of the Earth.

Like Mercury, Venus, Mars, and Pluto, the Earth is a *terrestrial planet*. Terrestrial planets are fairly dense, rocky bodies that have a solid surface. On the other hand, the giant planets, Jupiter, Saturn, Uranus, and

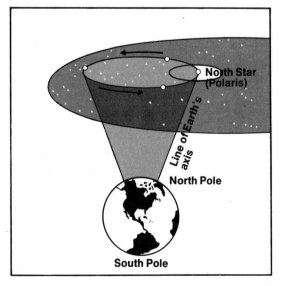

The Solar System. Each of the nine planets moves round the Sun in an elliptical orbit. The Earth's orbit is completed in just over 365 days. Mercury takes only 88 days, and Pluto takes 249 years. The Solar System is part of the Milky Way, a galaxy made up of thousands of millions of stars—of which the Sun is one.

The North Star. Polaris, the North Star, appears to us to be always in the same position because it is almost directly in line with the Earth's axis. But the Earth 'wobbles' as it rotates and its axis line describes a circle in the sky—taking 25,800 years to do so. Over thousands of years, various stars in turn become the North Star.

THE EARTH AND THE SOLAR SYSTEM

Heavenly body	Diameter in miles (and kilometres)		Average distance from Sun in millions of miles (and kilometres)		Period of orbital revolution	
The Sun	865,000	(1,392,000)	—		—	
Mercury	3,100	(4,989)	36.0	(57.9)	88	days
Venus	7,700	(12,392)	67.2	(108.1)	225	days
Earth	7,917	(12,741)	92.9	(149.5)	365	days
Mars	4,200	(6,759)	141.5	(227.7)	687	days
Jupiter	88,700	(142,749)	483.3	(777.8)	11.9	years
Saturn	71,500	(115,068)	886.1	(1,426.0)	29.5	years
Uranus	30,900	(49,729)	1,783.0	(2,869.4)	84.0	years
Neptune	28,000	(45,062)	2,793.0	(4,494.9)	164.8	years
Pluto	3,100-3,600	(4,989-5,794)	3,666.0	(5,899.9)	249.0	years

Neptune, are large, low-density bodies of gases, which rotate quite quickly on their axes.

The Earth moves in three main ways. It rotates around the Sun once every 365 days, 5 hours, 48 minutes, and 46 seconds at a speed of 66,000 mph (106,200 kph). Second, the Earth spins on its axis at a speed of 1,038 mph (1,670 kph) at the equator. This speed decreases towards the poles. Like other planets, the Earth's axis is tilted. The tilt of the Earth's axis is $23\frac{1}{2}°$ from a line perpendicular to its path around the Sun. The axes of all planets are tilted at less than 30° from the vertical, apart from Uranus, whose axis is tilted at slightly over 90°.

Together with the rest of the Solar System, the Earth moves in a third way. It is moving around the Milky Way at a speed of 43,000 mph (69,200 kph). The Milky Way is a galaxy (star system) which contains about 100,000 million stars. One of them is the Sun in our Solar System. The Solar System takes about 200 million years to rotate once around the centre of the galaxy. Millions of galaxies form the Universe, but no one knows just how big the Universe is. As a part of the Universe, the Earth may be moving in a fourth way, because many scientists believe that the Universe is expanding.

Like several other planets, the Earth has a satellite, the Moon. Recent studies of Moon rocks, obtained by Russian and American missions, have shown that the Moon is of the same age as the Earth (about 4.5 million years old). The Moon's diameter is 2,160 miles (3,476 km)—about one-fourth the size of the Earth. The Moon orbits the Earth at an average distance of 238,887 miles (384,400 km). The Moon rotates once on its axis and also completes one revolution around the Earth every $27\frac{1}{3}$ days. As a result, from the Earth we always see the same hemisphere of the Moon. The $27\frac{1}{3}$-day period is called the *sidereal month*.

Some scientists believe that the Moon was formed when part of the Earth was detached and flew away into orbit. Today, however, most scientists think that it was formed in the same way as the Earth and the other bodies in the Solar System. They consider that, about 5,000 million years ago, a large cloud of dust and gas collected. Part of this dust and gas concentrated into a dense core to form the Sun. The rest of the material rotated around the core. Smaller bodies formed in this rotating dust and gas and these bodies swept up more and more material. By 4,500 million years ago, these bodies had grown into planets.

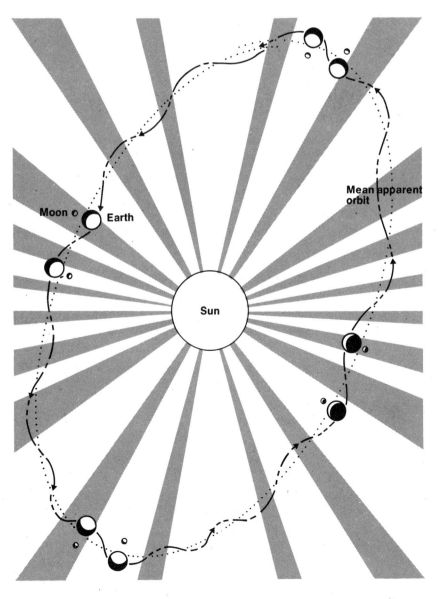

Earth and Moon Orbits. As the Earth and the Moon orbit the Sun, the Moon also revolves around the Earth, weaving in and out of the Earth's orbit. Only one side of the Moon is ever completely visible from the Earth because the Moon rotates once on its axis in the same time that it takes to revolve once around the Earth.

THE CHANGING YEAR

The Earth takes one *solar year* (365 days, 5 hours, 48 minutes, and 46 seconds) to complete one revolution round the Sun. Some early peoples, however, measured time in relation to the Moon. The month (or Moonth) was the period taken by the Moon to go from one new Moon to the next. This period is $29\frac{1}{2}$ days. This method of measuring caused much confusion because no number of 'Moon-ths' could be made to equal a solar year. Twelve months of $29\frac{1}{2}$ days is 354 days and 13 months totals $383\frac{1}{2}$ days.

In 1582, however, the Gregorian calendar was adopted, named after Pope Gregory XIII. This calendar is based on the idea of a seven-day week. A year of 52 such weeks totals 364 days, which is 1 day, 5 hours, 48 minutes, and 46 seconds less than a solar year. To eliminate the odd day, each year in the Gregorian calendar begins one day later in the week than the previous year. Leap years are fixed for every year that can be divided by four, such as 1976 and 1980. As a result, six hours are added on average to every year. This leaves a difference between the calendar and the solar year of 11 minutes and 14 seconds. Most of this error is eliminated by making century years that cannot be divided by 400—such as 1700, 1800, and 1900—*non*-leap years. But centuries, such as 2000, that can by divided by 400 are leap years. In this way, the difference between the Gregorian calendar and the solar year is reduced to 26.3 seconds. However, another error occurs because the solar year is slowly getting slightly shorter. This error amounts to 0.53 seconds every 100 years.

While the Earth rotates on its axis, half of the Earth is facing the Sun while the other half is facing away from the Sun. As a result, we experience day and night. The Earth takes 23 hours, 56 minutes, and 4.1 seconds to spin round once on its axis. This period of time, measured in relation to the stars, is called the *sidereal day.*

Because the Earth is divided into 360° of longitude and it completes one spin on its axis in a day (24 hours), we can see that time alters by one hour for every 15° of longitude (360÷24). Each place on Earth has its own *local time.* This can be measured forwards or backwards from the moment when the Sun reaches its highest point in the sky, which is noon or midday.

The Other Side of the World. Above: January in Europe—winter in the Dolomites. Below: January in the South Pacific, with summer sunshine.

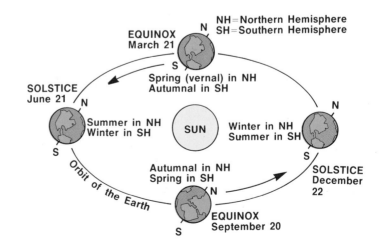

The Seasons. The Earth is tilted on its axis as it orbits the Sun. Each Hemisphere receives most sunlight when tilted towards the Sun.

Day and Night. While the Earth orbits the Sun, it also rotates on its axis. This rotation, towards the east, is responsible for day and night. Each point on the Earth's surface has 'day' when the Sun's rays can reach it, 'night' when it has spun 'round the back' out of reach.

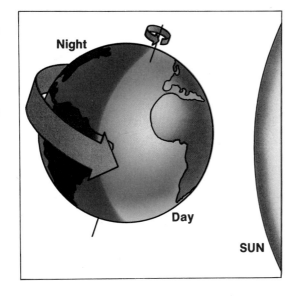

clock on by 12 hours. When they met on the *International Date Line*, there would be one day's difference between them. As a result, travellers crossing the International Date Line from east to west lose a day. Travellers crossing from west to east gain a day. The International Date Line, too, is an irregular line, which does not pass through land areas or island groups.

The term 'day' is sometimes used to refer to the hours of daylight as opposed to night. Hours of daylight vary because the Earth's axis is tilted. The North Pole is tilted towards the Sun for half of the year and so it enjoys continuous daylight. At the same time, the South Pole is in continuous darkness. The length of the day varies between the poles and the equator. The longest day in the Northern Hemisphere is usually June 21. The shortest is December 22. The reverse, of course, applies in the Southern Hemisphere.

The tilt of the Earth's axis is also responsible for seasons. Spring in the Northern Hemisphere begins on March 21, when the Sun is overhead at the equator. March 21 is called the *vernal equinox*. After March 21, the overhead Sun moves northwards until, on June 21, the *summer solstice*, it reaches its northernmost point. The overhead Sun then moves south back towards the equator, where it is again overhead on about September 20, the *autumnal equinox*. On December 22, the *winter solstice*, the overhead Sun reaches its southernmost point. In the Southern Hemisphere, the seasons are the reverse of those in the Northern Hemisphere.

But if every place on Earth could *choose* its own local time, there would be great confusion. To save this, the Earth is divided into *time zones*. The simplest time zones would be based on dividing the globe into equal segments of 15° and changing the time by one hour from one zone (segment) to the next. But this is unsatisfactory, because the boundary might pass through a city, giving different times in two sectors of the same city. Time zones, therefore, have irregular boundaries that make allowance for political frontiers.

The prime meridian, or 0° longitude, from which the time zones are measured, passes through Greenwich, England. A person travelling west of Greenwich for 180° of longitude would have to put his clock back by 12 hours. If another traveller were going east of Greenwich for 180°, he would put his

Standard Times Zones of the world showing how they differ from (fast or slow of) Greenwich Mean Time. The time in any particular zone is one hour earlier than that in the zone to its east. But at the International Date Line there is a variation of 24 hours.

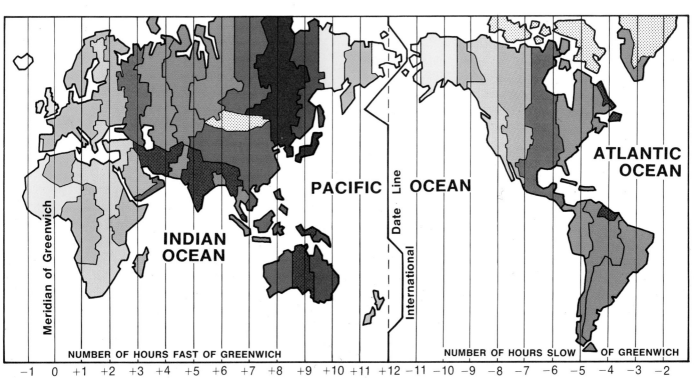

WEATHER AND CLIMATE

The climate of a country or other region is its average or typical weather over a period of time. Climate strongly affects our way of life. It influences, for example, the way we dress, the kinds of houses we live in, and the foods that farmers can grow. We can control the effects of climate in some ways. For example, central heating and air-conditioning make buildings more comfortable to live and work in. And scientists have succeeded in experiments to produce rain in dry regions by seeding clouds with solid carbon dioxide or other chemicals.

Climate is determined by several factors, one of which is latitude. The Sun's heat is concentrated at the equator. But at the poles much of the heat is absorbed by the atmosphere, and heat that does reach the ground is spread over a large area. Besides latitude, other factors, too, are important.

Mount Kilimanjaro, Africa's highest mountain, lies almost on the equator in Tanzania. This mountain is capped by snow and ice throughout the year. The reason is that, as one ascends a hill or mountain, the temperature falls by about 3.5°F for every 1,000 ft (6°C per 1,000 m). Mountains also affect rainfall. Onshore winds, bearing much moisture evaporated from the sea, are chilled as they rise over coastal mountain

JANUARY TEMPERATURES AT SEA-LEVEL
in degrees Centigrade

below −30 ☐ −30 to 0 ☐ 0 to 10 ☐ 10 to 20 ☐ 20 to 30 ☐ above 30 ☐

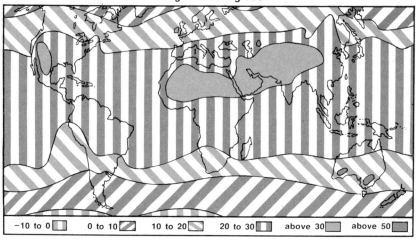

JULY TEMPERATURES AT SEA-LEVEL
in degrees Centigrade

−10 to 0 ☐ 0 to 10 ☐ 10 to 20 ☐ 20 to 30 ☐ above 30 ☐ above 50 ☐

ANNUAL PRECIPITATION

0 to 250 mm
(0 to 9.85 in)

250 to 500 mm
(9.85 to 19.7 in)

500 to 1,000 mm
(19.7 to 39.4 in)

1,000 to 1,500 mm
(39.4 to 59.1 in)

1,500 to 2,000 mm
(59.1 to 78.8 in)

more than 2,000 mm
(more than 78.8 in)

ranges. As air cools, it loses its capacity to hold moisture, which falls on the mountains as rain.

Oceans and seas influence climate. For example, the Gulf Stream is a warm ocean current which flows from the Gulf of Mexico across the Atlantic. One branch of this current flows past Norway. Onshore winds are warmed by the current and bring mild weather to Norway. In fact, the average January temperatures are 45°F (25°C) higher than is usual for Norway's latitude. Inland, beyond the moderating influence of the sea, conditions become more severe.

The definition of climatic regions is complicated by these and other factors. One simple classification that is widely used was made by a Russian, Vladimir Köppen. He divided the world into five main regions: tropical rainy climates with no cool season; dry climates; middle latitude rainy climates with severe winters; tundra climates; and polar climates. Each of these regions is subdivided according to rainfall and temperature, and also according to the distribution of rainfall season by season.

Weather is the day-to-day state of the atmosphere. The weather is determined by the temperature and pressure of the atmosphere, by the way the atmosphere moves, and by the amount of moisture it contains. Moisture may be present in the atmosphere as an invisible gas—water vapour—or it may be observable as clouds, fog, or dew. Clouds are the source of rain as well as of hail, sleet, and snow.

At weather stations on land and at sea, readings are taken of atmospheric conditions, usually every six hours. The readings from weather stations are sent by radio or teleprinter to forecast centres. At these centres, the reports are analysed. Weather maps, called *synoptic charts*, are drawn. These charts show temperatures, wind speeds, and direction and cloud amounts for each station. They are marked with lines, called *isobars*, that join points of equal atmospheric pressure.

Finally, cold and warm fronts are added to the charts. Fronts are features of depressions—complex air systems which form in temperate latitudes where the cold, polar air mass meets warm sub-tropical air. Depressions form when warm air forms a bulge in the cold air. As the warm air flows into the bulge, cold air moves in from the rear. This sets up an air system, in which winds circulate in an anti-clockwise direction in the Northern Hemisphere and clockwise in the Southern. The edges, or fronts, of the cold and warm air are zones where cloud and rain occur.

Weather forecasters examine the synoptic charts of the last 24 hours or so, and study information about conditions in the upper atmosphere. They then prepare a forecast, which is a summary of how they consider the weather is likely to develop in the next 12 to 24 hours. Today, forecasters often use computers to speed up their calculations.

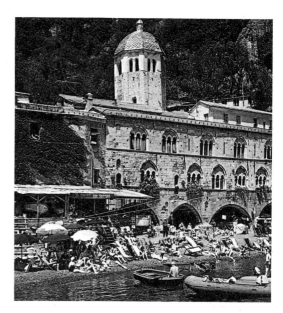

Weather and climate profoundly affect man's life. They largely determine where he can live, what shelter and clothes he needs, and what food is available to him. The chief elements of weather, apart from air pressure, are temperature, moisture, and wind.

Temperature is the most important factor in weather. The Earth's heat comes from the Sun. Because heat is trapped in the atmosphere, most parts of the Earth's surface have some warmth, even when on the side away from the Sun.

Moisture. About three-quarters of the Earth's surface is water. The heat of the Sun causes water to evaporate from the seas, from lakes and rivers, and from land and vegetation, forming water vapour in the atmosphere. Some of it condenses as clouds. More falls back to Earth as precipitation—rain, hail, snow, and dew.

Wind is air in movement. Some winds follow defined patterns across the surface of the Earth—the monsoons, for example, and the trade winds. Others are part of local weather conditions. Some winds are gentle, some destructive.

THE EARTH'S STRUCTURE

The Earth's crust averages 20 miles (32 km) in thickness beneath the continents and about six miles (9.7 km) under the oceans. The crust and the first land masses were formed in the earliest era of Earth history. But many changes have taken place since that time.

About 200 million years ago, the land masses were all grouped in one large continent, which geologists call *Pangaea*. This continent split into two parts, *Laurasia* and *Gondwanaland,* which drifted northwards. These two smaller continents later broke up and moved slowly to their present positions.

This movement, called *continental drift,* is still continuing today. The Earth's crust is cracked and split into a number of large, rigid *plates*. These plates are moved by powerful forces in the Earth's mantle. An example of movement occurs in the mountain ranges called *mid-oceanic ridges* that have been found in the oceans.

THE INTERIOR OF THE EARTH

CRUST

MANTLE

PLASTIC OUTER CORE

DENSER INNER CORE

FORMATION OF HIGHLANDS AND MOUNTAINS

Rocks found in highland and mountain areas are partly sedimentary in origin, partly igneous, and partly metamorphic. The sedimentary rocks originated as deposits in seas and lakes, and the igneous—such as basalt and granite—as molten magma rising from deep in the Earth. Metamorphic rocks were formed from sedimentary and igneous types by heat and pressure transformation. Some sedimentary rocks contain fossils, usually the remains of sea animals, showing that these rocks were once part of the sea bed.

Several theories have been advanced to explain these facts. One is that, in the course of millions of years, sediments—sand and mud—that were piled up on the sea bed by rivers formed layers so thick and heavy that the rocky crust supporting them broke and folded. The sedimentary layers were squeezed and pushed up by pressures from beneath, and in time became highlands—gradually to be worn down by erosion. Another theory suggests a continental rifting millions of years ago (A, above). An ocean filled the hollow between the segments as they spread apart, and sediments were deposited on the continental margins. Later a trench developed near one continental margin (B). As the continental segments moved together again, the ocean crust slid under the continent (C). In some places, magma rose to the surface through fissures. Eventually, the segments crashed back together once more (D).

These ridges are centres of volcanic activity and earthquakes. New crustal rock is being added to them by *magma* (molten rock) which wells up from the top of the mantle. This new rock is widening the oceans.

But the Earth is not growing in size. To

Quaternary   Tertiary   Mesozoic   Palaeozoic   Pre-Cambrian   Volcanic areas (of all ages)

balance the addition of new crustal rock, the crust in other places is being destroyed. This occurs when two plates are forced against each other. The edge of one plate is pushed beneath the other. As it descends, the crustal rock melts. Some of the melted rock returns to the surface through volcanoes and some returns to the Earth's interior. Sometimes, when two plates collide, sediments on the ocean floor between the plates are squeezed up into mountain ranges. For example, when the Indian sub-continent drifted against the Asian mainland, the ocean sediments between were forced up to form the Himalayas.

Rocks formed from molten magma, such as basalt and granite, are called *igneous rocks*. Rocks formed from sediments are called *sedimentary rocks.* Many sedimentary rocks are composed of fragments of eroded rocks, such as sand. These sediments pile up in layers in seas or lakes. They often contain the remains of living things as *fossils.* Limestones are sedimentary rocks formed mainly from the remains of sea creatures.

The third kind of rock is called *metamorphic.* Metamorphic rocks are igneous or sedimentary rocks, changed by heat, pressure, or chemical action within the Earth. For example, marble is metamorphosed limestone.

THE EARTH: FACTS AND FIGURES

Diameter: 7,926 miles (12,755 km) at the equator; 7,899 miles (12,712 km) at the poles

Circumference: 24,901 miles (40,074 km) at the equator; 24,857 miles (40,003 km) at the poles

Area: 196,940,400 sq. miles (510,073,270 sq km), of which 70.8 per cent. is covered by water

Highest point: Mount Everest 29,028 ft (8,848 m) above sea-level.

Lowest point: Challenger Deep (Pacific Ocean) 36,198 ft (11,033m) below sea-level

CONTINENTAL DRIFT
The idea that the continents of today are parts of one or two vast land masses that existed millions of years ago is an old one. It was given scientific form in 1910 by the German geologist and meteorologist Alfred Wegener. His maps, right, show (A) the continental masses in the late Carboniferous Period and (B) in the early Pleistocene Epoch.

Some of the evidence for continental drift relates to the shape of the continents: the eastern bulge of South America, for example, would fit fairly well into the concave western coast of Africa. But there is also more compelling evidence based on studies of geological structure, the past distribution of plants and animals, and past climatic changes. Modern plate tectonic theory and ocean studies tend to bear out the idea of continental drift.

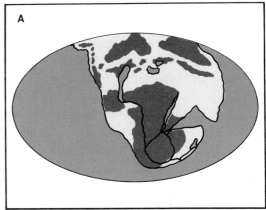

Dark blue: deep sea. Light blue: shallow sea. Brown: land.

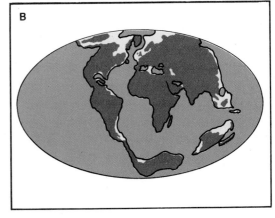

A broad outline of the Earth's history has been traced by *geologists* —scientists who study the rocks and other materials that make up the Earth's crust—and by *palaeontologists*—scientists who study the remains of ancient plants and animals preserved in rock. A fossil may consist of all or part of an animal's body; or it may be merely a rocky mould or cast showing the shape of a plant or animal that was pressed into the rock before it hardened, millions of years ago. Fossils give a clue to the age of rocks. The oldest rocks contain fossils of the simplest types of living things. More recent rocks have fossils of evolving plants and animals.

HISTORY OF THE EARTH

A time-scale of the Earth's history has six major divisions, called *Eras*, each of them representing many millions of years. They are, from the most distant to the most recent, the Azoic, Archaeozoic, Proterozoic, Palaeozoic, Mesozoic, and Cenozoic Eras. Often, the first three are grouped together as the Pre-Cambrian Era. Very little is known about the Earth in the Pre-Cambrian years. Eras are divided into *periods*, and periods further divided into *epochs*.

Million Years Ago
1
10
25
40
60
70
135
180
225
270
310
350
400
440
500
600

Geological Divisions		
CENOZOIC ERA	QUATERNARY PERIOD	RECENT EPOCH
		PLEISTOCENE EPOCH
	TERTIARY PERIOD	PLIOCENE EPOCH
		MIOCENE EPOCH
		OLIGOCENE EPOCH
		EOCENE EPOCH
		PALAEOCENE EPOCH
MESOZOIC ERA	CRETACEOUS PERIOD	
	JURASSIC PERIOD	
	TRIASSIC PERIOD	
PALAEOZOIC ERA	PERMIAN PERIOD	
	CARBONIFEROUS PERIOD	PENNSYLVANIAN PERIOD
		MISSISSIPPIAN PERIOD
	DEVONIAN PERIOD	
	SILURIAN PERIOD	
	ORDOVICIAN PERIOD	
	CAMBRIAN PERIOD	
PRE-CAMBRIAN ERA		

Conditions on Earth	Animal and Vegetable Life
Ice sheets decreased in size; glaciers melted. Sea levels rose. Land was eroded, and coastlines and landscapes were formed as we know them.	Forests spread. Marine and land animals developed as we know them. Man tamed other animals, learnt to grow food, and used various natural resources.
A time of great climatic changes. Ice sheets and glaciers covering much of Europe and North America melted four times. Land masses rose.	Many plants died out. An ape-like creature, identified as Man, appeared and discovered how to make stone implements. Mammoths and woolly rhinos died out.
Decrease in mountain building. The Earth's features took on the shapes we know. Climates cooler.	Animal life increasingly took on the forms we know. Man-like apes developed and became common.
The seas continued to retreat. Much mountain building and volcanic eruption.	Flowering plants developed much as today. Grazing mammals spread. Apes evolved further and spread.
The seas retreated from the land. There was extensive mountain building and volcanic eruption.	Cats, dogs, rodents, and bears evolved. Plant-eating animals became numerous. A primitive ape appeared.
Land subsidence caused the oceans to inundate coasts. The Atlantic Ocean was shaped much as today.	Grains and grasses appeared. Many fish, mammals, and insects began to take the form we know.
New mountain ranges continued to grow. In some parts of the world deep, fertile soil formed.	Flowering plants were dominant in vegetation. Dinosaurs had become extinct.
A major period of mountain building. Chalks formed in many regions. The seas continued to advance over the land. Coal swamps formed.	Flowering plants developed. Sea and land were still dominated by reptiles, many of them now giant and armoured. Mammals continued to evolve.
Much erosion of mountains, and much land inundated by the sea. The continents were cut up by swamps. Limestones as well as sandstones were formed.	Reptiles were dominant in the sea and on land. Archaeopteryx, the first bird-like creature, appeared. Dinosaurs grew to enormous sizes.
Much of the land was still arid. There was some volcanic activity, and marl and sandstone deposits formed. Copper and uranium deposits formed.	Conifers, ferns, and cycads were common. The first mammals appeared. Turtles and crocodiles appeared, as did dinosaurs—as yet only a few inches long.
A time of much Earth movement and mountain building. Melting glaciers left vast sedimentary deposits.	Deciduous plants appeared. Trilobites and eurypterids died out. Reptiles and insects were dominant.
The second period of the Coal Age. Land areas sank, producing vast swamps or regions of shallow sea. In other places, sea-beds rose and became dry land.	Huge evergreen trees and ferns grew. Fish were common, and large amphibians developed. Large insects evolved. Reptiles appeared on land.
The first period of the Coal Age: rotted vegetation became peat and, later, coal. Other minerals formed, including lead and zinc.	Mosses appeared. Amphibious creatures continued to develop. Large coral reefs began to form. Trilobites became less common.
The area of dry land increased. Much mountain building. Shales, slates, and sandstone formed.	Forests and ferns grew on land. Many fish evolved, including sharks. Amphibians and insects appeared.
Sea levels rose and fell. Mountain building in Europe. Limestones and sandstones were formed.	Leafless plants appeared on land. Eurypterids, animals that breathed air, developed in the sea.
Some mountain building. Boundaries between land and sea changed constantly. Great deposits of silt.	Still no life on land. Invertebrates were still common in the waters. Jawless fish appeared.
Seas covered much of the Earth's surface. Climates became milder. Some volcanic activity.	Invertebrate animals were common in the seas. The most advanced sea animal was the trilobite.
The Earth's crust formed. Vapour surrounding the Earth condensed to form water masses. Probably, much of the Earth's surface was desert.	No life on land. Algae and bacteria lived in the sea. Later, invertebrates—such as jellyfish, sponges, and worms—lived in the warmer waters.

LANGUAGES
- Caddo
- Maˋscoki
- Uto-Aztec
- Algonkin
- Iroquois
- Salish
- Eskimo
- Athabasca
- Maya
- Sioux
- Other linguistic groups

(Names of North American Indian tribes are printed in red)

EXPLORERS' ROUTES
- --- English
- —— Spanish
- —— Dutch
- ····· French

MAN EXPLORES THE WORLD

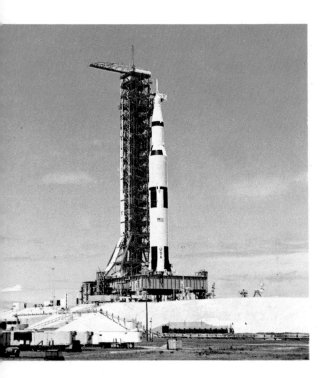

New Worlds to Explore. An American space rocket on the launch pad at Cape Kennedy (Cape Canaveral). By the 1900s, men had explored most corners of the Earth's surface. Explorers of a new type—scientists rather than adventurers—were turning their thoughts towards the other planets. In 1969, two American astronauts landed on the Moon, the first human beings to set foot on non-terrestrial dry land.

Throughout history, there have been periods when Man's desire for knowledge seems to have been particularly intense. At these times, science, art, and exploration have taken rapid strides forward. We are now living in such an age—an age when Man has taken his first steps into space.

The ancient Greeks lived through one of these periods of great intellectual activity. For example, as early as the 4th century B.C., Greek philosophers had learnt enough about the world to conclude that the Earth was a sphere. Greek knowledge of geography was summarized by Ptolemy, a scholar who worked in Alexandria in the A.D. 100s. The lands known to the Greeks included Europe, northern Africa, and part of Asia. Their knowledge of Asia owed much to Alexander the Great's march to the Indus Valley in the 300s B.C.

Ptolemy's biggest error was his underestimate, by nearly 40 per cent., of the circumference of the Earth. Ptolemy's miscalculation was still accepted 1,300 years later: when Columbus reached the West Indies, he thought that he must be in Asia. But discovery was not confined to the Europeans. While the Greeks, and later the Romans, explored the West, a great civilization was flourishing in China. Chinese traders extended their knowledge of geography over most of Asia.

After the Greek and Roman empires declined, Europe ceased for a time to be a centre of learning. Ptolemy's work was temporarily forgotten there and most people reverted to the idea that the Earth was flat. But, fortunately, the Arabs knew the works of Ptolemy and preserved Greek knowledge. In fact, during the Dark Ages of Europe, the Arab world became a centre of science and exploration. We owe our knowledge of the medieval empires of Africa to Arab traders and geographers who crossed the Sahara and recorded what they saw about the Negro empires in the south.

In the 800s, Norsemen began to sail the seas in search of plunder and lands to colonize. A Norseman, Leif Ericson, was probably the first European to land in North America. By the 1200s, the traders of Venice began to extend their horizons. Their search for trade culminated in the travels of Marco Polo through Asia (1271-95).

In the 1400s, another brilliant period of exploration began in Europe. It was linked with a development of science and art, called the Renaissance. Greek ideas were revived (often from Arab sources) and developed. In 1487, Bartholomeu Dias, a Portuguese captain, extended Greek knowledge of the world by sailing as far south as the Cape of Good Hope at the tip of Africa.

At this time, people who had believed the Earth to be flat were questioning this idea. One such man was Christopher Columbus, an Italian who interested the Spanish king and queen in the notion that he could reach Asia by sailing west. On October 12, 1492, Columbus reached San Salvador, one of the Bahamas Islands, but he believed that he

FICTION OR FACT?

Until a few hundred years ago, people knew very little about life in any continent except their own. But there were always many pretended travellers with tall stories to tell of the wonders they had seen in distant parts of the world. Everyone enjoyed such stories, but few really believed them.

One effect of the popularity of these accounts of fictional travels was that genuine travellers were disbelieved when they described the things they had seen. Marco Polo, who spent many years in China in the 1200s, was treated as a romancer when he returned to Italy and told about life in eastern Asia. Similarly, the explorations in Africa of the Portuguese priests Pedro Paez and Jerome Lobo in the early 1600s were for long treated as fictional, even by other travellers.

had reached Japan. Columbus continued his exploration of the Caribbean until 1504. Two other Italians, John and Sebastian Cabot, reached Canada in 1497, also thinking that they were in Asia.

But the Portuguese still wanted to reach Asia by sea to obtain gold and spices, so they sent Vasco Da Gama on a journey around Africa to India in 1497-98. This historic voyage was followed by another Portuguese achievement in 1519-21, when Ferdinand Magellan led an expedition that sailed around the world. Magellan was killed in the Philippines during this voyage.

The discovery of the New World led quickly to expeditions to conquer and explore the new lands. Between 1519 and 1521, the Spaniard Hernando Cortes conquered Mexico with a tiny force. He owed his success to the belief of the Aztec Indians that he was an invincible god. Another Spaniard, Francisco Pizarro, crushed the Incas in Peru in the 1530s. Meanwhile, in North America, an Italian, Giovanni da Verazzano, searched for a North-West passage in 1524. In the 1530s, a Frenchman, Jacques Cartier, explored the St. Lawrence River. In the 1540s, Spanish adventurers were exploring the southern parts of North America.

Rivalries developed in North America between various European powers, but exploration and settlement continued through the 1600s and 1700s. Between 1804 and 1806, two explorers, Meriwether Lewis and William Clark, finally made the long journey

EXPLORATION OF THE PACIFIC

— Cook (1st voyage—1769-70)
-- Cook (2nd voyage A—1773)
-·- Cook (2nd voyage B—1773-74)
···· Cook (2nd voyage C—1774)
— Fourneaux

Cook (3rd voyage—1776-78) ———
Return of Clerke, (1778-79) ------
after Cook's death
La Perouse (1785-88) ----
Baudin (1801) ———
Bardin (1802-03) ········

Forgotten Treasures. The ruins of Machu Picchu in Peru. Some of the more inaccessible spots on the Earth's surface are 'discovered' more than once. At Machu Picchu, 6,750 ft (2,000 m) high in the Andes Mountains, the ancient Incas built a fortress city that they later abandoned. For hundreds of years, the ruins remained untouched and forgotten, until the American explorer Hiram Bingham happened on them in 1911.

explore Lake Victoria and claim it as the source of the White Nile. The greatest European explorer in Africa was the missionary David Livingstone. He explored the River Zambezi and found the Victoria Falls. He later joined the search for the source of the Nile, and was still on his travels when he died in 1873. Henry Morton Stanley explored the Congo region (now Zaire) and enabled the king of Belgium to establish a colony there. The exploration of the desert interior of Australia also created many problems for explorers. Robert O'Hara Burke led the first north-south crossing of Australia, but on the return journey he and his deputy William Wills died of starvation.

In the late 1800s and early 1900s, serious exploration began in the polar regions. In 1893, Fridtjof Nansen, a Norwegian explorer, sailed into the icy Arctic Ocean. He spent two years in his ship before finally deciding to try to get to the North Pole by sled across the ice. He got within 272 miles of his target. However, it was an American, Robert E. Peary of the U.S. Navy, who first reached the North Pole, on April 6, 1909.

The South Pole lies at the heart of the bleak continent of Antarctica. On December

Man Explores the World

from the Missouri River to the Pacific coast.

Since the days of the ancient Greeks, there had been talk of a continent in the southern hemisphere. The first certain European landfall in Australia was made by a Dutch ship in 1606. However, Abel Tasman's voyage of 1642-43 was the peak of the Dutch exploration. He discovered Tasmania and New Zealand and, in 1644, he explored the north Australian coast. From 1768, the British Captain James Cook explored the Pacific region. He proved that Australia was a continent and explored Pacific islands before he was killed in Hawaii in 1779. By this time, the outline of the world was known and charted on maps.

By 1800, navigators and explorers had travelled along the coastlines of all the continents, except for that of Antarctica. The mainland of Antarctica was discovered in 1820, although the first landing was not made until 1821. In the 1800s, explorers travelled into the interiors of the unmapped continents. They were often dogged by disease and they suffered great hardships. In Africa, Mungo Park, a Scotsman, explored the River Niger in 1796 and in 1805-06. He was killed on this last expedition. Richard Burton and John Hanning Speke suffered many illnesses and mishaps before reaching Lake Tanganyika in 1858. Speke went on to

The Dark Continent. Mount Kenya, a huge, extinct volcano, is Africa's second highest mountain. Its existence was first recorded by a German missionary in 1849. Africa was for long called the 'Dark Continent' because so little was known about its interior. Much of the continent was not explored until the 1800s.

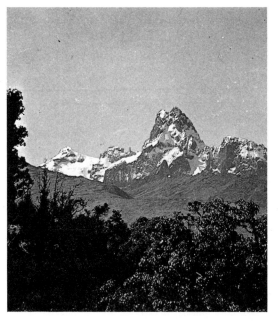

The Barrier. The Australian Alps, the highest mountains in Australia, rise at the southern end of the Great Dividing Range. For many years, the Range was an obstacle that prevented settlers on Australia's eastern coast discovering the great plains to the west. In 1813 the Range was penetrated by William Wentworth and other explorers.

14, 1911, the Norwegian Roald Amundsen reached the Pole. A month later, a British expedition under Captain Robert Falcon Scott also reached the South Pole, but the entire party perished on the return journey.

The development of aircraft made exploration far easier in the 1900s. The American Richard E. Byrd flew over the North Pole in 1926 and over the South Pole in 1929.

Even more dramatic than the arrival of the air age was the start of the space age in 1957, when Russia launched the first artificial satellite to orbit the Earth. Russia achieved another first in 1961, when cosmonaut Yuri Gagarin was sent into an orbit of the Earth. The United States Apollo project began in 1966 and reached its peak when Neil Armstrong stepped from his spacecraft onto the surface of the Moon on July 20, 1969. The Apollo project came to an end in 1975, when an Apollo craft linked up with a Russian Soyuz craft in orbit.

The space age has added to our knowledge of the Earth. Satellites circling the Earth send back photographs and other information which helps us in many ways. Satellites help us with weather forecasting, and aid navigators in fixing their position. Earth resources satellites help geologists to identify sources of valuable minerals. They also reveal areas that are in danger of pollution. Also, the recent study of the Moon has contributed to our knowledge of the history of our planet.

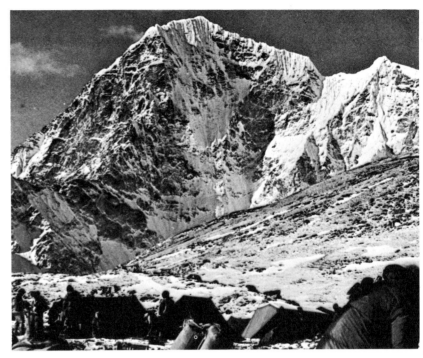

The Challenge of Height. Mount Everest in the Himalayas, the world's highest mountain. Many lives were lost in attempts to reach its summit. It was conquered in 1953.

FAMOUS EXPLORATIONS

Achievement	Explorer	Nationality
Possibly the first European landfall in North America (1003)	Ericson, Leif (?——?)	Norse
Travels in central Asia and China (1271-1295)	Polo, Marco (1254?-1324?)	Italian
Voyage around the Cape of Good Hope (1487)	Dias, Bartholomeu (1450-1500)	Portuguese
Exploration of the Caribbean (1492-1504)	Columbus, Christopher (1451-1506)	Italian
Landfall in Canada, with his brother Sebastian (1497)	Cabot, John (1450-1498)	Italian
First to sail around Africa to India (1497-1498)	Da Gama, Vasco (1469-1524)	Portuguese
First sea journey around the world (1519-1521)	Magellan, Ferdinand (1480-1521)	Portuguese
Conquest and exploration of Mexico (1519-1521)	Cortes, Hernando (1485-1547)	Spanish
Conquest and exploration of Peru (1530-1538)	Pizarro, Francisco (1478-1541)	Spanish
Discovery of Tasmania and New Zealand (1642)	Tasman, Abel Janszoon (1603-1659)	Dutch
Exploration of the South Pacific (1768-1779)	Cook, James (1728-1779)	English
Exploration of southern and central Africa (1841-1873)	Livingstone, David (1813-1873)	Scottish
First north-south crossing of Australia (1860-1861)	Burke, Robert O'Hara (1821-1861)	Irish
Exploration of central Africa (1874-1889)	Stanley, Henry Morton (1841-1904)	Welsh
First man to reach the North Pole (1909)	Peary, Robert E. (1856-1920)	American
First man to reach the South Pole (1911)	Amundsen, Roald (1872-1928)	Norwegian
First man to orbit the Earth (April, 1961)	Gagarin, Yuri (1934-1968)	Russian
First man to step onto the Moon (July, 1969)	Armstrong, Neil (1930-)	American

THE WORLD'S FEATURES

SURFACE FEATURES

Water covers 70.8 per cent. of the Earth's surface. The largest ocean, the Pacific, has an area of 63,800,000 sq miles (165,241,000 sq km). The world's greatest ocean depth of 36,198 ft (11,033 m) is in the Pacific. The Pacific contains many volcanic islands. In a sense, some of these islands are the highest mountains on Earth. For example Mauna Loa in Hawaii rises more than 30,000 ft (9,150 m) from the sea floor, although it is only 13,680 ft (4,170 m) above sea-level. Earthquakes and volcanoes are common, especially around the rim of the Pacific.

The Atlantic Ocean is the second largest ocean, with an area of 31,530,000 sq miles (81,660,000 sq km). Volcanic activity and earthquakes are common along the Atlantic mid-oceanic ridge, an underwater mountain range. Iceland is part of this ridge that juts above sea-level. The Atlantic is connected to the Mediterranean (and the Black Sea) by the narrow Strait of Gibraltar. The Baltic Sea is linked to the Atlantic by the Skagerrak. In the west, the Hudson Strait connects Hudson Bay with the Atlantic. In the tropics, the Caribbean Sea and the Gulf of Mexico are extensions of the Atlantic.

The third largest ocean is the Indian Ocean which covers 28,357,000 sq miles (73,444,000 sq km). It is linked to the Red Sea, the saltiest part of the world's oceans. The Arctic Ocean covers 5,440,000 sq miles (14,090,000 sq km). The Antarctic Ocean surrounds the continent of Antarctica. Some geographers consider the Antarctic Ocean as part of the Pacific, Indian, and Atlantic oceans.

The average salt content of the oceans is 35 per mille—35 parts of salts per 1,000 parts of water. In the Red Sea, salinity is increased by high evaporation to 41 parts per mille. Some inland seas have even higher salinities. The Dead Sea's salinity is 192 parts per mille, and the Great Salt Lake of Utah, in the United States, has a salinity of 203. There, the water is so dense that a swimmer cannot sink. Supplies of fresh water—from rain, melting ice, and rivers —reduce salinity. The world's largest inland body of water, the Caspian Sea, receives a lot of fresh water from rivers. Its salinity is 13 parts per mille.

Land areas occupy 29.2 per cent. of the

Above:
Fertile Plains. Rich farmland in Minnesota, in the United States. Some plains are deserts; others are the most fertile parts of the Earth's surface.

Left:
Voice of the Waters. The Iguaçu Falls on the Iguaçu River, on the border between Brazil and Argentina. Rivers cut valleys, irrigate the land, and, by spreading silt, make land fertile.

Mountain Grandeur. The Himalayas, the most magnificent and forbidding of the world's mountain ranges. Mountains occupy about one-fifth of the Earth's land surface.

GREAT MOUNTAINS		LARGEST ISLANDS		LONGEST RIVERS		
Mountain	Height in ft. (metres)	Island	Area in sq miles (sq km)	River	Length in miles (km)	Location
Everest	29,028 (8,848)	Greenland	840,000 (2,176,000)	Nile	4,160 (6,700)	North-east Africa
Godwin-Austen (K2)	28,250 (8,610)	New Guinea	316,856 (820,653)			
Kanchenjunga	28,168 (8,586)	Borneo	286,967 (743,241)	Amazon	3,900 (6,300)	Brazil
Mount Communism	24,590 (7,495)	Madagascar	230,000 (596,000)	Yangtze	3,100 (5,000)	China
Aconcagua	22,834 (6,960)			Zaire	2,718 (4,374)	Zaire
McKinley	20,320 (6,194)			Missouri	2,714 (4,368)	North America
Cotopaxi	19,344 (5,896)			Amur	2,700 (4,345)	China
Kilimanjaro	19,340 (5,895)			Hwang Ho	2,700 (4,345)	China
Mont Blanc	15,781 (4,810)			Lena	2,645 (4,257)	Siberia

Earth's surface. The nature of the land is largely determined by height. Three main height zones occur: plains near coasts; plateaux; and mountains.

Plains are good environments for Man. Gentle slopes make farming and travel easy, and the well-watered flood plains and deltas of tropical and temperate regions are among the most fertile places on Earth. The early civilizations of Mesopotamia, Egypt, and China grew up in river valleys with rich alluvial soil and plenty of water. But flooding sometimes causes tragedy. The Hwang Ho in China has burst its banks many times, drowning hundreds of thousands of people.

Some plateaux in tropical regions are attractive places for settlement, because they are cooler than the hot lowlands of the same latitude. But high plateaux are often hostile environments. All high regions have thinner air than land at sea-level, and as a result physical work becomes difficult. But plateau people adapt to the conditions in which they live.

Mountain areas are often bleak. Travel is difficult, and the mountains form barriers between peoples. There are three main kinds of mountains; volcanic mountains; fold mountains squeezed up by forces within the earth; and block mountains, which are blocks of land raised up by vertical Earth movement.

Dreamer's Paradise. The island of Morea, one of the Society Islands in the South Pacific.

Inhospitable Land. Life within the Arctic Circle, in Alaska.

FORCES OF CHANGE

The Earth's surface is constantly changing. The most dramatic and sudden changes are caused by forces originating within the Earth. These forces are linked with continental drift (*see* pages 16-17). They are responsible for volcanic activity, earthquakes, and mountain building. Other changes are caused by the forces of erosion which act on the Earth's surface.

Volcanic activity has been responsible for creating most of the rocks in the Earth's crust. Volcanoes are supplied with molten rock, or *magma*, in two main ways. First, heat sources in the Earth's mantle, which are probably radioactive, melt rock. The magma then rises through a conduit (channel) to the Earth's surface. Second, magma is created when one plate in the Earth's crust is forced beneath another. The descending rock is melted by friction and pressure.

Volcanic eruptions are of three main kinds: explosive, quiet, and intermediate. *Explosive eruptions* usually occur when explosive gases are trapped in the molten magma. When an explosive volcano erupts, the gases shatter the magma, which shoots out of the volcano as ash, as cinders, or sometimes in lumps called *volcanic bombs*. The greatest explosive eruption in history occurred in August 1883 at Krakatoa, a volcanic island between Java and Sumatra, in Indonesia. The eruption caused the 2,600-ft (792-m) high volcano to explode. Most of the island disappeared. Only three tiny islands, formerly part of the volcano's rim, were left. Ash and dust rose into the air and winds spread it around the world. The dust was seen in northern Europe three days later. The explosion also generated a destructive wave called a *tsunami*, which battered the coasts of Java and Sumatra, killing 36,000 people. Tsunamis are also caused by earthquakes.

In *quiet volcanoes*, such as Mauna Loa on Hawaii, the magma contains little gas or the gas can, at any rate, escape easily. As a result, great explosions do not occur. Instead, the magma flows through the vent of the volcano or through fissures (openings) in the ground as lava. *Intermediate volcanoes*, such as Vesuvius in Italy, combine features of explosive and quiet volcanoes.

Earthquakes occur around the edges of the plates in the Earth's crust. They are caused by movements along faults (cracks) in rocks. The major breaks in the Earth's crust are the plate edges themselves. For

Above:
River of Ice. A glacier flows slowly to the sea in Magdalena Fiord, in Svalbard. Glaciers tend to deepen the valleys through which they pass, laying bare the rocks.

Sea Power. Cape Reinga, at the northern tip of the North Island of New Zealand. The action of the waves continuously wears away the rocks, while at the same time building up the sandy beach.

example, the San Andreas Fault in California is 600 miles (960 km) long. It is the boundary between two plates that are moving in opposite directions. In 1906, a great earthquake destroyed the city of San Francisco. This earthquake occurred when the two plates along the San Andreas Fault jammed together, creating tension. The tension was released in a sudden jerk, making the ground shake. Near San Francisco, the shift along the fault was 15 ft (4.5 m).

Other earthquakes are caused when two plates are pushed against each other. The edge of one plate is forced down beneath the other. This is not a gradual movement, but one that proceeds in a series of jerks. Fold mountains are formed by a lateral pressure created when continental drift makes two plates collide. Sediments between the plates are compressed and squeezed up in large ripples called *folds*. The Alps, for example, are compressed sediments that accumulated on the sea bed. The formation of fold mountains is very slow. The Alps were formed as one plate, bearing the continent of Africa,

Above:
Pinnacles of Stone. The Dolomites, in the eastern Alps. Erosive forces have shaped the mountains into jagged peaks of spectacular beauty.

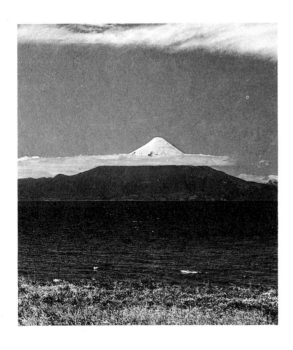

Spent Menace. The volcano of Osorno in southern Chile. It is in the Andes Mountains. Many of the high Andean peaks are volcanoes, some of them active.

drifted northwards and pressed against Europe. Geologists think that the Alps may still be rising. But this is difficult to prove, because as new mountains rise, the forces of erosion wear them down.

One erosive force acting on mountain slopes is called *weathering*. An example of this process is water freezing in a crack; it expands in volume and splits the rock. Sudden changes in temperature in hot regions also shatter rocks. Even the roots of plants that grow in rocky terrain can split rocks. Fragments of rock tumble down mountain slopes and pile up in heaps known as *scree*.

In dry regions, the wind is an important force. Armed with sand and dust, strong winds erode rock in a process rather like sand-blasting. They pick up the sand from dunes where it has piled up.

Water is a major agent of erosion. Some rainwater seeps into the ground; and rocks, such as limestone, that are soluble in water are dissolved. In this way, large caves are formed. Some rain, too, flows over the surface into streams and rivers. The rivers erode valleys, and carry eroded material to the sea. The Mississippi River in the United States, to take one example, carries 440 million tons of eroded material into the Gulf of Mexico in an average year.

Snow in mountain regions is compacted into ice and forms glaciers and ice sheets. Because of gravity, ice moves downhill. Frozen in the base of the ice are bits of rock. As a result, the moving ice is like a giant file. It scrapes against the ground, wearing out deep U-shaped valleys. At the base of glaciers, eroded material is dumped in *moraines*.

Natural erosion is a slow process and may hardly be noticeable in one person's lifetime. But Man sometimes speeds up erosion by misusing the land. When grasslands are ploughed and forests cut down, the soil is exposed to the wind and rain. Intensive farming robs the soil of its fertility and the soil becomes unstable. Unless it is protected and its fertility is restored, the soil is worn away and land that was once fertile can become as bare as a lunar landscape.

27

VEGETATION

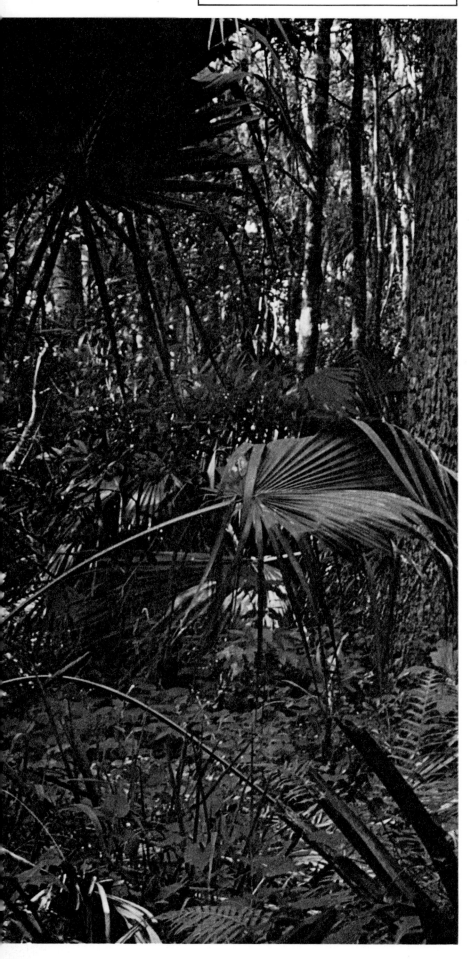

Whenever new land is formed on the Earth's surface, such as the volcanic island of Surtsey near Iceland, it is not long before plants start to grow. Their seeds may have been blown by the wind or carried by birds; or plants may have floated across the sea. Gradually, a stable community of plants develops, adapted to the soil and the climate. The final result is called the *climatic climax vegetation,* and this vegetation will flourish unless it is disturbed. However, in many areas Man has greatly changed the original vegetation. When we talk of *natural vegetation zones,* we often mean the original climatic climax vegetation rather than the actual vegetation, which is the result of Man's interference.

In general, belts of vegetation run E-W and are related to latitude. But there are many complications. For example, the height of the land in any particular place affects the climate and, therefore, also affects the vegetation.

Vegetation changes as one travels up a high mountain, because the temperature falls by 3.5°F for every 1,000 feet (6°C per 1,000 metres). In the Himalayas, the lower mountain slopes are covered by tropical forest. Gradually, the tropical forest gives way to deciduous forest—to trees that shed their leaves during one season of the year. Higher still, there are coniferous forests. Trees become more and more widely spaced and stunted as one nears the *tree line*—the upper limit of tree growth. Rich, grassy meadows which are located around the tree line finally give way to tundra, a region of grasses, lichens, and mosses. The tundra ends around 17,000 ft (5,000 m) above sea-level, where snow and ice begin.

This sequence of vegetation is broadly similar to the vegetation belts that encircle the Earth. Around the icy wastes of the North Pole is a region of treeless tundra, covering the far N of Europe, Canada, and Alaska. The soil just under the surface (the subsoil) is often frozen, and even in summer the temperature does not rise above 50°F (10°C). As one travels S, the temperatures gradually rise until sturdy trees appear. These trees become increasingly frequent until they form the boreal (northern) coniferous forest belt. Conifers, such as firs and pines, are specially adapted to cold conditions.

To the S of the boreal forests are the deciduous forests of the temperate zone. Deciduous trees include oaks, elms, and maples. Such forests need at least six months annually when the temperature exceeds the minimum necessary for tree growth, which stops entirely in winter.

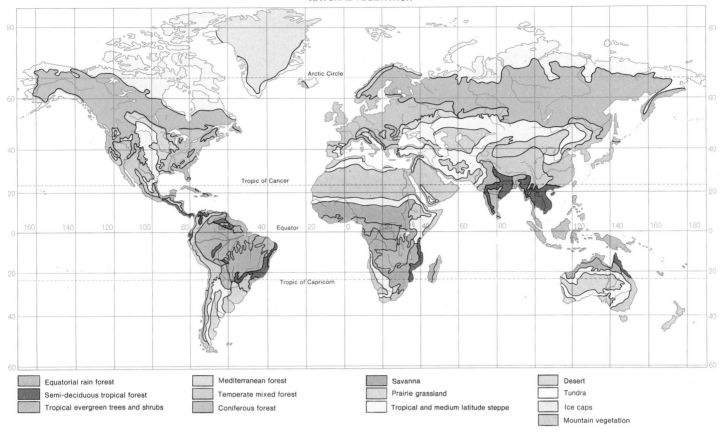

Equatorial rain forest
Semi-deciduous tropical forest
Tropical evergreen trees and shrubs

Mediterranean forest
Temperate mixed forest
Coniferous forest

Savanna
Prairie grassland
Tropical and medium latitude steppe

Desert
Tundra
Ice caps
Mountain vegetation

In some regions, the rainfall is too low for tree growth. Instead of deciduous forest, there are vast areas of grassland. These grasslands are called *prairies* in North America, *steppes* in Russia, *pampas* in Argentina, and *veld* or *velt* in Africa. The chief economic activity is livestock rearing, but large areas have been cultivated with such crops as cereals and cattle fodder.

Another forest belt occurs in warm Mediterranean regions. Mediterranean trees include evergreen oaks, corks, and such conifers as pines, firs, cypresses, and cedars. Such trees are adapted to summer drought and cool, moist winters.

The world's greatest desert, the Sahara, lies S of the Mediterranean zone. It covers about 3 million sq miles (7.8 million sq km). Other African deserts include the Namib and the Kalahari, which is really a semi-desert. The Middle East is largely desert, and desert extends into central Asia. The heart of Australia is desert, and deserts also occur in North and South America. The world's highest shade temperatures have been recorded in deserts, such as 136.4°F (58°C) at Azizia, in Libya, and 134°F (56.7°C) at Death Valley, California. Sand dunes build up in deserts, but large areas are bare, rocky wastes. Some plants survive these conditions. Cacti, for example, have spines instead of leaves to check water loss.

Bordering the deserts of Africa are vast tracts of savanna, which is tropical grassland with scattered trees. Savanna finally gives way to the hot, wet tropical forests around the equator. These forests cover Zaire and parts of West Africa, much of the Amazon and Orinoco basins of South America, and much of south-eastern Asia, including Indonesia. They consist of tall trees, often more than 100 ft (30 m) high. The foliage at the top may block out light from the forest floor. Trees in tropical forests include mahogany and rubber trees.

Opposite page:
Nature's Exuberance. The lush vegetation of a tropical rain forest.

Nature's Parsimony. The scant vegetation of the Kalahari Desert.

MAN CHANGES THE WORLD

AGRICULTURE

Before early men began to establish farming settlements and to plant crops and rear animals, they lived by hunting and gathering. This is still the way of life of such people as the Bushmen of the Kalahari in Africa. By studying them, we find that they use up most of their energy in the struggle to stay alive.

Farming began about 8,000 years ago when people discovered that some of the plant species that they collected yielded more food than others. By planting the seeds of the best food crops, they increased their food supplies. They also stopped moving around from one place to another, and built permanent homes. The earliest farm settlements were probably in Turkey and Palestine. From these areas, farming spread to the valleys of the Tigris-Euphrates Rivers in Mesopotamia and of the Nile in Egypt. There, people learned how to irrigate the land with water from the rivers. Irrigation was important in lands where the rainfall was light and infrequent. Around 5,000 years ago, the iron plough was in use in Egypt and Mesopotamia. Farming also began at an early date in China and, later, the lands of the Roman Empire became highly-developed agriculturally, and crop rotation was introduced to conserve the fertility of the soil.

In farming societies, food supplies were

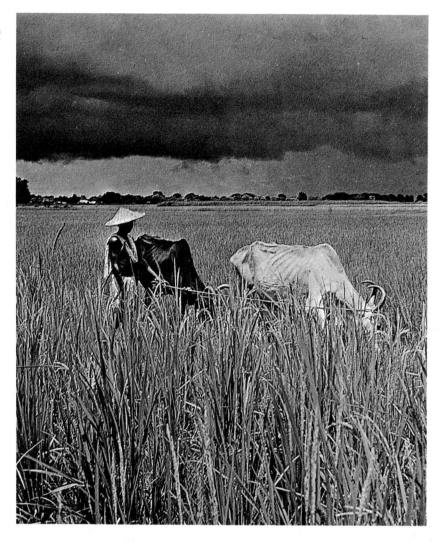

The Staff of Life. Rice cultivation in South-East Asia. Rice is the world's chief food crop. Probably mankind's biggest single step in the process of civilization was learning how to grow food crops. Primitive men had to exist on whatever food they could find; if they could not kill animals or find edible plants within reach of their caves or forest shelters, they had to move elsewhere. Once they had discovered how to grow their own food, they were able to build settlements and, eventually, cities. But in most parts of the world, crop growing involves hard, unremitting labour.

The Useful Plant. Harvesting hemp. Not all crops are grown for food, some—such as hemp, cotton, and flax—are used in making textiles, others have varied uses. Soya-beans, for example, are a valuable food, and can be used also in the manufacture of paints, adhesives, linoleum, and many other products.

reliable and people were relieved of the constant pressure for survival. As a result, civilizations developed, with advances in science and the arts. This led to improvements in farming techniques, the discovery of new plant hybrids, which yielded more food than the old varieties, and the breeding of better quality livestock.

Farming involves changing the environment. It has led to the destruction of much of the natural vegetation in many areas. For example, most of the original deciduous forests of temperate zones have been cleared to create farmland. This happened slowly and gradually in Europe; but in the United States, after European settlement began in the 1500s, it happened quickly. In the last 300 years, most of North America's deciduous forest has disappeared and wild animal populations have dwindled. Fast development has its dangers, such as soil erosion.

Crop farming may be divided into two types: crops which must be planted regularly; and shrub and tree crops that are more permanent. Most of the food grown for man and livestock—as well as materials to make fibres, oils, and plastics—comes from only about 50 crops.

Climate and soils are the factors that

determine what the chief crops in any area will be. For example, rice is the world's principal food crop, in terms of the number of people who depend on it. The main producers are China and India; and it flourishes throughout the wet, hot tropical lands of eastern Asia. Wheat, barley, and rye are more important cereals in cooler regions. The world's leading barley producers, for example, are the U.S.S.R., France, and the United Kingdom. Sugar is an important food. It is produced from two plants: sugar cane which grows in the tropics, and sugar beet which grows well in temperate regions.

Crops of grass, lucerne, and clover are produced as food for animals. Cattle are raised for their milk and other dairy products, and for meat and hides. Huge cattle ranches are located on the great grasslands of North and South America and Australia. Dairy farming requires rich grass. Other important farm animals are pigs, sheep, goats, and poultry.

Developments in agriculture have become especially important as the world's population has increased in the 1900s. In 1925, the world's population was less than 2,000 million. By 1970, it was more than 3,600 million and it was expected to rise to 4,457 million by 1980.

The most important recent development in agriculture has been the 'Green Revolution'. This revolution of farming techniques and plant hybrids has been particularly important in the countries where much of the population expansion is taking place.

Nature's Luxuries. Vines growing on a sunny slope below the ancient walls of Monteriggioni, near Siena in Italy. In addition to grapes—for eating and for making wine—luxury crops include tea, coffee, spices, and tobacco.

Scientists have produced new hybrids of such important food crops as maize, rice, and wheat—hybrids that yield bigger harvests. Agricultural experts have also been advising farmers on improved techniques and the use of fertilizers. In the 1940s Mexico—to take an example of what can be done—had to import wheat, but by 1964 it was exporting half a million tons a year. The new varieties of rice introduced in the Philippines yield about 15 times more rice than old varieties.

Another important source of food is the sea. Today fishing techniques are rapidly improving. Radar, spotter planes, and even satellites are used to locate fish.

Harvest of the Sea. A Polynesian islander carries a fish he has harpooned. People living in coastal areas or on islands where there is little or no agricultural land rely heavily on fish as a source of protein. To others, fish may be one item in a varied diet.

Man Changes the World

SHELTER

Food, clothing, and shelter are three essentials of life. Numerous archaeological finds have shown that prehistoric men often found shelter in caves. But once men began to farm the land, they built shelters for themselves and their livestock in the most suitable places.

The types of houses men build vary from place to place, according to the climate and the building materials available. For example, houses in forest regions are usually made predominantly of timber; and stone houses are common where easy-to-work stone is quarried. The Eskimo's igloo consists of blocks of compacted snow; the Mexicans use sun-baked bricks called *adobe*; and many Africans weave grass huts.

Climate affects the design of houses. In hot, dry regions, houses often have thick walls to keep out the heat, as well as flat roofs where the infrequent rainfall can be collected and where people can sleep at night. In cold regions, houses have to be sturdy, too, to keep out the cold, but they have steeply-inclined roofs to carry away heavy rainfall and snow.

In regions where earthquakes are common, people often live in extremely light structures which do little harm to their inhabitants if they cave in. In constructing larger buildings, architects have to take special precautions. They establish extremely secure foundations and use reinforced materials that will sway but not collapse when the ground shakes.

Modern technology has contributed much to comfortable living. Electric lighting, air-conditioning, central heating, and refrigeration have all helped to make life comfortable for people living in areas where the weather outside is harsh.

The distribution of population over the Earth's surface is very uneven. More than two-thirds of the world's people live in areas which total less than 8 per cent. of the available land. About one-third of the world's people live in urban settlements—ranging from towns with a few thousand inhabitants to cities of several millions. The proportion of people living in towns as opposed to rural areas varies, of course, from country to country. For example, about 90 per cent. of the people of Britain now live in urban areas, although about 150 years ago 80 per cent. were country-dwellers. In large parts of Africa, less than 10 per cent. of the people live in urban settlements today; and Bangladesh, China, India, and Indonesia have enormous rural populations—80 to 85 per cent. In Russia and Japan, just over half the people live in urban areas; this is a low proportion by comparison with most European countries and the United States.

The first cities grew up some time after

Top: Shelter for Millions. The skyline of Manhattan, in New York City. As civilization developed, certain points on the Earth's surface were found to have particular advantages for defence, communications, trade, or industry. In these spots, people grouped their shelters, forming villages, then towns and cities. The biggest cities grew upwards as well as outwards. Today, many people would like to see the size of cities limited.

Shelter for a Few. Above left: Aborigines in a desert region of Australia shelter in their rough tent. Left: A village on a Samoan island, in the Pacific Ocean. Half of the world's people live in small villages.

man began to practise agriculture. Farming of the land led to a surplus of food and materials, enabling some people to turn their energies to making goods and providing services. Some of the earliest towns were built in Mesopotamia in the 5000s and 4000s B.C. At a later date, the great city of Babylon probably had a population of 80,000.

Early towns may have been primarily trading centres, markets through which materials that were unavailable locally could be imported. Later, towns assumed an administrative role and handicraft industries were established. Writing and record-keeping were essential to carry on town business, and such developments stimulated the rapid growth of civilization.

From Mesopotamia, town development spread to Egypt, the Indus Valley, and China, and later to the Mediterranean. Greek and Roman towns were small by modern standards. Ancient Athens reached a peak population of about 150,000. Ancient Rome probably had a population of 200,000, but both Athens and Rome were great administrative capitals and centres of learning.

In the Middle Ages, increasing trade furthered the growth of cities. The Industrial Revolution from the 1700s led to an acceleration of city growth and to a relative decline in rural populations. Canals and railways made it possible for food and raw materials to be carried swiftly into cities, and manufactured products were exported. Towns tended to be divided into distinct zones. Administrative and business areas developed, and industrial zones were often separated from residential areas. Cities differed considerably in character. Some were centres of trade or industry. Others were cultural and art centres, or centres of government.

Cities today face many problems. Shortage of building space has led to the construction of tall buildings. Many residents of these buildings complain that they have no feeling of living in a genuine community. As cities grow in size, so traffic problems and certain types of crime increase. Some city authorities anticipate that they will soon be facing other severe problems if growth continues, such as water shortages or insufficient electricity supplies. Furthermore, in a nuclear war, the concentration of people in cities makes them extremely vulnerable to attack. Because of the problems of large cities, many countries build 'new towns' as a means of preventing existing cities growing larger. In China, the government encourages people to leave the cities and return to the land.

A City in the Sand. The old settlement of Ahwaz, in Iran, was restored to importance in modern times by the discovery of petroleum in its neighbourhood. Cities have developed in many unexpected places: in jungles, in deserts, or cut into cliffs. But, always, there is a reason for their existence. One, for example, may be sited at a point where jungle trails once crossed, another where a caravan route crosses a river.

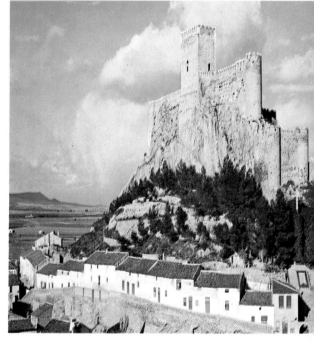

Protection. The castle of Almanza, in Spain, stands guard over its clustering village. In the past, such villages were usually self-contained communities, able to meet most of their own needs. Their life revolved around the church and the castle of the local nobleman.

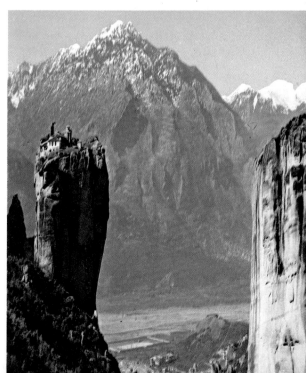

Protection. The Meteora monasteries near Kalabaka, in Greece. Built in the Middle Ages on the top of natural rock pinnacles 1,000 ft (300 m) high, the monasteries were cut off from the world and felt themselves safe from invaders. Access was by means of steps cut in the rock and baskets that were hauled up with ropes.

Man Changes the World

COMMUNICATIONS

Transport and communications are essential to the existence of all societies except the simplest subsistence economies. The people of early civilizations, such as the Chinese, Incas, and Romans, were mostly great road-builders. Along their roads, animals and human labourers hauled the goods they needed.

Animals are still used for transport today; camels in the Middle East, for example, and

So Many Roads. Approach roads to the George Washington Bridge across the Hudson River, in the United States—part of a highway system used by thousands of New York commuters daily.

The Iron Horse. A railway in the Andes Mountains, between Mendoza in Argentina and Santiago in Chile. Railways were the first major communications achievement of the machine age.

water buffaloes in southern Asia. But modern technology has changed the face of transport in the last 100 years.

Land Transport includes roads, railways, and pipelines. Road transport is the only means of conveying goods all the way from door to door. The construction of motorways and freeways has speeded it up, and it is of vital importance in most of the world's industrial countries. The United States has about one-third of the world total of commercial vehicles, and Japan and western Europe combined account for another third. The United States also has about half of the world's passenger cars.

The U.S.S.R., however, relies more on railways. In tons per kilometre, the U.S.S.R. handles about 40 per cent. of the world's total rail trade. This is partly explained by the long haulages necessary in that vast country. Railways are also important in eastern European countries. Railways developed in the 1800s and they have played a major role in opening up the interiors of Africa, North America, and the U.S.S.R. For example, the Trans-Siberian Railway, which runs 5,800 miles (9,400 km) from Moscow to Vladivostok, has contributed greatly to Siberia's growing prosperity. One disadvantage of railways is that trains cannot travel up steep slopes. But engineers can build tunnels that enable railways to go almost anywhere. Underground railways are important in many cities.

Pipelines serve areas where petroleum and natural gas are mined or distributed. Pipelines also transport such essential things as water and gas.

Water Transport includes coastal and lake shipping, river and canal transport, and ocean shipping. Coastal shipping is important in many countries where bulky raw materials, such as coal and ores, must be transported from mines to industrial areas. Haulage is slow but costs are low. Lake shipping is well-developed in such areas as the Great Lakes of North America and the lakes of Africa. Large rivers, such as the Rhine and the Danube, are major transport arteries, and canals remain important in some areas. But canal usage has declined in many countries because of competition with faster forms of travel.

The great canals, however—chiefly the Panama Canal and the Suez Canal (which was reopened in 1975 after having been closed for eight years)—play a major role in international shipping. These canals are short cuts on ocean routes, saving much time for cargo ships.

Shipping has been important in trade since ancient times, progressing from sails to steam engines. Today most ships are powered by oil-burning engines. Giant oil tankers are especially important. Some of them can carry 300,000 tons of crude oil. A major recent advance has been the introduction of container ships. These vessels are

Removing the Obstruction. The Panama Canal cuts through the Isthmus of Panama, between the continents of North America and South America. It connects the Atlantic and Pacific Oceans, and has enabled ships to avoid the long and hazardous journey around Cape Horn, at the southern tip of South America. Another great waterway, the Suez Canal, connects the Mediterranean Sea with the Red Sea and the Indian Ocean. It cuts through the narrow strip of land joining Asia to Africa.

designed to transport large metal boxes, called containers, which are packed by the sender and loaded complete into the ships. This method simplifies the handling of goods and saves time.

Air Transport. Ocean travel is a relaxing means of travel for passengers, but is comparatively slow. Jet aircraft can circle the world in 26 hours, and supersonic jets travel at speeds of 1,400 mph (2,250 kph). Freight traffic by air is, however, expensive compared with sea transport, though the use of helicopters may reduce the cost, and these aircraft may have a bright future in the transport of freight. Modern technological developments suggest that the world will continue to 'shrink' as faster and faster methods of travel are devised.

The effects of communications technology on the environment cause increasing concern. Motorways speed traffic, but add to the concrete jungle. Aircraft pollute the atmosphere, and the noise they make may often be almost intolerable.

Bridging the Gap. The Golden Gate Bridge connects the peninsula of San Francisco to northern California. It crosses a sea channel, but most bridges have been constructed to carry roads and railways across rivers.

Man Changes the World

<table>
<tr><td>

TRADE AND INDUSTRY

</td></tr>
</table>

Trade is an essential part of contemporary life and has a major influence on the standards of living of peoples throughout the world. Trading involves the buying and selling—or the exchange—of all kinds of products—foods and other agricultural goods, minerals, and manufactured items.

About half of the world's people work on farms. But in industrialized countries only a small proportion are engaged in agriculture. In Britain, for example, only about four out of every 100 people work on the land; whereas in such countries as Nepal and Tanzania more than 90 per cent. of the population are farm workers. British farming is highly efficient; but it does not produce

Technological Landscape. Industrial plant on Delaware Bay, in the United States. Though refineries, factories, and storage depots have added little to the beauty of the world, they have made an immense contribution to mankind's comfort and health.

Making the Desert Blossom. An oil well in Libya. Petroleum is the world's largest single source of energy.

all the food that the country needs and so food must be imported. However, despite this fact, the British people enjoy a far higher standard of living than either the Nepalese or the Tanzanians. The reason for Britain's superiority in standards of living is a reflection of two factors: trade and industry.

The most valuable items involved in world trade are not farm products, but minerals and the products manufactured from them. As a result, countries where most people work on farms find it very difficult to achieve high standards of living, even if their agriculture is very efficient and they export farm produce. But countries can raise their standards of living either by industrialization, which is very costly, or by exploiting their mineral reserves, provided that they have such reserves.

Mineral extraction industries employ

comparatively few people, but they are very important to the economies of many countries. Minerals fall into two categories: those that are used to supply energy; and those, such as metallic ores, that are used in manufacturing industries.

Minerals used to produce energy are petroleum, natural gas, and coal and lignite. In the late 1960s, about 40 per cent. of the world's energy came from petroleum products, 18 per cent. from natural gas, and 38 per cent. from coal and lignite. The remaining four per cent. came from hydro-electricity, nuclear power, wood, and other sources. Hydro-electricity is important in regions where mineral fuels are scarce.

Petroleum plays an extremely important part in world trade. Since the 1930s, world production of petroleum has increased by more than seven times. Some countries, such as the United States and the U.S.S.R., use most of their own petroleum resources at home, but many petroleum producers sell most of their production to other countries. About four-tenths of all petroleum is exported. In the 1970s, many petroleum-producing nations decided to raise the price and restrict the supply of this fuel, so as to conserve their resources. This action created many problems in countries that relied on imported petroleum. In some, coal mining, which had been declining because of competition with petroleum, suddenly assumed a new importance. However, coal remains far less important in world trade than petroleum, although both the United States and the U.S.S.R. are coal exporters.

Other minerals, especially such metallic ores as copper and bauxite (aluminium ore), are costly and are, consequently, very important economically. Some developing countries rely on the production of one or two minerals for their export income. For example, the export of copper is vital to Zambia and bauxite exports are extremely important to Surinam and Guyana.

Most industrialized countries have to import minerals. The process of industrialization, which has changed the world, began in Britain in the late 1700s. During the 1800s, it spread to other countries of western Europe, and to the United States, Canada, Japan, Russia, and Sweden. In the early stages, industries were based mainly on local supplies of minerals. But as world trade developed, so minerals were extracted in some countries and sold to other, industrialized, countries. The industrialized countries then sold manufactured products, such as machinery, on the world market.

Broadly, the countries of the world today can be divided into two groups: the indus-

Industrial Power. Heavy industry at Pittsburgh, in the United States. Pittsburgh, on the Ohio River, is America's biggest inland port and the world's chief steel city. Many great manufacturing cities have grown up in places where there are rich mineral deposits or where there are favourable circumstances for the import of raw materials.

Production. Above: An atomic-power plant in France. Below: Mass production in a Japanese factory.

trialized nations and the non-industrialized or 'developing' nations. North America, Europe, the U.S.S.R., Australia, and Japan combined have only about 30 per cent. of the world's population, but they produce about 85 per cent. of the world's industrial goods. The rest of the world has some industrial regions (for example, in Brazil, China, and South Africa). But despite these pockets of industry, the rest of the world has 70 per cent. of the world's peoples and makes only 15 per cent. of the world's industrial products. In the industrialized world, North America is the chief region. With seven per cent. of the world's population, it manufactures 35 per cent. of the total industrial production.

The chief group of trading nations is western Europe. Its prominence is partly due to the fact that western Europe consists of small nations that trade between themselves. Western Europe and North America combined handle about two-thirds of the world's exports. Japan, the U.S.S.R., and eastern Europe are also important; consequently, it can be seen that the industrialized nations are also the chief trading nations.

Dictionary of Geography

Abrasion The wearing away of rock surfaces or other surfaces by the rubbing action of materials carried by wind, running water, or moving ice. *See* ATTRITION; CORRASION; DENUDATION; EROSION; WEATHERING.

Accretion The gradual building up of water-borne material such as mud and sand in an area that is regularly flooded.

Adobe Unburnt bricks that have been dried in the sun. Usually the bricks are made of clay mixed with grass or straw.

Advection A horizontal movement of air or a liquid. The term is often used in connexion with the movement of warm air from tropical to temperate latitudes or from sea to land. *See* CONVECTION.

Aerology The science that deals with the atmosphere.

Affluent *See* TRIBUTARY.

Agglomerate A mass of broken rocks or lumps of lava thrown out by a volcano and cemented in ash. *See* LAVA.

Agriculture The science and practice of cultivating the land in order to produce crops. The term is sometimes used in a wider sense to include the rearing of livestock.

Agronomy The scientific study of farm methods in order to produce a higher yield of crops. The agronomist is concerned with such matters as drainage, plant diseases, pest control, soil conservation and quality, and the potential of various types of seeds.

Air *See* ATMOSPHERE.

Alpine peaks enclose the Engadine, part of the valley of the River Inn, in Switzerland. The village of St. Moritz in the valley is a popular resort.

Air mass A mass of air having a uniform temperature and humidity. Its boundaries are called *fronts*. An air mass develops in a high-pressure area when the air over a broad land or sea surface remains still long enough to adopt the surface temperature and humidity. It retains these qualities when it moves elsewhere. Air masses are described as *tropical or polar* (hot or cold); *maritime or continental* (moist or dry). The four main types are tropical maritime, tropical continental, polar maritime, and polar continental. *See* FRONT.

Air-stream A breeze or wind; any moving current of air.

Ait, Eyot A small island, especially in a river.

Alfalfa *See* LUCERNE.

Algonkian Name given to a system of North American rocks. They are Pre-Cambrian and sedimentary. The name is taken from the Algonkian Indians who lived between Hudson Bay and the Great Lakes. *See* PRE-CAMBRIAN ERA; PROTEROZOIC PERIOD; SEDIMENT.

Alluvium Solid material carried by a river, and eventually deposited on the river bed, on nearby flooded land, or in lakes or deltas. It may include silt, sand, gravel, and organic material.

Alp A mountain pasture, especially in Switzerland, formed of a gentle slope high up in the mountains. It is usually covered with snow in winter, but provides rich grazing for farm animals in summer.

Alpine Having to do with the Alps, the largest mountain range in Europe. The word is also used (with small initial *a*) to describe any high mountains, or their climate, plants, and wildlife. Strictly speaking, an alpine region is one that lies above the timber line and below the snow line.

Altiplano A high plain in the Andes Mountains. Specifically it refers to a barren plateau 13,000 ft (3,960 m) above sea level in Bolivia.

Altitude The height of any object, or the vertical distance above mean sea level. The altitude of any feature on the earth's surface, such as a mountain, is usually measured in feet or metres. *See* MEAN SEA LEVEL.

Altocumulus A cloud formation made up of small, round, white or grey clouds massed closely together at heights of from 8,000 to 20,000 ft (2,500 to 6,000 m). *See* CLOUD; CUMULUS.

Altostratus A type of cloud that forms a thick greyish-blue sheet or veil from about 6,500 to 20,000 ft (2,000 to 6,000 m) up. Sometimes it produces continuous heavy rain. *See* CLOUD; STRATUS.

Amphibole A group of dark-coloured minerals containing silica (a common whitish substance that occurs as quartz, flint, etc.). The most important of them is hornblende, a mineral made up mostly of calcium, magnesium silicate, and iron.

Anaerobic A term describing animals that can live in the absence of free oxygen. It is also used of soil that is in an airless condition, as when waterlogged.

Anemometer An instrument for measuring the speed and direction of the wind.

Aneroid barometer An instrument that measures the pressure of air without using mercury or any other fluid. It is made up of a hollow metallic box from which almost all the air has been removed. Changes in outside air pressure cause the sides to bulge out or cave in. These movements are registered by a spring and a needle on a graded scale. *See* ATMOSPHERIC PRESSURE; BAROGRAPH; BAROMETER.

Antarctic As an adjective, the term describes the South Polar regions. As a noun, it refers to that part of the earth's surface that lies within the Antarctic Circle (latitude 66° 32′ S).

Anticline An arch-like fold in the layers of rock of the earth's crust, formed when pressure from each side forced a section of rock upwards in a bulge. *See* FOLD.

Anticyclone A region of high atmospheric pressure, with the highest pressure at its centre. A spiral flow of air moves clockwise round the centre in the Northern Hemisphere; anti-clockwise in the Southern Hemisphere. An anticyclone is usually accompanied by settled weather. *See* CYCLONE.

Antipodes Points that are opposite to each other on a diameter of the earth (180° of longitude apart). People in Britain sometimes call Australia and New Zealand 'the Antipodes' because they are 'at the opposite end' of the earth.

Anti-trades Winds that blow in a generally westerly direction in the upper air above the trade winds. The trade winds blow in the reverse direction close to the earth's surface. *See* TRADE WINDS.

Anvil cloud A cumulonimbus cloud (thunder-cloud) whose top is flattened out, in rather the shape of a blacksmith's anvil. *See* CUMULONIMBUS.

Aphelion The point in its orbit at which a heavenly body in the solar system is farthest from the sun. The earth at this point (4 July) is about 94 million miles (151 million km) from the sun. *See* PERIHELION.

Apogee The point in the orbit of any planet or the moon when it is at its farthest from the earth. *See* PERIGEE.

Appleton Layer A layer in the ionosphere, about 150 miles (241 km) high, that reflects radio waves back to earth. It is named after Sir Edward Appleton, the physicist who discovered it. *See* IONOSPHERE.

Arable land Land that is used for growing crops, or is suitable for ploughing and cultivation; as distinct from pasture, woodland, vineyards, and house gardens.

Arc of the meridian A measurement made along a meridian (line of longitude) to discover accurately the shape and size of the earth.

Archaeozoic The second of the three Pre-Cambrian Periods, after the Azoic and before the Proterozoic.

Archipelago A group of islands in a sea or other broad expanse of water.

Arctic The region of the North Pole and that near the Pole. Strictly speaking, the Arctic is the region within the Arctic Circle (latitude 66° 32′ N).

Artesian basin A basin-shaped layer of porous rock in the earth's crust that is trapped between layers of non-porous rock, thus also trapping water. *See* ARTESIAN WELL.

Artesian well A deep, narrow well sunk into an artesian basin and producing a continuous stream of water at the surface. The water is forced upwards because the outlet of the well is below the level of the water's source. *See* ARTESIAN BASIN.

Atlas A book of maps.

Atmosphere Air; the envelope of mixed gases that surrounds the earth, up to a height of some 300 miles (about 480 km). It consists of about 78% nitrogen and about 21% oxygen. Argon, carbon dioxide, helium, and some other gases make up the remaining 1%. Often, the atmosphere also contains water vapour. The atmosphere is tasteless, odourless, and colourless. Its characteristics vary at different altitudes. For this reason it is usually described as having four layers —the troposphere, stratosphere, ionosphere, and exosphere. *See* ATMOSPHERIC PRESSURE; EXOSPHERE; IONOSPHERE; STRATOSPHERE; TROPOSPHERE.

Atmospheric pressure The pressure of the atmosphere on the surface of the earth as a result of its weight of about 6,000 million million tons. The average atmospheric pressure at sea level is 14.7 pounds per sq inch (1,033 g per sq cm). Atmospheric pressure decreases with altitude. *See* ATMOSPHERE; BAROMETER.

Atoll A circular or horseshoe-shaped coral reef that encloses a lagoon. *See* CORAL REEF; LAGOON.

Attrition The constant wearing down of pieces of rock into ever finer particles as they are carried along by wind, water, or ice. It

Anticline. This enormous fold of rock in Wyoming, in the United States, has been carved into sharp ridges by erosion.

differs from *abrasion*, the wearing away of rock surfaces. *See* ABRASION; CORRASION; DENUDATION; EROSION; WEATHERING.

Aureole (i) The zone of rocks immediately surrounding a mass of igneous rock and metamorphosed as a result. (ii) A corona or halo round the sun or moon. *See* IGNEOUS ROCK; METAMORPHIC ROCK.

Aurora Spectacular coloured lights seen in the sky at night. They are believed to be the result of electrical discharges from the sun, and occur between 50 and 600 miles (80 and 1,000 km) up from the earth. Those seen in the Northern Hemisphere are called the *Aurora Borealis* or *Northern Lights*; those in the Southern Hemisphere, the *Aurora Australis* or *Southern Lights*.

Autumn The season of the year between summer and winter. *See* SEASON; SUMMER; WINTER.

Avalanche The fall down a mountainside of masses of snow, ice, rocks, or earth. *See* LANDSLIDE.

Axis of the earth An imaginary line through the earth, joining the North and South Poles, round which the earth rotates once in every 24 hours.

Azimuth (i) A bearing read clockwise in degrees from north. (ii) The arc of the horizon between the longitudinal plane of the observer and a vertical plane that passes through the observer and a particular heavenly body. *See* BEARING.

Azimuthal or Zenithal projections A group of map projections in which the surface of the globe of the earth is represented on a flat piece of paper as if the centre of the paper touched the globe; all points on the map have their true compass directions from this centre. *See* PROJECTION.

Azoic Meaning literally 'without life', the term refers to the earliest part of the Pre-Cambrian Era, when there was no animal life on earth. *See* PRE-CAMBRIAN ERA.

Backing An anti-clockwise change in the direction of the wind; for example, from N to NW. *See* VEERING.

Badlands Any region where drought and very porous soils occur together, and where erosion by wind and water has resulted in deep ravines and steep hills. The name was originally given to typical areas of South Dakota and western Nebraska in the United States. *See* EROSION.

Balloon-sonde, Ballon-sonde A balloon carrying instruments for checking the weather. It is released into the upper atmosphere where it eventually bursts. The instruments, with their recorded information, float safely down to earth by parachute. Some sondes carry radio transmitters.

Bar (i) A ridge of material such as mud, stones, or sand that forms across a bay or the mouth of a river. (ii) A unit of atmospheric pressure equal to one million dynes per sq cm. It is equivalent to 29.53 inches (750 mm) of mercury at 0° C in latitude 45°. Usually the *millibar* is used in measuring atmospheric pressure. *See* MILLIBAR.

Barograph An aneroid barometer that records atmospheric pressure on a moving drum. A pen shows pressure changes hour by hour as a continuous line drawn on graph paper fastened round the drum. *See* ANEROID BAROMETER; ATMOSPHERIC PRESSURE; BAROMETER.

39

Barometer An instrument for measuring the pressure of the atmosphere. A common type of barometer, the *mercury barometer*, indicates atmospheric pressure by balancing the weight of a column of mercury against the weight of a column of air. *See* ANEROID BAROMETER; ATMOSPHERIC PRESSURE; BAROGRAPH.

Barrier reef A coral reef lying parallel to the shore and separated from it by a wide stretch of deep water. Usually there are many openings in the reef. *See* CORAL REEF.

Barysphere (i) The core of the earth. (ii) The whole of the earth's interior, including mantle and core. *See* CORE; EARTH; MANTLE.

Basalt A dark, fine-grained igneous rock. *See* IGNEOUS ROCK.

Basin (i) A hollow in the earth's crust; if filled with water it may be a *lake basin* or an *ocean basin*. (ii) The area drained by a river system. *See* ARTESIAN BASIN.

Bay A wide, open, curving arm of the sea, or a lake within the coastline or shoreline. *See* BIGHT.

Bayou A marshy creek found in flat country, especially a creek along the lower reaches of the Mississippi River in the United States.

Beach The shore of a sea, consisting of a strip of pebbles, sand, or mud lying between low and high water marks.

Bearing The direction of one object from another; the horizontal angle between the meridian and an object seen by the observer, measured in degrees clockwise from N. *See* MERIDIAN.

Beaufort Wind Scale An internationally accepted series of numbers devised by Admiral Sir Francis Beaufort in 1805 to indicate different wind strengths.

Beaufort No.	Description	Speed (mph)
0	Calm	Less than 1
1	Light air	1–3
2	Light breeze	4–7
3	Gentle breeze	8–12
4	Moderate breeze	13–18
5	Fresh breeze	19–24
6	Strong breeze	25–31
7	Near gale	32–38
8	Fresh gale	39–46
9	Strong gale	47–54
10	Whole gale	55–63
11	Storm	64–75
12	Hurricane	above 75

Bedrock Solid, unweathered rock lying beneath the soil and subsoil with which most of the earth is covered.

Belt A zone of vegetation or rock outcrops that has its own definite characteristics, making it distinct from the rest of the region.

Bight A wide bay. *See* BAY.

Bill A long, narrow cape or headland jutting out to sea. *See* CAPE; HEADLAND; PENINSULA; POINT.

Billabong Australian aboriginal term for a pool or a small *cul-de-sac* stream flowing out from a river. Usually, billabongs are created by heavy rains in areas that are normally hot and dry. After a period they dry up.

The Great Barrier Reef, off the coast of Queensland, in Australia, stretches for about 1,250 miles (2,000 km).

Biogeography The study of the geographical distribution of plants and animals. *See* PHYTOGEOGRAPHY; ZOOGEOGRAPHY.

Biosphere The surface zone of the earth where living things can exist. It extends a short distance into the earth's crust and upwards into that part of the atmosphere that contains sufficient oxygen to support life.

Bitter lake A lake whose waters hold large amounts of chemical salts in solution.

Blizzard A blinding snowstorm with strong icy winds. In temperate zones, blizzards are caused when cold air masses move in and meet warm air masses.

Blow-hole Hole resulting from the partial collapse of the roof of a coastal cave. Spray shoots through the hole at high tide and can be seen above the cliff-top level.

Bluff A steep headland or cliff. Specifically, the term refers to the remains of a spur that has been cut back by the wearing away of a meandering river valley. *See* SPUR.

Bocage Farming country region of northwestern France, made up of small fields separated by hedges and trees. In some places, dry stone walls take the place of hedges.

Bog An area of waterlogged, spongy ground with rotting vegetation lying on the surface. Eventually the vegetation turns into a layer of peat. *See* PEAT; MARSH.

Bore A wall of foaming water that travels at speed up a river. It occurs when a tidal wave enters an estuary and breaks as it proceeds up river. *See* ESTUARY; TIDAL WAVE.

Boreal (i) The *boreal zone* is a climatic zone in which winters are snowy and summers short. (ii) The *boreal era* was a dry climatic era in Eurasia with warm summers and cold winters (7500–5500 B.C.). The word *boreal* means 'pertaining to the north'.

Boulder clay Thick clay with stones of all sizes formed of rock that has been crushed and deposited by glaciers and ice sheets. *See* CLAY.

Bourne A stream that flows along a normally dry valley and that appears only when the water table rises. *See* WATER TABLE.

Brae Scottish for a bank or hillside. *See* BROW.

Brake A dense mass of trees or brushwood; a thicket.

Breaker A wave that breaks as it reaches shallow water near the shore.

Bay. A tranquil inlet on the Dalmatian coast of Yugoslavia, with the village of Sveti Stefan.

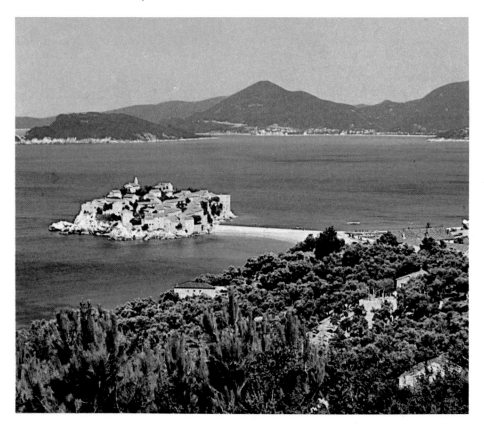

Breakwater A man-made barrier built out from the sea coast to break the force of the waves and so protect the shore.

Broad A lake-like sheet of water, usually part of or close to the estuary of a slow-moving river. Broads are common in parts of East Anglia, in Engand.

Brow The edge or upper part of a hill. *See* BRAE.

Bush An area of wild and uncleared vegetation, shrubby or wooded. The term is commonly used in southern Africa and in Australia.

Butte A flat-topped hill of hard rock in arid areas of the United States that has resisted total wearing away by flash floods (fierce but short-lived torrents). *See* MESA.

Buttress A mass of rock that juts out of a mountainside.

Cainozoic *See* CENOZOIC ERA.

Cairn, Carn A man-made pile of stones, constructed as a monument or built as a landmark.

Caledonian Term used to describe ancient mountain ranges that run through western Scandinavia, Scotland, and north-western Ireland. The term also describes the Silurian-Devonian Periods when these mountains were being formed.

Cambrian Period A Period in the earth's history, believed to have started about 600 milion years ago. It was the first Period of the Palaeozoic Era. At this Period large areas of today's continents were covered by sea. *See* PALAEOZOIC ERA.

Canal A man-made inland waterway used for transport. Sometimes a river that is too narrow or too shallow for navigation is 'canalized' to enable it to take boats. Small canals are also constructed to irrigate arid land. *See* IRRIGATION.

Cancer, Tropic of A line of latitude at 23° 32′ N of the equator where the sun shines directly overhead on June 21. It marks the northern boundary of the Tropical Zone. *See* CAPRICORN, TROPIC OF.

Canyon A steep-sided, usually deep gorge cut by a river through soft rock in a dry region. The Grand Canyon of the Colorado River in the United States is the largest and best-known example.

Cape A large, rocky headland jutting out to sea. Often carved by the waves into a point, capes help to shelter adjoining coasts. Also called *headland, mull,* and *promontory. See* FORELAND; POINT.

Capricorn, Tropic of A line of latitude at 23° 32′ S of the equator where the sun shines directly overhead on December 21. It marks the southern boundary of the Tropical Zone. *See* CANCER, TROPIC OF.

Carboniferous Period The fifth Period of the Palaeozoic Era, when much of the low land was covered repeatedly by the sea, and decaying vegetation in the swamps dried out, becoming compressed into coal. The Period began about 350 million years ago and lasted about 80 million years. It is also called the *Coal Age. See* PALAEOZOIC ERA.

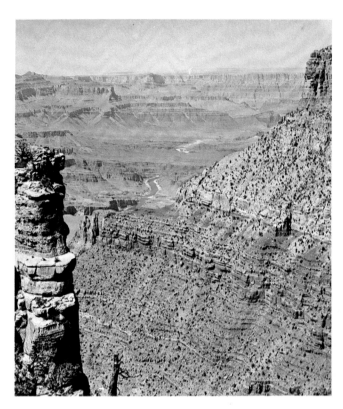

The Grand Canyon in Arizona, in the United States, was cut by the Colorado River. It extends for more than 200 miles (320 km) and is more than one mile (1.6 km) deep in places.

Cardinal points The four main points of the compass: north, south, east, and west. *See* COMPASS.

Cartography The art of drawing maps and charts. *See* CHART; MAP; MAP READING.

Cascade A small waterfall. Sometimes cascades are constructed in parks and gardens in such a way that the water flows down a series of steps, tinkling as it falls. *See* WATERFALL.

Cataract A large waterfall or a series of rapids, as on the Nile. *See* RAPIDS; WATERFALL.

Catchment area The area from which a river draws its water supply; limited by the watershed—the ridge on the other side of which the water flows away from the area. Also called *drainage area. See* DRAINAGE; RIVER; WATERSHED.

Cave, Cavern A hole in the earth's crust, usually running horizontally or almost horizontally. Sea caves may be made by the waves beating against a cliff face, or by the varying air pressures produced by the waves in a crack in the rock. Inland caves may be formed in limestone regions when rock underground is dissolved by streams of water containing carbon dioxide.

Cay, Key, or Kay A small, low island of sand and coral, such as the Florida Keys.

Ceiling (i) A certain level in the atmosphere, usually the lowest layer of thick cloud. (ii) The effective height to which any man-made object (aircraft, balloon, rocket, etc.) can climb.

Celsius scale A scale used on thermometers, an alternative name for the *centigrade scale.* Anders Celsius (1704-1744), a Swedish astronomer, invented the scale in 1742 by dividing the interval between the freezing and boiling points of water into 100 parts. Freezing point =0°, boiling point =100°. (On the original scale these markings were reversed.) To convert Celsius into Fahrenheit,

multiply the C reading by 9/5 and add 32. Thus F=9C/5+32. *See* FAHRENHEIT.

Centigrade scale *See* CELSIUS SCALE.

Cenozoic, Cainozoic Era The most recent Era in the earth's history, coming down to the present. Sometimes it is called the *Tertiary Era* or the *Age of Mammals,* because mammals first began to appear at the beginning of the Era—about 70 million years ago. It is divided into the Tertiary and Quaternary Periods.

Chalk A white, fine-grained type of limestone, easily broken. It is made up mainly of shells and skeletons of tiny animals. Most of it was formed during the Cretaceous Period, which began about 135 million years ago. *See* CRETACEOUS PERIOD; LIMESTONE.

Channel (i) The deep navigable part of a harbour. (ii) The deepest part of a river bed. (iii) A stretch of sea between two land masses. (iv) An irrigation ditch. (v) A tapering estuary.

Chaparral Vegetation made up of low evergreen trees, shrubs, and herbs growing on dry slopes, specifically in the southwestern United States and western Mexico. The term comes from a Spanish word meaning 'evergreen oak'.

Chart (i) A map of the sea and coastline, drawn up for use by navigators. (ii) A weather map. (iii) Sometimes, loosely, any map. *See* MAP.

Chimney (i) In England, especially in the Lake District, a narrow, vertical split in a rock face used by climbers in their ascent. (ii) In the United States, an upright pillar of rock.

Chinook A warm, dry SW wind that blows down the eastern slopes of the Rocky Mountains in the United States and Canada in winter and early spring. It quickly raises the temperature, melts the snow, and exposes grass that can be used for grazing. Consequently, it is highly important to the farmer.

The white cliffs of England. Steep chalk cliffs rise above the waters of the stormy English Channel.

Cirrocumulus A type of cloud that produces a 'mackerel sky'. It is formed of very high sheets or patches broken into globules or soft, rounded masses. *See* CLOUD.

Cirrostratus A type of cloud that looks like a very high, whitish, translucent veil. It often produces a halo round the sun or moon. *See* CLOUD.

Cirrus A type of high cloud with wispy and fibrous-looking bands and streamers. It is sometimes called 'mares'-tails' or 'banners'. *See* CLOUD.

City (i) A large and important town, where thousands of people live and work. (ii) An ancient town, which may be small, with a cathedral and bishop. (iii) In some countries, a town so designated by royal charter or letters patent.

Clay A type of smooth, slightly oily earth. It is fine-textured and consists of sedimentary rock. When wet, it turns into slippery, sticky mud. Many clays have a large content of mineral fragments. *See* SEDIMENT.

Cliff A steep rock face. Sea-coast cliffs are formed by the wearing action of the waves. Canyon cliffs are cut by deep rivers. Glaciers also grind out cliffs along valley walls.

Climate The general weather conditions over a fairly large area of the earth's surface, observed during a period of years. Sunshine, rainfall, altitude, and geographical location all affect climate. *See* WEATHER.

Climatology The scientific study of the earth's various climates—their distribution and the way they affect human, animal, and plant life. *See* CLIMATE; METEOROLOGY; WEATHER.

Cloud A mass of water vapour floating in the air; it is made up of condensed droplets of water or ice crystals. Clouds usually form high above the earth's surface and adopt many different shapes. They are classified according to their height and shape. *See* ALTOCUMULUS; ALTOSTRATUS; CIRROCUMULUS; CIRROSTRATUS; CIRRUS; CONDENSATION; CUMULONIMBUS; CUMULUS; NIMBOSTRATUS; RAIN; STRATOCUMULUS; STRATUS.

Coal A soft brown or black rock, found in layers, and formed hundreds of millions of years ago of carbonized plant remains. It burns well and is a very important source of energy.

Coastline The shoreline of the continents and other land masses. Also, specifically, the highwater mark of medium tides or, sometimes, the mark reached by the biggest storm waves.

Col A high pass in the mountains; a saddle-like gap between two peaks or across a ridge. It is formed when two corries (glacier-formed, half-moon depressions at the head of mountain valleys) eat into opposite sides of a ridge and meet.

Cold desert (i) A term for the polar regions and tundra, where plant life is limited or non-existent because of snow and ice. (ii) A high inland plateau such as the Gobi Desert, in Asia, whose altitude and distance from the sea make it very cold and dry. *See* DESERT; TUNDRA.

Cold front The boundary between a mass of warm air and a mass of colder air that overtakes it and burrows underneath it. The moving boundary is marked by a rise in pressure, a drop in temperature, and (in the Northern Hemisphere) a veering of the wind. Such fronts are indicated on weather charts by a series of black triangles on the leading edge of the line marking the front. *See* FRONT; OCCLUDED FRONT; WARM FRONT.

The crater of Mount Ararat in Turkey. The volcano last erupted in 1840, causing great oss of life and property.

Compass An instrument for finding directions. It has a magnetized needle that turns freely on a pivot and always points to the magnetic North and South Poles. The needle is mounted on a card marked in degrees and showing the cardinal points. *See* CARDINAL POINTS; GYRO-COMPASS.

Condensation Water droplets formed when the air has more water vapour than it can hold. The amount it can hold depends on its temperature. Condensation occurs when the temperature of the saturated air drops below its *dew point.* Clouds and rain result from the condensation of water vapour in the atmosphere. *See* DEW POINT; SATURATION.

Conical projections Projections designed for mapping limited areas, such as specific countries or parts of continents, with reasonable accuracy. An imaginary paper cone is placed over a globe with its apex on the globe's axis. The surface features of the globe are then projected onto the cone, and

the cone unrolled and laid out flat. Meridians now appear as straight lines radiating from the apex; lines of latitude are arcs of concentric circles. The line of latitude at the point where the cone touched the globe is called the map's *standard parallel.* On the standard parallel, the scale is the same as on the globe. At other parallels, there is distortion. Some conical projections lessen the distortion, for example by having two standard parallels. *See* PROJECTION.

Coniferous forest A forest whose trees are mostly evergreen and cone-bearing, with needle-like leaves. Such forests grow in a variety of climates from the subtropical to the subarctic, and yield valuable softwood timber.

Conservation The preservation for future generations of natural surroundings, including soil, plants, and animals. Conservationists try to guard against waste and pollution and help to maintain the balance of nature. *See* ECOLOGY.

Continent One of the seven major unbroken land masses of the earth: Asia, Europe, Africa, North America, South America, Australia, and Antarctica. *See* CONTINENTAL DRIFT.

Continental drift The theory that the continents, during millions of years, have shifted their positions by 'drifting'. It suggests that originally there was only one huge continent and that in time it broke into several pieces. These pieces later drifted into the present positions of the continents. They drifted or 'floated' because being made of *sial* (rocks composed largely of SIlica and ALumina) they were lighter than the underlying *sima* (rocks consisting chiefly of SIlica and MAgnesia). Some modern studies have tended to support the theory. *See* CONTINENT.

Continental shelf Shelf round the coasts of the continents. It slopes gently, and is covered by shallow water. Its approximate edge is at the depth 100 fathoms (about 600 ft or 182 m). In places, the shelf extends outwards from the shore for several hundred miles. Beyond the continental shelf is the continental slope. *See* CONTINENT; CONTINENTAL SLOPE.

Continental slope Slope leading from the outer edge of the continental shelf to the deep ocean floor. Its slope is steeper than that of the shelf. *See* CONTINENT; CONTINENTAL SHELF.

Contour line Line on a map joining all places at the same height above (or depth below) sea level.

Convection The upward flow of air that has been heated by contact with the earth's surface. As it warms, it expands and rises. Cold air takes its place at the surface, and is in turn heated and caused to rise. *See* ADVECTION.

Coral reef A ridge of rock in the sea that is composed chiefly of the skeletons of small animals called *reef-building coral polyps*. The upper part of such reefs is usually at or near the surface. Coral reefs occur only in warm seas. *See* ATOLL.

Cordillera A group of mountain ranges, specifically the Andes in South America and other great ranges in North America. The word comes from the Spanish word meaning 'chain'.

Core The dense, innermost part of the earth, with a diameter of about 4,320 miles (6,950 km). There is evidence of a molten *outer core* and a solid *inner core*. *See* CRUST; MANTLE.

Coriolis Force An effect, caused by the rotation of the earth, on objects moving over the earth's surface. Such objects appear to be pushed to the right in the Northern Hemisphere and to the left in the Southern Hemisphere.

Corn Belt A region in the United States where three-quarters of the country's corn (maize) is grown; it lies to the S and SW of the Great Lakes.

Corrasion The wearing away of a river's bed by the abrasive action of solid materials carried in the water. *See* ABRASION; EROSION.

Corrosion The wearing away of rocks as the result of a chemical process. An example is the corrosion of limestone rocks by carbonic acid in solution. *See* EROSION.

Cosmic A term that describes anything occurring or located in outer space; for example, *cosmic rays*.

Cotton Belt A region of the south-eastern United States that grows more cotton than any other area. It includes 14 states: Alabama, Arizona, Arkansas, California, Georgia, Louisiana, Mississippi, Missouri, New Mexico, North Carolina, Oklahoma, South Carolina, Tennessee, and Texas.

Crater (i) The funnel-shaped depression at the top of a volcano. (ii) A hollow in the earth's surface where a meteorite has struck. *See* VOLCANO.

Crater lake A lake formed in the crater of a volcano. *See* CRATER; LAKE; VOLCANO.

Creek (i) A small stream. (ii) A small inlet or tidal estuary of a river.

Cretaceous Period A Period in the earth's history, the third and most recent Period of the Mesozoic Era. It began about 135 million years ago. *See* MESOZOIC ERA.

Crust The outer shell of the earth, about 10 to 30 miles (16 to 50 km) thick. It is also called the *lithosphere*. *See* CORE; MANTLE.

Crystal A solid body with smooth, flat surfaces and sharp edges and corners, whose atoms are arranged in regular patterns—for example, a diamond or a grain of salt.

Cumulonimbus A type of dense thundercloud, often with a very dark base, and sometimes with an anvil-shaped top. *See* ANVIL CLOUD; CLOUD.

Cumulus A white, detached, billowing type of cloud formed in upward, convection currents of air. *See* CLOUD; CONVECTION.

Current (i) The upward movement of an air mass. (ii) The distinct flow of water in the channel of a river or stream. (iii) The driving force of an ocean. Ocean currents are caused by the wind and the rotation of the earth, and flow both in deep and in shallow waters. They may be warm or cold, and have a marked influence on climate.

Cycle A series of stages or occurrences that are repeated in the same order. Specifically it refers to the *cycle of erosion*—the series of changes in a landscape from youth to age. The young, sharply-peaked mountains gradually wear away until only a smooth, slightly-undulating peneplain is left. *See* EROSION; PENEPLAIN; WATER CYCLE.

Cyclone A region of low atmospheric pressure in which winds spiral inwards towards the centre of lowest pressure. In the Northern Hemisphere, the winds blow anticlockwise; in the Southern Hemisphere, clockwise. In temperate regions cyclones are called *depressions*; in tropical regions they are more violent, and are called *typhoons* or *hurricanes*. *See* DEPRESSION; HURRICANE; TYPHOON.

Cylindrical projections Map projections based on the idea of rolling a paper cylinder or tube round a globe along the equator. The surface features of the globe are projected onto the cylinder, and the cylinder is then rolled out flat. On the map so formed, meridians and lines of latitude become straight lines that cross each other at right angles. As a result, all land lying near the Poles is much exaggerated in shape and size. *See* MERCATOR'S PROJECTION; PROJECTION.

Dam A barrier built across a river to hold back and control the flow of water. It may be made of earth, rock, or solid masonry. The water is usually used for irrigation or to produce electricity.

Dead reckoning Navigation of a ship solely by calculating its direction and the distance it has travelled.

Debris Loose surface material from the break up of rocks that has been moved from its original site; also called *detritus*.

Deciduous forest A cool temperate forest made up of broad-leaved trees such as oak, elm, and beech that shed their leaves each winter and renew them the following spring. In monsoon forests, trees shed their leaves during the hot season.

Declination (i) In astronomy, the angular distance of a heavenly body, such as a star or a planet, N or S of the equator. (ii) Magnetic declination is the angle between magnetic north (to which a magnetic compass needle points) and true north.

Deep *See* TRENCH.

Deep space Space beyond the boundaries of the Solar System. *See* SOLAR SYSTEM.

Degree (i) A measurement unit of temperature. It has different values, according to whether the scale used is Celsius or Fahrenheit. (ii) A unit of angular measurement equal to 1/360th part of a circle, used to measure latitude and longitude. *See* CELSIUS SCALE; FAHRENHEIT SCALE; LATITUDE; LONGITUDE; REAUMUR SCALE; TEMPERATURE; THERMOMETER.

Delta A fan-shaped area of land where a river enters a body of standing water. It is built up from alluvial material carried by the river and deposited at its mouth. It is named after *delta* (Δ), the fourth letter of the Greek alphabet, but not all deltas have this triangular shape. In time, a delta becomes crisscrossed by a network of small channels. *See* ALLUVIUM.

Denudation The wearing away of a land surface by natural outside forces such as wind, water, and ice. *See* ABRASION; ATTRITION; CORRASION; EROSION; WEATHERING.

Deposition The raising of the level of the land by the gradual building up of rock particles carried by wind, water, or ice. *See* EROSION.

Depression (i) A hollow or low-lying area in the earth's surface. (ii) A centre of low pressure, also called a *cyclone*, producing unsettled weather. *See* CYCLONE.

Desert A large area of land that lacks water. A desert region is sometimes defined as one having less than 10 inches (250 mm) of rain a year. Deserts include *hot deserts*, such as the Sahara, and *cold deserts*, such as the Gobi. *See* COLD DESERT.

Desiccation The gradual drying up of a region, either because of changes in climate or as a result of man's interference.

The Kariba Dam on the River Zambezi in Africa provides hydro-electric power for Rhodesia and Zambia. Lake Kariba, formed by the dam, is 175 miles (280 km) long.

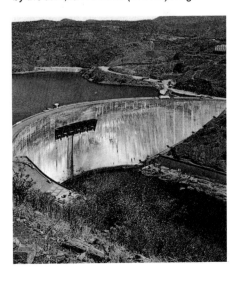

Devonian Period A Period in the earth's history sometimes called the *Age of Fishes.* The oceans covered the continents, leaving behind a thick layer of mud that eventually turned into rock. The Devonian was the fourth Period of the Palaeozoic Era. It began about 400 million years ago and lasted about 50 million years. *See* PALAEOZOIC ERA.

Dew point The temperature below which water vapour in the air begins to condense into droplets of water. *See* CONDENSATION; SATURATION.

Dike *See* DYKE.

Divide *See* WATERSHED.

Doldrums A low-pressure belt round the equator with light surface winds or calms. The doldrums are hot and often wet. Sailing ships avoided them because of the danger of being becalmed.

Down An upland tract of gently rolling pasture land, especially the chalk hills of southern England. *Downland* sometimes refers specifically to the grasslands of Australia and New Zealand. *See* GRASSLAND; PASTURE.

Dune. Shifting wastes of sand are fringed by steppe lands and hills in the Gobi Desert of Mongolia and China.

Drainage (i) The process of water loss from an area by means of streams that flow into a river. (ii) The drying-out of wet land in order to make it suitable for agriculture or building. Surplus water is led away in ditches or drainpipes. *See* BASIN; CATCHMENT AREA.

Drought An unusually long period of dry weather in a normally fertile area. As a result of a prolonged drought, crops may wither away and animals die.

Drowned valley A valley that has been filled with water from the sea, a river, or a lake. This may be the result of a raising of the water level, a lowering of the land level, or deliberate flooding to form a reservoir. *See* RESERVOIR.

Dune A hill or mound of sand formed by the wind blowing over beaches or deserts. Crescent or half-moon-shaped dunes are called *barkhans;* long dunes in parallel lines are called *seif-dunes.*

Dust Minute solid particles in the air, so small and light that they settle only when the air is still.

Dust bowl An area of drought or semi-drought and dust storms where the wind has blown or is blowing the topsoil away. Often sheltering plants have been destroyed by over-grazing or similar misuse of the land. Originally the name was given to a part of the Great Plains region of the central United States. *See* DROUGHT; DUST STORM.

Dust storm A storm in which a strong wind sweeps up the topsoil from the land, producing choking clouds of dust. *See* DUST BOWL.

Dyke, Dike (i) A ditch. (ii) A bank or wall of earth to protect low-lying land from being flooded. (iii) A sheet of igneous rock in molten form that cuts across layers of other rocks, forces its way to the surface through a crack, and becomes solid. *See* IGNEOUS ROCK; MAGMA; VOLCANO.

Earth A general term for the planet on which we live, the solid material of which it is made, or the loose material (soil) that lies on the surface over solid rock. *See* CORE; CRUST; MANTLE; SOLAR SYSTEM.

Earthquake A sudden rolling, shaking, or trembling of the earth's crust, caused by the breaking and movement of rocks or by volcanic action. Most earthquakes take place along belts where new mountains and volcanoes have been formed. The longest of these belts is the *circum-Pacific belt.* Another major belt is the *Alpine belt,* which runs through southern Europe and Asia. The majority of earthquakes occur under the ocean. *See* VOLCANO.

Easting Distance that a point on a map is eastwards from a given meridian, usually the meridian that is the origin of the map grid. *See* GRID; MERIDIAN; NORTHING.

Echo-sounder An instrument used at sea to measure the depth of water, or to locate shoals of fish or other underwater objects. It operates by sending sonic or supersonic vibrations through the water and recording the time that it takes an echo to bounce back.

Eclipse The obscuring of the whole or part of a heavenly body by another. A *lunar eclipse* occurs when the earth lies between the sun and the moon; a *solar eclipse* occurs when the moon comes between the earth and the sun.

Ecology The study of the relationship between living things (plants and animals) and their environment. It may be subdivided into human, animal, plant, and bio-ecology (the effect of animal and plant life on each other).

Economic geography A branch of human geography that is concerned with trade and commerce, industry, agriculture, and mineral and other resources. *See* GEOGRAPHY; HUMAN GEOGRAPHY.

E-Layer *See* KENNELLY-HEAVISIDE LAYER.

Environment The sum total of the conditions affecting the lives of people, animals, and plants. These conditions may be internal, such as body temperature and bacteria; or external, such as material surroundings and climate.

Eocene Epoch The second Epoch of the Tertiary Period in the history of the earth. It began about 50–60 million years ago at a time when the first types of modern mammals appeared. *See* TERTIARY PERIOD.

Epicentre A point or line on the earth's surface directly above the point of origin of an earthquake. *See* EARTHQUAKE.

Epoch A subdivision of a Period in geological time. Only the Tertiary and Quaternary Periods are subdivided in this way. *See* GEOLOGICAL TIME; TERTIARY PERIOD; QUATERNARY PERIOD.

Equator The (imaginary) great circle round the earth at right angles to the earth's axis, at latitude 0°, and midway between the North and South Poles. *See* AXIS OF THE EARTH; GREAT CIRCLE.

Equatorial rain-forest Dense, evergreen forest that grows in a region within about 7° N and S of the equator. This is an area of heavy rainfall, high temperatures, and almost no difference between the seasons. The result is a mass of many kinds of trees and plants all struggling for space and light. Trees appear to grow at three or four different layers or levels. Most trees are hardwoods. The forests of the Amazon and Congo (Zaire) basins are typical.

Equinox The time of the year when the sun is directly over the equator at noon. The word means 'equal night' and is a time when days and nights are of equal length in all parts of the world. The sun 'crosses the equator' twice a year so there are two equinoxes in a year: the *spring* or *vernal* equinox occurs about March 21, and the *autumn* or *autumnal* equinox about September 20.

Era A main division of geological time, subdivided into a number of Periods. The Eras are the *Pre-Cambrian, Palaeozoic, Mesozoic,* and *Cenozoic.* Each of them covers an immense number of years. *See* GEOLOGICAL TIME.

Erg The sandy part of the Sahara.

Erosion The wearing down of earth and rock, and the carrying off of the debris by wind, water, or ice. A similar process is called *weathering. See* ABRASION; ATTRITION; CORRASION; CORROSION; CYCLE; DENUDATION; WEATHERING.

Escarpment A steep slope or inland cliff, produced usually by erosion.

Estuary The broad V-shaped mouth of a river, formed generally by the sinking or washing away of the coastal land. When the tide flows in, river and sea water mix. *See* BORE; DELTA; DROWNED VALLEY.

Evaporation The change of a liquid into a vapour (gas). The water vapour in the atmosphere results from the evaporation of surface water on the earth. The rate of evaporation depends on the temperature of the air, its movement, and the amount of water vapour it already holds. *See* CONDENSATION; DEW POINT.

Everglade A swampy region with tall grass, a few trees, and decayed plant life underfoot. The name is applied particularly to an area of more than 5,000 sq miles (12,950 sq km) of such vegetation in southern Florida, United States.

Exosphere The fourth and topmost layer of the atmosphere. It starts about 300 miles (480 km) up from the earth and has almost no air molecules. *See* ATMOSPHERE; ATMOSPHERIC PRESSURE; IONOSPHERE; STRATOSPHERE; TROPOSPHERE.

Fahrenheit scale A scale used on thermometers, the freezing point of water being 32° and the boiling point 212°. To convert Fahrenheit into Celsius, first subtract 32 from the Fahrenheit reading, then multiply the result by 5/9. *See* CELSIUS SCALE; RÉAUMUR SCALE; TEMPERATURE; THERMOMETER.

Fathom A unit of length used mainly to measure water depth; equal to 6 ft (1.8 m).

Fault An enormous crack or break in the rock strata of the earth's crust, caused by horizontal or vertical movement.

Fauna The animal life of a specific geological period or of a region of the earth. The corresponding term for plant life is *flora*.

Fertilizer Plant food, either natural or artificial, used as an additive to increase the fertility of the soil. Fertilizers contain nitrogen, phosphorus, potassium, or other elements necessary for plant growth.

Fiord, Fjord A long, narrow inlet of the sea, usually enclosed by steep cliffs. The fiords of southern Norway are famous for their beauty. Probably, they are the result of erosion by glaciers moving seawards. *See* GLACIER.

Flint A form of silica, a hard, dark grey mineral. It is usually found in chalk or limestone. In the Stone Age, men shaped flints into implements and weapons by chipping them. Flint was also once valued as a 'sparking stone' for making fire.

Floe A horizontal sheet of floating ice that has become detached from the polar ice mass. *See* ICEBERG.

Flood The inundation by water of an area that is normally dry. It may be caused by a sudden rise in the sea level, or by a change in the water level of a river or lake because of abnormal rains or the melting of snow.

Flood plain Low-lying land by the side of a river that is regularly flooded when the river overflows its banks. The river deposits alluvium and gradually builds up the flood plain. *See* ALLUVIUM; FLOOD.

Flora The plant life of a specific geological period or of a region of the earth. The corresponding term for animal life is *fauna*.

Fluvial Of or belonging to a river.

Fog Low cloud that forms near the surface of the land or on a body of water. It occurs when warm, moist air passes over a cold surface, causing the moisture to condense. *Advection fogs* form when warm air touches cold areas of land or water. *Radiation fogs* occur when ground warmth is released quickly into the upper atmosphere and the air close to the ground is chilled as a result. In the International Meteorological Code, a fog exists when objects 1 kilometre away can no longer be seen. *See* CLOUD; CONDENSATION.

Föhn A warm and dry wind that blows down the slope of a mountain on the side away from windward. It is common in the valleys of the northern Alps. In North America it is called a *chinook*.

Fold, Folding The bending of rock strata, usually molten strata, by pressures beneath the earth's crust. Folding usually takes place along lines of weakness.

Foreland A headland jutting out into the sea. *See* CAPE; POINT.

Forest A large area of unbroken woodland. *See* CONIFEROUS FOREST; DECIDUOUS FOREST; EQUATORIAL RAIN-FOREST.

Fossil The impression or remains of an animal or plant that lived during some earlier age, preserved in sedimentary rocks, in ice, or in amber. A fossil may be the hard remains, or a replacement by minerals of the substance of the animal or plant, retaining the original shape. Or the fossil may be a mould of, for example, a footprint or the outline of a leaf.

Freezing-point The temperature at which a liquid becomes cold enough to turn into a solid. Each liquid has its own freezing-point. Water freezes at 0°C (32°F). *See* CELSIUS SCALE; FAHRENHEIT SCALE; FROST; ICE; RÉAUMUR SCALE; TEMPERATURE.

Frigid zone The climatic zone of cold winters and cool summers in polar regions. *See* PERMAFROST; TEMPERATE ZONE; TORRID ZONE.

Front The boundary between cold and warm air masses at the earth's surface. *See* COLD FRONT; OCCLUDED FRONT; WARM FRONT.

Frontier A narrow tract of country on either side of the boundary line between two nations. Sometimes it is confused with *boundary*, which is the actual line of demarcation.

Erosion. Ayers Rock, a giant eroded outcrop of sandstone in central Australia, rises 1,110 ft (335 m) above the plain. Depending on the light, its colour varies from purple to red or orange.

Frost Thousands of tiny crystals of frozen moisture that form on solid objects when the temperature falls to 0°C (32°F) or below. *See* FREEZING-POINT; TEMPERATURE.

Fumarole A hole or vent in the ground in volcanic regions through which fumes or gases escape.

Gale A strong wind that blows at speeds of between 39 and 63 mph (62–101 kph)—numbers 8, 9, and 10 on the Beaufort Wind Scale. *See* BEAUFORT WIND SCALE; WIND.

Gap A break in a ridge through which a river sometimes flows. A dry gap is known as a wind-gap. *See* GATE.

Gate (i) A wide gap in a ridge. (ii) An opening or valley in a low range of hills. (iii) An entrance to a bay between headlands.

Geest An area of heath-covered coarse sand and gravel, found in the northern lowlands of Germany and neighbouring regions of Poland, the Netherlands, and Denmark.

A rocky foreland, part of the bleak Cliffs of Moher on the coast of Clare, in Ireland.

The Great Geysir, north of Reykjavik in Iceland, has given its name to all other geysers. Today, its activity is reduced, but the dozens of similar hot springs nearby include the largest geyser in the world.

Geodesy A branch of mathematics that deals with the shape, size, weight, and curvature of the earth. Geodetic surveys determine exact points on the earth's surface, and measure areas large enough to take the curvature of the earth into account.

Geography The study of the earth's surface as the home of mankind. It describes its physical features, resources, climate, soils, plants, animals, and peoples and their distribution. It draws largely from the natural sciences and social sciences, such as geology, meteorology, botany, zoology, economics, and history. *See* BIOGEOGRAPHY; ECONOMIC GEOGRAPHY; HUMAN GEOGRAPHY; PHYSICAL GEOGRAPHY; PHYSIOGRAPHY; PHYTOGEOGRAPHY; REGIONAL GEOGRAPHY; SOCIAL GEOGRAPHY; ZOOGEOGRAPHY.

Geoid A term—meaning 'earth-shaped body'—that refers to the general shape of the earth. Although the earth is spherical, it is not a true sphere (globe or ball) because it bulges slightly towards the equator.

Geological time The earth's history described chronologically from the Pre-Cambrian Era, through the Palaeozoic and Mesozoic Eras, to the Cenozoic Era. The chief divisions of geological time are Eras, Periods, and Epochs. Broadly speaking, each Era ended in a time of orogeny—active mountain building. *See* EPOCH; ERA; OROGENY; PERIOD.

Geology Earth science; the study of the history, build-up, and materials of the earth and of the forces that act upon it. More specifically, it deals with the history and structure of the earth's crust. *See* GEOPHYSICS.

Geomorphology *See* PHYSIOGRAPHY.

Geophysics The physics of the earth. It is sometimes thought of as including *seismology, geodesy, meteorology,* and other sciences. *See* GEODESY; METEOROLOGY; SEISMOLOGY.

Geosyncline An enormous downward curve or inverted arch in the earth's crust, generally many miles wide and hundreds of miles long—a *syncline* on a very large scale. Marine sediments that settled at the bottom of the curve have resulted in the formation of thick deposits of sedimentary rock. *See* SYNCLINE.

Geyser A hot spring that periodically throws steam and hot water up into the air with explosive force. Geysers are common in the north island of New Zealand, in Iceland, and in Yellowstone National Park in the United States. *See* HOT SPRING.

Glaciation The covering of part of the earth's surface by an ice sheet or a glacier. Major glaciation occurred in northern Europe and northern North America during the Pleistocene Epoch. *See* ICE AGE.

Glacier A very slow-moving river of ice that creeps down a valley from a snowfield above the snow line. It moves under the influence of gravity.

Globe Another name for the earth; a sphere with a map on it, representing the earth.

Gnomonic projection A type of azimuthal map projection used for ocean charts, radio work, and plotting earthquakes. It is also used for star maps. It is made by projecting a part of the globe onto a plane that is tangent to it. Great circles become straight lines, but shapes are badly distorted. *See* AZIMUTHAL PROJECTIONS; PROJECTION.

Gradient (i) A slope whose steepness is expressed either in degrees as an angle to the horizontal, or as a proportion between the vertical interval and the horizontal. (ii) The slope of the bed of a stream, expressed usually in ft per mile or cm per km.

Granite A hard, coarse-grained, light-coloured igneous rock made up of feldspar, quartz, and mica or hornblende. *See* IGNEOUS ROCK; QUARTZ.

Grassland A plain where grass is the natural and main vegetation. Grasslands do not receive enough rainfall for extensive tree growth, but are wetter than deserts. *See* DOWN; PAMPAS; PLAIN; PRAIRIE; STEPPE.

Graticule The network of meridians and parallels (lines of latitude) drawn on a map. *See* GRID; MERIDIAN; PARALLEL OF LATITUDE.

Gravel Deposit of small rounded stones often mixed with clay or sand. The stones are water-worn because they were laid down a long time ago by streams, glaciers, or the action of the sea.

Gravity, Gravitation The force that attracts masses of matter to each other. Gravitation holds the universe together. The moon's and sun's gravity causes the tides on earth. Newton's *law of gravitation* states that all bodies attract each other, and that the force of their attraction depends on their size and the distance between them.

Great Circle An imaginary line drawn right round the earth, with its plane passing through the earth's centre. The equator is a great circle. Each two corresponding meridians together form a great circle.

Green belt A strip of open country surrounding a town or lying between towns, with the aim of preventing over-building. Green belts include parks, playing fields, and unspoilt countryside, and are kept free of houses, factories, and other development.

Greensand A kind of sand or sandstone that contains glauconite—a greenish mineral formed in the sea.

Greenwich Mean Time (G.M.T.) A reckoning of time that is known as *standard time* in the British Isles and Western Europe. The sun's crossing of the meridian 0° is taken as 12 noon. *See* GREENWICH MERIDIAN; STANDARD TIME.

Greenwich meridian The meridian 0° that passes through Greenwich in England, the former site of the Royal Observatory. Longitude is expressed in degrees E or W of this meridian. *See* GREENWICH MEAN TIME; STANDARD TIME.

Grid Horizontal and vertical lines drawn on a map to form a pattern of squares. Each line is numbered, and any point on the map can be referred to by giving its *co-ordinates*—its position in relation to nearby lines. The vertical lines are called *eastings*, and the horizontal lines *northings*. *See* EASTING; GRATICULE; NORTHING.

Groyne A barrier or fence built out into the sea from the shore in order to stop the drift of sand or shingle.

Glacier. Above: Monte Rosa, on the Italian border. Below: Lauterhorn, in Switzerland.

Gulf A large, deep sea inlet extending far inland. *See* BAY.

Guyot A round, flat-topped mountain on the bed of the ocean. Guyots are probably of volcanic origin, and rise from the sea bed to within half a mile (0.8 km) of the surface. Many of them have been identified in the Pacific Ocean.

Gyro-compass A compass based on the use of a spinning gyroscope. The axle of the gyroscope stands parallel to the earth's axis, thus indicating true north. Unlike an ordinary compass, the gyro-compass makes no use of the earth's magnetism. *See* COMPASS.

Habitat The natural environment of any particular animal or plant.

Hail, Hailstones Hard globules of ice that fall from cumulonimbus clouds, often during thunderstorms. As moist air rises and freezes, ice globules form and increase in size as more water vapour freezes onto them. When too heavy to be sustained by upward air currents, they fall as a hailstorm. *See* CUMULONIMBUS.

Hamada The name of large areas of the Sahara, especially in Libya, where the surface is bare and rocky. *See* DESERT.

Harmattan A strong, hot, dusty wind that blows over parts of western Africa from the Sahara. *See* WIND.

Haze A cloud of dust, smoke, salt, or other particles that reduces visibility close to the earth's surface. A haze is said to exist when visibility is less than 1.25 miles (2 km) but more than 0.6 miles (1 km). *See* FOG.

Headland *See* CAPE.

Headwater The source of a river or a river system.

Heath, Heathland A stretch of open, uncultivated country covered with low shrubs such as ling.

Heat wave A spell of weather that is unusually hot and dry for the area in which it occurs.

Heaviside-Kennelly Layer *See* KENNELLY-HEAVISIDE LAYER.

Hemisphere One half of a sphere or globe. On the earth's surface, the half north of the equator is called the *Northern Hemisphere;* the half south of the equator, the *Southern Hemisphere.*

Highland Another term for a high upland. Sometimes the word is used in the plural as a proper name, as in the *Highlands of Scotland. See* UPLAND.

High seas The open sea or ocean outside any country's territorial waters.

High water The highest point reached by the tide; the time at which the tide is highest. *See* TIDE.

Hinterland Region lying inland from a coast or port.

Horizon The line where earth and sky seem to meet, and beyond which the observer cannot see because of the curvature of the earth.

Horse Latitudes Subtropical zones of high pressure on either side of the equator close to latitudes 30°N and 30°S. They lie between the trade winds and the westerlies and are marked by light winds or calms, and light rainfall. *See* TRADE WINDS; WESTERLIES.

Hot spring A continuous stream of hot water that bubbles to the surface from underground in volcanic regions. It differs from a geyser in that the water does not come out with explosive force. Such a spring is also called a *thermal spring. See* GEYSER.

Human geography Geographical studies concerned with man and his activities in relation to the earth—including race, settlement, migration, distribution, work, and commerce. *See* ECONOMIC GEOGRAPHY; GEOGRAPHY.

Humidity The amount of water vapour in the air. The air can hold only a certain amount of water vapour at any given temperature. The warmer it is, the more it can hold. When it can hold no more it is said to be *saturated.* The amount of water vapour in the air compared with the amount the air would hold when saturated is the *relative humidity. See* CONDENSATION; HYGROMETER; SATURATION.

Hurricane (i) A severe tropical cyclone that blows occasionally over the West Indies and the Gulf of Mexico. Hurricanes originate in mid-Atlantic, and generally move first westwards and then north-eastwards. (ii) A wind *Force 12* or more on the Beaufort Wind Scale. *See* BEAUFORT WIND SCALE; CYCLONE; DEPRESSION; TYPHOON; WIND.

Hydro-electric power Electricity produced by water power. Waterfalls, natural or man-made, turn turbines and dynamos, thus generating electricity. *See* DAM.

Hydrography The science that deals with the surface water of the earth, including the charting and sounding of the seas, the investigation of tides and ocean currents, and the surveying of coastlines. It makes important contributions to navigation.

Hydrological cycle *See* WATER CYCLE.

Hydrosphere The waters of the earth, which cover two-thirds of its surface. The hydrosphere includes seas, lakes, rivers, ice and snow fields, and the water vapour in the atmosphere.

Hygrometer An instrument that measures relative humidity. One kind is based on the use of some material—such as human hair —that expands and contracts as humidity changes. Another kind has two thermometers—a dry-bulb that records the air temperature, and a wet-bulb that is cooled by evaporation and records a lower temperature. Relative humidity is worked out from the difference between these two readings. *See* HUMIDITY; THERMOMETER.

Ice The solid state of water, formed when the air temperature falls to freezing-point or below. Ice floats on water because it is less dense. *See* FREEZING-POINT; FROST.

Ice Age A time when much of the earth's surface was covered with ice. The term commonly refers to the latest Ice Age, the Pleistocene Epoch, which began about 1 million years ago. During this Epoch much of North America and northern Europe was covered with ice. *See* GLACIATION; PLEISTOCENE EPOCH.

Iceberg A huge mass of ice floating in the sea after having broken off from an ice barrier or the end of a glacier. Only about one-ninth of an iceberg can be seen above the water. *See* FLOE.

Ice-cap A mass of ice covering land in polar regions. Huge ice-caps, such as those that cover Antarctica and much of Greenland, are sometimes called *ice-sheets.*

Ice-field A large continuous sheet of ice; a very large floe. *See* FLOE.

Igneous rock Rock that has solidified from molten rock or *magma.* It is one of the three main kinds of rocks that make up the earth's crust. *See* MAGMA; ROCK.

IGY (International Geophysical Year) A period during which scientists from 66 nations worked together on various projects connected with the earth and its surround-

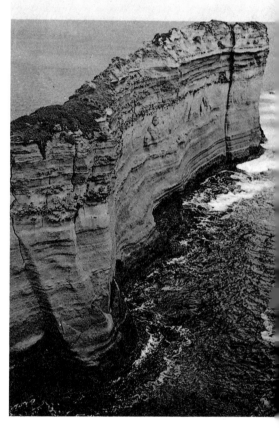

Granite. A towering granite 'peninsula' on the coast of Victoria in Australia.

ings. The IGY began officially on July 1 1957, and ended on December 31 1958.

Impermeable rock Non-porous rock that therefore does not allow fluids to seep through it.

Inland sea A sheet of water, larger than a lake, that is not connected with the open sea—for example, the Dead Sea. *See* LAKE.

Inlet A narrow bay. *See* BAY; GULF.

Insolation Radiant energy from the sun received by the earth and its atmosphere in a spectrum of radiation. About half is absorbed by the earth; the remainder is reflected back or absorbed by the atmosphere. *See* RADIATION.

Island. Above: A 'dream island' in Polynesia, in the Pacific. Below: An island in the River Rhine, at Kaub in Germany. The toll-collecting castle of Pfalzgrafenstein was built in 1327.

International Date-Line An imaginary line that roughly follows the 180° meridian in the Pacific Ocean. E and W of the Date-Line the date differs by one day. The line zigzags in places to avoid land areas. Each 15° longitude means one hour's difference in time (the earth turns through 380° in 24 hours). To make up for the total difference of 24 hours between one side of the line and the other, ships sailing E (or planes flying E) alter the date to gain a day, and those sailing W lose a day.

Inversion of temperature An increase of air temperature with increase in altitude, unlike the usual decrease in temperature with increasing height. Inversion sometimes occurs in a valley on a cold night. The air loses heat rapidly by radiation. It becomes dense and sinks to the valley floor. On the slopes above, the air is still warm. Temperature inversions also occur high in the atmosphere. *See* RADIATION; TEMPERATURE.

Ionosphere The layer of the atmosphere that lies between the stratosphere and the exosphere; it extends from about 50 miles to 300 miles (80 km to 480 km) above the earth. *See* ATMOSPHERE.

Irrigation The artificial channelling of water to dry land in order to make it suitable for agriculture. This may be done by controlling and diverting flood water, as on the banks of the Nile; or by damming a river and drawing water when necessary from the artificial lake thus created. *See* DAM.

Island A piece of land, smaller than a continent, entirely surrounded by water.

Isthmus A narrow bridge of land joining two large land areas—for example, the Isthmus of Panama, which joins South America to Central America.

Jet stream A westerly current of air moving in the upper atmosphere at speeds of between 60 and 270 mph (100 and 430 kph). Such air streams occur at heights of about 40,000 ft (12,000 m).

Jungle Originally a term meaning wild, rough land. Now the word popularly denotes any wild tangle of vegetation, especially a tropical rain-forest.

Jurassic Period The middle Period of the Mesozoic Era, lasting from about 180 million to 135 million years ago. It is known as the *Age of Great Reptiles. See* MESOZOIC ERA.

Karst A rugged type of landscape made up of barren limestone, such as occurs in north-western Yugoslavia, the northern Pennines in England, and the Yucatan region of Mexico. Its main features are dry valleys, sink-holes, caverns, and underground streams.

Kennelly-Heaviside Layer The lowest section of the ionosphere, lying 60 to 75 miles (96 to 120 km) above the earth; also called the *E-layer*. It reflects long-wave radio signals back to earth. *See* IONOSPHERE.

Khamsin A hot, dry, and generally dusty wind that blows across parts of Egypt and the south-east Mediterranean. Elsewhere, the wind is often called the *sirocco. See* SIROCCO.

Kilometre A measure of length. It equals 1,000 metres or 0.621 statute miles.

Kopje An Afrikaans word for a small isolated hill in southern Africa.

Lagoon (i) A shallow stretch of water separated completely or partly from the sea by a narrow strip of land. (ii) A stretch of water enclosed by an atoll. (iii) A stretch of water lying between a coral reef and the mainland. *See* ATOLL; CORAL REEF.

Lake A body of water surrounded by land. Very large lakes are called *inland seas*. The hollows that fill with water to form lakes may be caused by erosion, deposition, volcanic action, or by faults in the earth's crust. *See* CRATER LAKE; DEPOSITION; EROSION; INLAND SEA.

Lake dwelling A prehistoric house built on stilts in a lake or other standing water in various parts of Europe. The dwelling stood on a wooden platform which in turn was supported by wooden piles driven into the mud.

Land breeze A breeze blowing from land to sea at night as a result of the fact that land heats and cools more rapidly than the sea. Faster cooling of the land at night causes higher pressures over the land, so that air flows out to the sea.

Landslide, Landslip The slipping of large masses of rock and earth down the side of a mountain or hill. It may result from the action of water seeping into the earth, earthquakes, waves undermining a sea coast, or cracks in the rocks. *See* AVALANCHE.

Laterite A reddish, porous rock, rich in oxides of iron and aluminium. It is soft and plastic when quarried, but hardens when exposed to the air. It occurs in parts of South America, eastern and western Africa, Sri Lanka, and southern India.

Latitude Angular distance in degrees N or S of the equator of any point on the earth's surface. Latitude is measured along an imaginary north-south line called a *meridian*. The latitude of the equator is 0°. *See* MERIDIAN; PARALLEL OF LATITUDE.

Lava White-hot molten rock or magma thrown out by a volcano. *See* MAGMA; VOLCANO.

Lee, Leeward The side away from or protected from the wind.

Levee A ridge or bank along the edges of a river's channel. When a river floods, it deposits material on its banks. This ridge of material grows higher with each flooding and is raised above the level of the surrounding flood-plain. Sometimes, artificial levees are constructed of earth and sandbags to contain a river during flooding.

Lightning Static electricity in the sky. It is an enormous electric spark caused by a discharge between positively-charged and negatively-charged areas within clouds, between clouds, or between a cloud and the earth. *See* THUNDERSTORM.

Light-year The distance light travels in one year—a measure often used in astronomy. It is equal to 5,880,000 million miles (9,463,000 million km).

Lignite A soft, brownish-black coal with more moisture and less carbon in its make-up than domestic coal. The formation of lignite is a stage in the process from vegetable matter to (black) coal.

Limestone A sedimentary rock consisting of at least 50% calcium carbonate ($CaCO_3$). Some types of limestone formed from the hard remains of tiny sea creatures. Others formed when minerals held in solution were deposited as the water evaporated. *See* CHALK.

Lithosphere *See* CRUST.

Littoral The seashore, or the stretch of land between high and low water marks.

Llanos A Spanish term meaning 'level lands'; specifically the huge tropical grasslands of the Orinoco basin, in Venezuela and Colombia. *See* GRASSLAND.

Loch A Scottish lake, or a long, narrow arm of the sea such as is common along the coast of north-western Scotland.

Lode A thick vein of mineral ore in rock, or a number of closely parallel veins.

Loess Fine, clinging, yellowish dust that has been deposited as a soil by the wind. It comes from deserts and from areas that were once covered by ice sheets. Loess soils are highly-valued for agriculture because they are fertile and well-drained.

Longitude The angular distance measured in degrees east or west of the prime meridian—0°, the Greenwich meridian. *See* GREENWICH MERIDIAN; LATITUDE; MERIDIAN; PARALLEL OF LATITUDE.

Long-range forecast A weather forecast for a period longer than 5 days. In some countries, such forecasts are issued each month for the month ahead. *See* WEATHER.

Lough The Irish name for a lake or an arm of the sea; it corresponds to the Scottish *loch*. *See* LOCH.

Low, Atmospheric *See* DEPRESSION.

Lucerne A perennial fodder plant of the pea family. Its ability to withstand heat, cold, and drought and its very high yield make it a valuable food for farm animals. It is also called *alfalfa*.

Lunar day (i) The period of time taken by the earth to rotate once in relation to the moon. This is about 24 hours 50 minutes; because while the earth rotates, the moon is also orbiting about the centre of gravity of the earth and the moon. (ii) A day on the moon, which is about 336 hours long.

Maelstrom (i) A whirlpool. (ii) A powerful current that sweeps between two of the Lofoten Islands off the north-western coast of Norway.

Magma Molten rock at a very high temperature and pressure lying under the surface of the earth. Igneous rocks are formed from magma. Magma loses much of its water and gas when thrown out by volcanoes as *lava*. *See* IGNEOUS ROCK; LAVA; VOLCANO.

Magnetic Pole One of two poles of the earth's magnetic field to which the free-swinging magnetic needle of a compass points. The North magnetic Pole is near Prince of Wales Island, in northern Canada. The South magnetic Pole is in Wilkes Land, in Antarctica. The magnetic poles are not true N and S. The North magnetic Pole is about 1,000 miles (1,600 km) from the North Pole; the South magnetic Pole about 1,500 miles (2,400 km) from the South Pole. *See* COMPASS; POLE.

Mallee Low eucalyptus scrub that covers much of subtropical south-eastern and south-western Australia.

Mantle A dense layer of rock lying between the earth's crust and core. It is about 1,800 miles (2,900 km) thick. *See* CORE; CRUST.

Map A picture on a flat surface of part or all of the earth's surface, drawn to a definite scale. Certain features (political, physical, etc.) will be shown and others left out, according to the purpose of the map. *See* MAP READING; PROJECTION.

Map reading The art of 'reading' the signs and symbols of a map and, with the help of a compass, of finding one's way from one place to another.

Maquis A low, shrubby, scented kind of vegetation that grows in the Mediterranean region. It is made up of such plants as myrtle, lavender, and heath.

Maritime climate Climate common on islands and on the coasts, particularly the western coasts, of the continents. The moderating influence of the sea gives a relatively mild winter and a warm summer. Rainfall is fairly heavy.

Marsh A low-lying region of soft, wet land, treeless and partly or completely under water. *See* BOG; SWAMP.

Massif A central mountainous mass, such as the Massif Central of France.

Meadow An area of grassland mown for hay. A *water-meadow* lies on the flood plain of a river and is regularly flooded. *See* PASTURE.

Mean sea level The average level of the sea. It is calculated by taking readings at various places over a long period.

Mediterranean climate A climate marked by hot, dry summers and moist, warm winters. It occurs on the western edges of continents, latitudes 30° to 40°.

Mercator's projection A type of cylindrical projection in which the distances between parallels of latitude widen as they approach the poles, but the distances between meridians remain the same. It is useful for navigators because it shows true directions. It was invented in the 1560s by Gerhardus Mercator, a Flemish geographer. *See* CYLINDRICAL PROJECTIONS; PROJECTION.

Meridian A line of longitude; a half circle of the earth's surface from the North Pole to the South Pole. *See* LONGITUDE.

Mesa A steep-sided, flat-topped upland area that is found in arid regions, specifically in the western United States. It is the remains of a plateau that has been denuded. The Spanish word *mesa* means 'table'. *See* PLATEAU.

Mesozoic Era The time in the earth's history between the Palaeozoic Era and the Cenozoic Era, also called the *Age of Reptiles*. It occurred between 225 million and 70 million years ago. It is divided into the Triassic, Jurassic, and Cretaceous Periods.

Lake. Above: The 'dark lake', Scutari, between Albania and Montenegro in Yugoslavia. Top: An artificial lake in Austria, formed by the building of dams.

Lava. The beautiful but fearsome sight of a river of lava, destroying everything that will burn.

National Park. Above: A lion ignores intruders as it enjoys a meal in a national park in Kenya. Below: Peace and beauty in a national park in Yugoslavia.

Metamorphic rock One of the three main kinds of rocks that make up the earth's crust. Originally igneous or sedimentary rock, it has been metamorphosed (transformed) by pressure, heat, and chemical action. *See* IGNEOUS ROCK; ROCK; SEDIMENT.

Meteor A metallic or stony body that enters the earth's atmosphere from outer space. Meteors become visible when friction with the air makes them glow. They are also called *shooting stars. See* METEORITE.

Meteorite A meteor that hits the earth instead of burning up in the atmosphere. *See* METEOR.

Meteorology The scientific study of the earth's atmosphere and weather. Among other things it enables the experts to forecast the weather. *See* WEATHER.

Mid-latitudes *See* TEMPERATE ZONE.

Midnight Sun The sun that is visible at midnight during the summer months in latitudes greater than 63.5°. It never sinks below the horizon, and shines 24 hours a day.

Migration The mass movement from one region to another of certain birds and other animals, or people. Some animals (chiefly birds) migrate seasonally.

Mile A measure of length. (i) *Statute mile* equals 1,760 yards or 1.609 kilometres. (ii) *Geographical mile* equals one-sixtieth of a degree or one minute of latitude. This is

taken to mean about 1,850 metres at the equator.

Millibar A unit of pressure used by meteorologists when preparing their weather charts; 1,000 millibars=1 bar, or 29.53 inches (750 mm) as indicated on a barometer. *See* BAR.

Mineral (i) An inorganic substance that occurs in nature and that possesses distinct chemical and physical properties. Most minerals are crystalline in structure. Rocks consist of various mineral particles. (ii) Any substance—such as coal, gold, or oil—that is mined.

Mineral water Water that holds some mineral substance in solution, often drunk as a medicine or tonic.

Minute A unit of measurement. It may be one-sixtieth of an hour, or one-sixtieth of a degree of angle. In measurements on the earth's surface, a minute is a distance corresponding to one-sixtieth of a degree of latitude or longitude.

Miocene Epoch An Epoch in the earth's history from about 25 million to about 10 million years ago. It was the fourth Epoch of the Cenozoic Era. During it, there was much earth movement and erosion. *See* CENOZOIC ERA; TERTIARY PERIOD.

Mirage A type of optical illusion caused by air conditions. In hot deserts, for example, the hot air near the ground is less dense than air above it. Refraction of light rays through this layer can give the distant viewer an impression of a lake. Mirage can also cause the apparent transposition of distant objects.

Mistral A very cold wind that sweeps down the Rhône Valley from a high pressure area over the central plateau of France to a low pressure area over the Mediterranean.

Mole A breakwater or stone pier designed to shelter a harbour. *See* BREAKWATER.

Monsoon A wind system that reverses its direction from season to season. More specifically, the term applies to the seasonal winds over the Indian Ocean, blowing from the SW from April to October and from the NE from October to April. The SW winds bring torrential rains to India and neighbouring areas.

Moon (i) The earth's only natural satellite, which travels round the earth once every lunar month. It is a dead world without water

or atmosphere, and its distance from the earth varies between 221,000 and 253,000 miles (356,000 and 407,000 km). Its mass is less than one-eighteenth that of the earth and its diameter about one quarter (2,160 miles; 3,476 km). Man first set foot on it in 1969. (ii) In a general sense, the satellite of any planet.

Moorland A wild, open, upland area of land, often covered with grass and bracken.

Moraine Large quantities of rock debris carried along and deposited by a glacier. The material deposited at either side is called a *lateral moraine*; that at the bottom of the valley forming the glacier's course, a *ground moraine*; and that at the melting front of the glacier, a *terminal moraine. See* GLACIER.

Mountain A high hill, or any defined mass of land that rises above the surrounding region, and that has steep sides.

Mulga Scrub composed mainly of species of acacia, which is the characteristic vegetation on the edges of the deserts of central Australia. *See* SCRUB.

Nadir The point in space opposite the zenith, directly downward from the zenith. *See* ZENITH.

Narrows The narrow part of a river, a pass, a strait, or a valley.

Nation A large group of people united by attributes they have in common—language, origin, culture, or history. A nation may be an independent state, such as Italy or Britain. But some national groups are not independent. The Basques, for example, live in France and in Spain.

National park An area of natural beauty or other special interest that has been set aside for protection. Some national parks have the object of preserving wild life. Usually, national parks are open to the public.

Navigation The art of finding one's way from place to place, keeping on a chosen course, and knowing where one is all the time. Expert navigation is especially important for ships and aircraft. *See* CHART; COMPASS; DEAD RECKONING; MAP READING; SEXTANT.

Neap tide The tide with the smallest range between low water and high water. It occurs when the gravitational pull of the sun is at right-angles to that of the moon. *See* SPRING TIDE.

Neogene The Miocene and Pliocene Epochs.

Neolithic A period of time in the history of man commonly known as the *New Stone Age*. Men still used stone tools, but they began to farm, to weave, and to make pottery. It started about 6000 B.C. and lasted until the discovery of metals in about 3000 B.C.

Nimbostratus A type of low cloud that forms dark grey, ragged sheets. It often brings heavy rain or snow. *See* CLOUD.

Nimbus A term that means a *rain cloud*. It forms part of the composite names of some clouds to indicate that they are rain-bearing—for example, *cumulonimbus. See* CLOUD.

Northing The horizontal lines of a map grid, indicating the northerly progression from the point of origin of the grid. *See* EASTING; GRID.

Nullah In India, the normally dry bed of a stream that temporarily fills with water after heavy rainfall. *See* WADI.

Oasis A spot in a desert where there is water, and as a result vegetation grows and permanent settlement is possible. The water may be from a spring or there may be a stream. Oases are most numerous in areas of artesian basins. Desert routes lead from one oasis to the next. *See* ARTESIAN BASIN; DESERT.

Obsidian A glassy volcanic rock.

Occluded front A front formed in a depression when a cold front overtakes a warm front and lifts it off the earth's surface. The horizontal movement of the warm front is occluded (shut off). *See* COLD FRONT; DEPRESSION; WARM FRONT.

Ocean One of the great sheets of salt water that surround the continents and make up more than 70% of the earth's surface. The five oceans are: the Pacific, Atlantic, Indian, Arctic, and Antarctic or Southern oceans. *See* SEA.

Oceanography The scientific study of the oceans. *See* OCEAN.

Oligocene Epoch An Epoch in the earth's history that lasted from about 40 million to 25 million years ago. It was the third Epoch of the Cenozoic Era. *See* CENOZOIC ERA; TERTIARY PERIOD.

Orbit of the earth The path taken by the earth round the sun. It is elliptical. On 4 July (called *aphelion*) the earth is about 94 million miles (151 million km) from the sun; on 3 January (*perihelion*), about 91 million miles (147 million km). The earth takes 1 year to complete its orbit at a speed of about 66,000 mph (106,000 kph).

Ordovician Period A division of time in the earth's history, the second Period of the Palaeozoic Era. It lasted from about 500 million to 440 million years ago. *See* PALAEOZOIC ERA.

Oolitic A type of limestone rock made up of small rounded grains called *oolites*, which look rather like fish roe.

Ore A mineral deposit that contains enough metal to make it economically valuable. *See* MINERAL.

Orientation Aligning a N—S line on a map with an imaginary N—S line on the ground in order to identify the features shown on the map.

Orogeny A major phase of mountain building. It involves the formation of mountain ranges as a result of rocks being thrown up into folds or blocks. *See* FOLD; GEOLOGICAL TIME.

Orographic rainfall Rain caused by the rising of moisture-laden air over a mountain range. The air is cooled and releases some or all of its moisture as rain.

Orrery A working model that shows the movements of the planets. It is made up of a central ball representing the sun, and lesser globes to represent the planets. The device is run by clockwork.

Orthographic projection A type of azimuthal map projection showing the globe as viewed from a distance. This means that only one hemisphere can be seen. The scale is accurate only in the centre of the projection. It is used mainly for astronomical charts. *See* AZIMUTHAL PROJECTIONS.

Orthomorphic projection A type of map projection in which the shape of a small area on the earth is faithfully copied on the map. It is also called *conformal projection. See* PROJECTION.

Outcrop An exposed piece of bedrock.

Pack ice Large masses of ice floating on the sea, the result of the breaking up of an ice-field by wind and wave. *See* ICE-FIELD.

Padang In South-East Asia, a treeless stretch of scrubland.

Padi, Paddy Rice plant or rice grain in the husk. Rice is grown in a *paddy-field.*

Palaeocene Epoch An Epoch in the earth's history that lasted from about 70 to 60 million years ago. It was the first Epoch of the Cenozoic Era. *See* CENOZOIC ERA; TERTIARY PERIOD.

Palaeogene The Palaeocene, Eocene, and Oligocene Epochs.

Palaeozoic Era The first Era of geological time after the Pre-Cambrian. The rocks that date from this Era are the first to show traces of organic remains. It occurred between 600 million and 220 million years ago, and is divided into the Cambrian, Ordovician, Silurian, Devonian, Carboniferous, and Permian Periods.

Pampas. A gaucho, a cowboy of the pampas, prepares his midday meal.

Pampas The vast grasslands that surround the estuary of the Rio de la Plata in South America. *See* GRASSLAND; PRAIRIE; STEPPE.

Panhandle A strip of land belonging to one state sticking out like the handle of a pan into the territory of another state—for example, the panhandle of Oklahoma in the United States.

Parallel of latitude A line on a map that is parallel to the equator and that joins all points the same distance N or S of the equator. *See* LATITUDE.

Pass A narrow way through or over a mountain barrier. It is usually the result of erosion. *See* COL.

Pasture Grassland on which domestic animals graze. *See* MEADOW.

Peak A mountain summit, usually formed by the erosion of surrounding rocks. *See* MOUNTAIN.

Peat A dark brown or black organic mass found in wet, marshy ground. It consists of partly rotted vegetation that has collected under waterlogged conditions over a long time. In some parts of the world it is used as fuel.

Pediment A gentle, smooth, sloping rockface at the base of a steeply eroded upland. Pediments were possibly formed by flood waters.

Peneplain A region made almost level by the eroding force of rivers and the rain. *See* CYCLE; EROSION; PLAIN.

Peninsula A piece of land projecting into the sea or into a lake.

Perigee The point in the orbit of the moon or a planet when it is nearest the earth. *See* APOGEE.

Perihelion The point in its orbit at which a heavenly body in the Solar System is nearest the sun; the earth at this point (3 January) approaches to about 91 million miles (147 million km) of the sun. *See* APHELION.

Period A division of geological time that is shorter than an Era but longer than an Epoch. *See* EPOCH; ERA; GEOLOGICAL TIME.

Permafrost Subsoil and bedrock that is permanently frozen, as in polar regions. Even though the topsoil may thaw temporarily, the permafrost sometimes goes down as far as 2,000 ft (610 m). *See* FRIGID ZONE; TUNDRA.

Permeable rock Rock that will let fluids seep through it, either because it is porous (coarse-grained, with open texture) or pervious (cracked and fissured).

Permian Period The last Period of the Palaeozoic Era, from about 270 to 225 million years ago. *See* PALAEOZOIC ERA.

Polder in the region of IJsselmeer, formerly the Zuider Zee, in the Netherlands. Many of the windmills that are a feature of the Dutch landscape are water pumps, used for draining the land.

Petroleum A mixture of hydrocarbons lying in the earth's crust, valuable as a source of energy. Geologists believe that the remains of plants and animals buried in muddy ocean beds millions of years ago, and then pressed into sedimentary rock by the weight of sand and mud above them, gradually changed into oil and natural gas.

Physical geography That aspect of geography that is concerned with the physical environment—including the form and composition of the land, the oceans, natural vegetation, and the nature of the atmosphere. *See* GEOGRAPHY; HUMAN GEOGRAPHY; PHYSIOGRAPHY.

Physiography The study of the natural features of the earth's surface and their relationship to one another. It is also called *geomorphology*.

The port of Luanda in Angola. Ports are vital to the economies of countries that live by trade.

Phytogeography The geography of plant distribution. *See* BIOGEOGRAPHY.

Piedmont A region of foothills.

Pipe (i) The vent of a volcano. (ii) A cylindrically-shaped mass of mineral ore. (iii) Vertical joints in chalk, filled with gravel and sand.

Plain A broad, flat area of land or one that is almost featureless. Generally, such areas are at low elevation. *See* GRASSLAND; PAMPAS; PRAIRIE; STEPPE.

Planet A heavenly body that revolves round the sun in a path called an *orbit*. The nine planets that, with the sun, make up the Solar System are Mercury, Venus, Earth, Mars, Jupiter, Saturn, Uranus, Neptune, and Pluto. *See* SOLAR SYSTEM.

Planetary winds The general pattern of wind systems near the earth's surface—such as the *doldrums, trade winds,* and *westerlies.* *See* DOLDRUMS; TRADE WINDS; WESTERLIES.

Planimeter An instrument used to measure areas on maps and charts. A pointer traces the perimeter of the territory to be measured, and the instrument then indicates the corresponding area.

Plankton A mass of tiny animal and plant life that floats in seas and oceans and provides valuable food for many marine animals.

Plantation A large estate, usually in tropical or subtropical regions, where a crop such as sugar or coffee is grown for profit.

Plateau A large, level area of land that, generally, rises above the surrounding land. It may have been formed by the upward thrusting of a plain, or by layers of hard rock resisting denudation. Some plateaux, how-

ever, lie between higher fold mountain ranges. Plateaux of this type are described as *intermontane.*

Pleistocene Epoch An Epoch in the earth's history when ice sheets covered large parts of Europe and North America. It began about 1 million years ago and was the sixth Epoch of the Cenozoic Era. It is also called the *Ice Age. See* CENOZOIC ERA; ICE AGE; QUATERNARY PERIOD.

Pliocene Epoch An Epoch in the earth's history that began about 10 million years ago and lasted for 9 million years. It was the fifth Epoch of the Cenozoic Era. During it, there was great volcanic activity and climates grew cooler. *See* CENOZOIC ERA; TERTIARY PERIOD.

Pluvial Rainy, or accompanied by rain.

Point A narrow piece of land that juts out into the sea; a cape. *See* CAPE.

Polder A term used in the Netherlands to describe a stretch of land, at or below sea level, that has been reclaimed from the sea by the building of dykes.

Pole The N or S end of the earth's axis. *See* AXIS OF THE EARTH; MAGNETIC POLE.

Port A place on the coast or on a river where ships can tie up and load and unload passengers and cargo. The term is also used in relation to aircraft.

Pot-hole A circular hole cut in the bed of a river by the action of pebbles carried by the swirling waters.

Prairie Huge treeless plains of tall grasses found in the interior of North America. They correspond to the *pampas* of South America and the *steppes* of Europe and Asia. *See* PAMPAS; STEPPE.

Pre-Cambrian Era The time in the earth's history before the Cambrian Period; the first of the four great geological Eras. During it, the earliest forms of life appeared. Rocks from this Era are exposed at the earth's surface in some places, and have not undergone changes as other rocks have. The Era ended about 600 million years ago.

Precipice The vertical or very steep face of a mountain, cliff, or rock.

Precipitation Moisture such as rain, snow, or hail that falls on the earth's surface as a result of condensation in the atmosphere. *See* HAIL; RAIN; SNOW.

Prevailing wind The wind direction most common in any particular place.

Projection A representation on a flat map of the surface of a geographical globe with its network of meridians and parallels. All projections are inaccurate in one respect or another, and each kind aims at accuracy in some particular aspect, such as directions, shapes, or areas. *See* AZIMUTHAL PROJECTIONS; CYLINDRICAL PROJECTIONS; CONICAL PROJECTIONS; GNOMONIC PROJECTION; MERCATOR'S PROJECTION; ORTHOGRAPHIC PROJECTION; ORTHOMORPHIC PROJECTION.

Proterozoic Period The last Period of the Pre-Cambrian Era. *See* PRE-CAMBRIAN ERA.

Pumice A light-coloured, volcanic rock that has solidified while gases were bubbling out of it. It is very light, floats on water, and is used as an abrasive.

Puszta The flat, temperate grasslands of the Hungarian plains.

Quadrant An instrument formerly used in astronomy and navigation for measuring angles and elevations; now replaced by the sextant. *See* SEXTANT.

Quagmire An area of soft, wet land; a bog or marsh that quakes underfoot. *See* BOG; MARSH.

Quarry A pit or open site where commercially valuable stone is cut or excavated.

Quartz The hardest of common minerals. It is found in many rocks, and looks like pieces of broken glass. It has many industrial uses, and some kinds of quartz—such as amethyst—are valued as gemstones. Chemically, it is a crystalline form of silicon dioxide.

Quaternary Period The second and most recent Period of the Cenozoic Era, made up of the Pleistocene and Recent Epochs. This latest Period included the last great *Ice Age* and the first appearance of man on earth. *See* CENOZOIC ERA; ICE AGE.

Quicksand A deep mass of loose, fine-grained sand that, when wet, will not support the weight of men or animals and becomes a deadly trap.

Race (i) One of the main divisions of mankind, consisting of people that have a recognizable combination of physical traits handed on from one generation to another. (ii) A fast tidal current flowing through a narrow channel.

Radiation The process whereby radiant energy is emitted by a body in the form of short-waves or long-waves. The sun's emission of radiant energy is called *solar radiation*. The loss of heat from the earth's surface at night by long-wave radiation is called *terrestrial radiation*. *See* INSOLATION.

Radioactive decay The continuous and increasing breakdown of certain radioactive elements in nature. The decay of radioactive elements in rocks makes it possible for geologists to determine the age of rock strata. *See* RADIOCARBON DATING.

Radiocarbon dating A method of finding out the ages of long-dead organisms or substances derived from them. All living things contain (as well as carbon) carbon-14, a radioactive isotope of carbon. As soon as an animal or plant dies, it begins to lose its carbon-14 at the rate of 50 per cent every 5,730 years. By measuring the proportions of C-14 and carbon left in a piece of peat, say, or the bone of a dinosaur, scientists can calculate the object's age. The method is accurate up to 20,000 or 30,000 years.

Rain Water droplets that fall to the earth as a result of the condensation of water vapour in saturated clouds. *See* CLOUD; CONDENSATION; RAIN GAUGE; RAINFALL; SATURATION.

Rain gauge An instrument used to measure the amount of rainfall in a particular area in any given period of time. The simplest kind of gauge is a funnel fitted into a glass jar, which holds the water. Rainfall is usually measured in millimetres or inches. *See* RAIN; RAINFALL.

Rainbow A bow or arch of 7 colours seen in the sky when sunshine occurs during rain. The raindrops refract and reflect the sun's rays, each drop acting as a prism. The 7 colours are: violet, indigo, blue, green, yellow, orange, and red.

Rainfall The total amount of water that falls on a given region during a certain period of time, as recorded by a rain gauge. Melted snow and hail are also included in this total. *See* PRECIPITATION; RAIN; RAIN GAUGE.

Rain-forest *See* EQUATORIAL RAIN-FOREST.

Rand An Afrikaaner word for a ridge of hills in southern Africa. Specifically the *Rand* refers to the *Witwatersrand*, a rich goldfield near Johannesburg.

Range (i) A chain of mountains. (ii) In some countries, grazing land for cattle, goats, and sheep.

Rapids Part of a river where the current flows with great swiftness, and the water is broken. It is caused by a sudden steepening of the stream or by a series of hard rocky outcrops that break up the flow in midstream. *See* WATERFALL.

Ravine A long, deep, narrow valley.

Raw materials Natural substances that are manufactured into consumer goods. They may be raw substances such as iron ore, or partly-processed substances such as wholemeal flour.

Reach Any specified section of a river, or the section of a canal between locks.

Réaumur scale A thermometer scale in which the freezing point of water is marked 0° and the boiling point 80°. It is little used today.

Reef (i) A ridge of rock at or near the surface of the sea, specifically one made of coral. (ii) In mining, a vein of gold, or some other metal or metal ore. *See* ATOLL; BARRIER REEF; CORAL REEF.

Region An area or division of the earth's land surface without definite boundaries but with distinctive characteristics.

Regional geography The geographical study of any region that differs sufficiently from the surrounding area to be distinctive. *See* GEOGRAPHY.

Relief map A map that shows the *relief* —differences in height—of any part of the earth's surface. The physical landscape may be shown by various means, including contour lines, hachures, and tinting. *See* CONTOUR.

Representative fraction The ratio of the distance between any two points on a map and the actual distance between them on the ground. It indicates the *scale* of the map. For example, the representative fraction of 1 inch to 1 mile is 1/63,360, because there are 63,360 inches in 1 mile. *See* MAP; SCALE.

Reservoir A place where a large volume of water is stored for hydro-electric or irrigation purposes, or to form the water supply of a populated area. The reservoir may be a natural store, such as a lake, or may be man-made, perhaps by damming a river. *See* DAM.

Ridge of high pressure A long, narrow area of high atmospheric pressure wedged in between two areas of low pressure. It is often marked by a temporary period of fine weather. *See* ATMOSPHERIC PRESSURE; TROUGH.

Rift valley A long, narrow depression formed when the land sinks between two parallel faults. The most famous is the Great Rift Valley that stretches some 3,000 miles (4,830 km) from Syria to eastern Africa, and is only 20 to 30 miles (30 to 50 km) wide. *See* FAULT.

Rip A stretch of broken or agitated water caused by the ebb of waves piling up on the shore or by tidal streams.

River A freshwater stream that flows in a definite channel to the sea, a lake, or another river. The point where it originates is called its *source*. Lesser rivers that contribute water to it are called *tributaries*. *See* WATERSHED.

Roaring Forties A southerly region of the oceans, located between latitudes 40° S and 60° S, where there are no land masses and where, as a result, the prevailing NW to W winds blow fiercely and regularly. *See* WESTERLIES.

River. Above: Some comparatively minor rivers carry a great volume of water. The vast Lake of the Ozarks in Missouri, in the United States, was formed by the damming of the Osage, a tributary of the Missouri. Below: Budapest on the Danube. Most of the world's great cities are built on rivers.

Salt Flat. A thick deposit of salt covers the desert in the Danakil Depression, in northern Ethiopia. The Depression, some 400 ft (120 m) below the level of the nearby Red Sea, once held a salt lake that has now evaporated.

Rock The hard mass of mineral matter that forms most of the earth's crust. *See* IGNEOUS ROCK; METAMORPHIC ROCK; SEDIMENT.

Rotation of the earth The spinning of the earth on its axis from W to E. One rotation takes approximately 24 hours. The speed of rotation at the equator is about 1,050 mph (1,690 kph). *See* AXIS OF THE EARTH.

Saddle A type of wide, level col located in a ridge between two peaks. *See* COL.

Salinity The relative saltiness of sea-water. The amount of salt in the earth's seas as a whole averages 34.5 parts per thousand.

Salt flat The bed of a former salt lake, now dried up and forming a flat salt-crust, often many miles long.

Sand A mass of tiny rock particles, mostly quartz. Sand is coarser than silt or dust, but finer than gravel. *See* DUST; GRAVEL; SILT.

Sandbank A band of sand in the sea or in a river; formed by tides and currents, and sometimes visible at low water. A low sandbank is called a *shoal*.

Saturation (i) The condition of the atmosphere when it can hold no more water vapour. (ii) The state of rocks whose pores are filled with water. *See* CONDENSATION; WATER TABLE.

Savanna Tropical grassland with some scattered trees. The word is also spelt *savannah* or *savana*. *See* GRASSLAND.

Scale The ratio between distance on the ground and the length that represents it on a map. It may be stated in words, shown as a graduated line, or given as a representative fraction. *See* REPRESENTATIVE FRACTION.

Scrub The vegetation that grows in semi-arid conditions, on poor, stony soils. It is made up largely of coarse grasses, bushes, and low, stunted trees. *See* MULGA.

Sea A general term for the mass of salt water that covers most of the earth's surface. Specifically, it refers to lesser divisions of the oceans such as the North Sea and the China Sea. *See* OCEAN.

Sea level *See* MEAN SEA LEVEL.

Season One of the four major periods of the year—spring, summer, autumn, and winter. Each season is marked by differences in the temperature and the lengths of night and day. These differences are caused by the earth's annual journey round the sun and the angle at which it is tilted on its axis. *See* AUTUMN; SPRING; SUMMER; WINTER.

Seaway A ship-canal or inland waterway that is navigable by seagoing ships. A famous example is the St Lawrence Seaway in North America.

Second In the measurement of angles, one-sixtieth of a minute. Also, one-sixtieth of a minute of time.

Sediment Material that has been carried by wind, water, or ice and eventually deposited. Hence *sedimentary rock* is a type of rock made up of sediments that have been laid down in layers and compacted or cemented together. *See* ROCK; SILT.

Seismology The scientific study and measurement of earthquakes and other earth tremors. *See* EARTHQUAKE.

Selva Specifically, a Brazilian word for the equatorial forest of the Amazon basin; generally, any region of similar vegetation. *See* EQUATORIAL RAIN-FOREST.

Shooting star *See* METEOR.

Shore *See* COASTLINE.

Sial *See* CONTINENTAL DRIFT.

Sierra A Spanish term for a mountain range with jagged, saw-like peaks.

Silt Fine muddy soil that is carried by rivers and streams and deposited as sediment in lakes, deltas, and river valleys. *See* DELTA; SEDIMENT.

Silurian Period The third Period of the Palaeozoic Era. It occurred between 440 million and 400 million years ago. *See* PALAEOZOIC ERA.

Sima *See* CONTINENTAL DRIFT.

Simoon An Arab term for an extremely hot, dry, sandy, swirling wind that blows over the northern Sahara.

Sirocco, Scirocco A warm, dusty, southerly wind that blows from the Sahara across northern Africa and on to Sicily and southern Italy. It normally lasts for one or two days, and, though dry over Africa, may be very humid by the time it reaches Sicily.

Slate A metamorphic rock that is finely layered and splits easily into sheets. It has several commercial uses. *See* METAMORPHIC ROCK.

Smog A word coined from 'fog' and 'smoke' to indicate a condition of fog polluted with smoke. It is a particular nuisance in industrial areas.

Soil Erosion. By contour ploughing, following the natural contours of the land, a farmer can prevent soil being washed away and improve irrigation.

Sextant An instrument used for measuring angles, especially the angle of the sun or some other star. In this way a navigator at sea or in a plane is able to fix his position in terms of latitude and longitude.

Shale Sedimentary rock made up of compressed mud or clay. It splits easily into thin layers. *See* SEDIMENT.

Shelf A platform or ledge under the sea. *See* CONTINENTAL SHELF.

Shield A large area of very old, crystalline, Pre-Cambrian rock, exposed at the earth's surface. Well-known examples are the Canadian Shield and the Deccan in India. *See* PRE-CAMBRIAN ERA.

Shoal (i) A shallow region in a sea, river, or lake. (ii) The ridge of sand or rocks that causes it. (iii) A very large group of fish.

Snow Precipitation in the form of delicate, feathery ice crystals. It occurs when water vapour condenses at a temperature below freezing-point and so passes directly from a gas to a solid. All snowflakes have the same basic six-sided shape. *See* CONDENSATION; PRECIPITATION.

Social geography A division of human geography that is particularly concerned with the study of populations and their settlement. *See* GEOGRAPHY; HUMAN GEOGRAPHY.

Soil The mixture of rock particles, decomposed organic matter, aqueous solution of chemicals, and gases that makes up the

Swamp. Cypress Gardens, Florida, in the United States, a swamp that has been turned into a pleasure ground. Oaks and cypresses rise above quiet waterways and rare and beautiful plants.

loose surface-layer of the earth. Agriculturally, the most important part of the soil is the thin layer of fertile topsoil.

Soil erosion The removal, mainly by wind and water, of the topsoil. Man often speeds this process by over-grazing, burning, or deforestation. The soil is lost at a faster rate than it can be replaced, and disastrous conditions such as *dust-bowls* result. *See* DUST-BOWL; EROSION.

Solar System The system to which the earth belongs. It is made up of the sun together with the nine planets, the asteroids, the meteoroids, the comets, and the various particles that revolve round it in almost the same plane. It is part of the galaxy called the *Milky Way*. *See* PLANET; SUN.

Solar year A period of slightly more than 365 days, the time taken by the earth to make one complete orbit of the sun. *See* ORBIT OF THE EARTH; YEAR.

Solstice The time of year when the difference between lengths of days and lengths of nights is at its greatest. At this time the sun reaches its maximum declination from the equator. The summer solstice in the Northern Hemisphere is about June 21, and the winter solstice about December 22. In the Southern Hemisphere these dates are reversed.

Sound (i) A stretch of water, usually broader than a strait, that links two larger areas of water. (ii) An inlet of the sea. *See* INLET; STRAIT.

Spit A narrow, low-lying tongue of sand, shingle, or gravel sticking out into the sea from the coast.

Spring (i) One of the four seasons of the year, between winter and summer. In the Northern Hemisphere it begins about 21 March; in the Southern Hemisphere, about 23 September. (ii) A natural flow of water gushing from the ground, usually because the water table cuts across ground level. *See* HOT SPRING; WATER TABLE.

Spring tide A very high and very low tide that occurs twice a month. It occurs when the earth, sun, and moon are in the same, almost straight line—at the time of the new moon and full moon. The gravitational pull of the sun is then added to that of the moon. *See* NEAP TIDE; TIDE.

Spur A small ridge jutting out from a mountain or a hill.

Stalactite An icicle-shaped deposit hanging from the roof of a cave. It is formed by drops of water containing calcium carbonate dripping from the roof and evaporating. *See* STALAGMITE.

Stalagmite A stumpy, icicle-shaped mineral deposit rising from the floor of a cave. Like the stalactite, it forms from a solution of calcium carbonate dripping from the roof and evaporating. It grows immediately below the stalactite, with the result that eventually both 'icicles' join to form a pillar. *See* STALAC-TITE.

Standard time The time adopted in a particular longitudinal zone with reference to the mean time of a standard meridian. Most standard times differ from Greenwich Mean Time by an exact number of hours or half hours. Standard time is also called *zone time*. *See* GREENWICH MEAN TIME; INTERNATIONAL DATE LINE.

Steppe Temperate grassland, as a rule level and treeless. Specifically, steppes are found in the mid-latitudes of Europe and Asia. They correspond to the *prairies* of North America and the *pampas* of South America. *See* PAMPAS; PRAIRIE.

Storm (i) Force 11 on the Beaufort Wind Scale. (ii) Any violent disturbance of the atmosphere accompanied by wind, rain, snow, thunder, or other manifestation. *See* BEAUFORT WIND SCALE.

Strait, Straits A narrow stretch of water that connects two larger bodies of water. *See* SOUND.

Stratification The layering of sedimentary rocks. *See* SEDIMENT.

Stratocumulus A type of low cloud formation that looks like a greyish-white serrated blanket across the sky. *See* CLOUD.

Stratosphere The layer of the atmosphere that lies between 10 and 50 miles (16 and 80 km) above the earth. It is above the troposphere. *See* ATMOSPHERE.

Stratum A single, generally distinct, horizontal layer or bed of sedimentary rock (plural *strata*). *See* SEDIMENT.

Stratus A type of low cloud that stretches in a grey, unbroken layer. *See* CLOUD.

Subsidence Sinking of the earth's crust. It may be a small local settling of the earth produced perhaps by the collapse of old mine workings, or a huge depression across part of a continent, such as the Great Rift Valley. *See* RIFT VALLEY.

Summer One of the four seasons of the year, between spring and autumn. In the Northern Hemisphere it begins about 21 June; in the Southern Hemisphere, about 21 December. *See* SEASON.

Sun The star round which the earth orbits and from which the earth receives its heat and light. It is about 93 million miles (150 million km) from the earth. Its mass is about 330,000 times greater than the earth's mass. Scientists believe that the sun has a liquid centre, which is surrounded by flaming gas. All life in the Solar System depends for its existence on the sun's radiation. *See* RADIATION; SOLAR SYSTEM.

Survey The precise measuring and estimating of the location, size, shape, and features of any part of the earth's surface in order to map it.

Swamp An area of permanently waterlogged soil and vegetation. *See* BOG; MARSH.

Syncline A trough or downfold in a layer of rock. *See* ANTICLINE; GEOSYNCLINE.

Tableland A flat-topped upland or plateau with steep sides.

Taiga Coniferous forests in the high latitudes of the Northern Hemisphere. Tundra lies to the north and steppes to the south.

Temperate Zone A climatic zone lying between the Torrid Zone and the Frigid Zone. More accurately known as the *mid-latitudes*, there are really two such zones, one between the Tropic of Cancer and the Arctic Circle and the other between the Tropic of Capricorn and the Antarctic Circle. The seasons are well-marked and distinct, and the sun's rays are never directly overhead.

Temperature As an element of climate, the degree of heat in the atmosphere, in the land, or in the sea. It is measured with a thermometer. Various factors affect the temperature of any particular place, such as latitude, altitude, wind direction, and distance from the sea. *See* CLIMATE; THERMOMETER.

Terrain The physical characteristics and features of any stretch of country.

Tertiary Period The first Period of the Cenozoic Era. It lasted from about 70 million to about 1 million years ago. It is divided into the Palaeocene, Eocene, Oligocene, Miocene, and Pliocene Epochs. *See* CENOZOIC ERA.

Taiga. The seemingly limitless coniferous forests of Siberia, in the U.S.S.R.

Valley. Typical valley scenery in the high mountain ranges of southern Africa, with their thick vegetation, stark cliffs, and deeply-cut canyons.

Theodolite An instrument used in surveying for measuring angles and directions. It is mounted on a tripod and is made up of a telescope and a spirit-level with graduated horizontal and vertical scales. *See* AZIMUTH; SURVEY.

Thermal A rising current of warm air, caused when an area of the earth's surface is heated by the sun's rays.

Thermometer An instrument for measuring temperature. It consists of a glass tube marked with a graduated scale. The base of the tube is a bulb containing mercury, alcohol, or some other liquid that expands or contracts with changes in temperature. The resultant changes in the level of liquid in the tube are read off on the scale. *See* CELSIUS SCALE; FAHRENHEIT SCALE; TEMPERATURE.

Thunderstorm A violent storm in the atmosphere accompanied by lightning, thunder, rain, and sometimes hail. Such storms occur when pockets of warm, moist air rise into layers of colder air. The moisture in the air condenses and falls as heavy showers of rain or hail. At the same time, large positive and negative charges of electricity build up, and eventually discharge as lightning. The lightning heats the air, causing it to expand at great speed, sending out the sound waves called thunder. *See* CONDENSATION; LIGHTNING.

Tidal wave The rocking motion of the water body round the edges of oceans and seas, produced by tides. Under certain conditions, at high tide, a wall of tidal water enters an estuary and advances up a river. Such tidal waves are called *bores*. Often, the term tidal wave is mistakenly used for the phenomenon properly called a *tsunami*. *See* BORE; TIDE; TSUNAMI.

Tide The rise and fall in the level of the oceans and seas caused chiefly by the gravitational pull of the moon and the sun. *See* NEAP TIDE; SPRING TIDE.

Timber line, Tree line The upper limit of tree growth on a mountainside, or a limit of latitude beyond which trees do not grow. The timber line varies from place to place, depending on climate, soil, slope, species of trees, and other factors.

Topography A detailed description of the physical features of an area, or their representation on a map, called a *topographical map*.

Tornado A small, extremely violent whirlwind moving destructively across country at 20 to 40 mph (30 to 60 kph). At its centre, the wind has a velocity of more than 200 mph (320 kph). The tornado forms a heavy black cloud from which a twisting funnel stretches down to the ground. Tornadoes occur most frequently in the Middle Western United States. *See* HURRICANE: WATERSPOUT: WHIRLWIND.

Torrid Zone The climatic zone lying between the Tropic of Cancer and the Tropic of Capricorn, where the weather is almost always hot and the sun shines directly overhead twice each year. *See* FRIGID ZONE; TEMPERATE ZONE.

Town *See* CITY.

Trade route A land, sea, or air route that has become an established route for commerce over the years because of the existence of suitable ports and distribution facilities, proximity to sources of supply, or other factors.

Trade winds Fresh, dry, steady winds that blow from high pressure areas in subtropical regions towards the low pressure region of the equator. In the Northern Hemisphere they blow from the NE and in the Southern Hemisphere from the SE. The trade wind belts vary a little in latitude during the course of the year. *See* ANTI-TRADES.

Volcano. Mount Etna, on the eastern coast of Sicily, is one of the few active volcanoes in Europe. It is known to have erupted more than 80 times. Its snow-clad upper slopes overlook shady woods, vineyards, and orange groves.

Tree line *See* TIMBER LINE.

Trench One of the deepest parts of the ocean, usually a steep-sided depression in the ocean floor. The Marianas Trench in the western Pacific has the greatest known ocean depth—36,198 ft (11,033 m). Such ocean valleys are also called *deeps* or *troughs*.

Triangulation A method of surveying large areas based on the mathematical principle that if at least one side and two angles of a triangle are known, the other two sides and the third angle can be calculated. The angles

are measured by a theodolite. *See* SURVEY; THEODOLITE.

Triassic Period A time in the history of the earth, the first Period of the Mesozoic Era. It lasted from about 225 million to 180 million years ago. *See* MESOZOIC ERA.

Tributary A stream or river that flows into another, also called an *affluent*. *See* RIVER.

Tropic *See* CANCER, TROPIC OF; CAPRICORN, TROPIC OF.

Troposphere The layer of the atmosphere that extends from sea level to about 10 miles (16 km) above the surface of the earth. Most weather phenomena occur in the troposphere. *See* ATMOSPHERE.

Trough (i) A trench or deep in the sea bed. (ii) A U-shaped valley, sometimes made by a glacier. (iii) A long narrow region of low pressure in the atmosphere sandwiched between regions of higher pressure. (iv) At sea, a hollow between two successive waves. *See* TRENCH.

Tsunami A great sea wave caused by a submarine earthquake. In the open sea, it may be only a few feet high, but it is usually several hundred miles long and may travel at up to 600 mph (970 kph). When it reaches shallow coastal waters it sometimes rears up to great heights before sweeping inland, where it sometimes causes much destruction. It is often mistakenly called a *tidal wave*. The word *tsunami* is Japanese. *See* WAVE.

Tufa A spongy, porous, chalky deposit produced by springs that are rich in lime; also called *travertine*.

Tundra A treeless Arctic plain that has no corresponding region in the Southern Hemisphere. It lies between the northern polar ice and the northern coniferous forests. Part of the ground is frozen throughout the year and much is waterlogged. Vegetation is made up mostly of sparse grasses, mosses, and lichens. *See* PERMAFROST.

Typhoon A violent tropical storm that occurs in the western Pacific Ocean and the China Sea. It is accompanied by revolving winds that exceed 100 mph (160 kph), thunderstorms, and very heavy rain. It is also called a *tropical cyclone*, and is the same kind of storm as a hurricane. *See* CYCLONE; HURRICANE.

Upland Land that is higher than the surrounding region. *See* HIGHLAND.

Valley A long narrow hollow in the earth's surface formed usually by a glacier or river but sometimes by folding and faulting. Often a river flows through it. *See* FAULT; FOLD.

Van Allen Radiation Belt A belt of radiation surrounding the earth. It begins about 500 miles (800 km) out from the equator and extends some 40,000 miles (64,000 km). The belt probably consists of electrically charged particles trapped in the earth's magnetic field. *See* RADIATION.

Veering A clockwise change in the direction of the wind—for example from N to NE. The opposite of *backing*. *See* BACKING.

Vegetation The plant life or flora of the earth. *See* FLORA.

Vein A long narrow crack in a rock mass containing deposits of crystalline matter. Veins are often the source of metallic ores. *See* LODE.

Veld A Dutch term meaning 'field'. In South Africa it refers to open grassland.

Vent The pipe or the opening in the earth's crust through which magma rises in a volcano. *See* MAGMA; VOLCANO.

Visibility The distance it is possible to see clearly. Visibility depends partly on the observer's height above sea level, the clarity of the atmosphere, the time of day, and the physical features of the surrounding land. *See* HORIZON.

Volcano An opening in the earth's crust through which magma is forced up to the surface. The magma that flows out is called *lava*. As the volcanic materials pile up around the vent, a typical conical hill is formed. An *active* volcano is one that erupts regularly; an *extinct* volcano is one that has not erupted since the beginning of recorded history, and is consequently regarded as dead; a *dormant* volcano is one that has been quiet for a long period but may still be active. *See* LAVA; MAGMA; VENT.

Wadi An Arabic term for a steep-sided rocky stream bed in a desert. *See* NULLAH.

Warm front The boundary line between a mass of cold air and an overtaking mass of warm air. The approach of a warm front is marked by rising temperatures and an increase in cloud formations. *See* COLD FRONT; FRONT; OCCLUDED FRONT.

Water cycle The continuous circulation of water from the earth's surface into the atmosphere, and back to the surface. Water evaporates from oceans, lakes, rivers, land, and vegetation. In the atmosphere, condensation takes place, followed by precipitation of the water as rain or snow. The water that has returned to the surface subsequently evaporates again. *See* CONDENSATION; EVAPORATION; PRECIPITATION.

Waterfall A sudden steep fall of the water in a river because of an abrupt drop in the river's bed. This may be the result of differences in the hardness of the rock forming the bed, or have been caused by a sudden uplift or landslide, or volcanic activity. *See* CASCADE; CATARACT; RAPIDS.

Watershed High land that separates two or more river systems; the crest of the surrounding uplands that form the boundary of a river basin. *See* BASIN; CATCHMENT AREA; DRAINAGE; RIVER.

Waterspout A tornado over the sea. A funnel of cloud spirals down to the surface of

the sea and scoops up a twisting column of water. *See* TORNADO.

Water Table The junction between saturated and non-saturated layers of soil. Water trickles through most soils until the lower layers become saturated. After heavy rain the water table rises; after drought it sinks.

Wave, Ocean A surging motion of the ocean surface caused by the friction of wind blowing over it. The water moves in a kind of circle from trough up to crest and down and back into trough again. It does not actually travel forward.

Weather The day-to-day or hour-to-hour condition of the atmosphere at any particular place. The factors to be taken into account include temperature, precipitation, clouds, sunshine, wind, and visibility. *See* CLIMATE.

Weathering The breaking down of parts of the earth's crust by exposure to the atmosphere. It is one of the principal processes of denudation. *Mechanical* or *physical weathering* is caused by changes in temperature and by frost. *Chemical weathering* results from natural chemical processes such as oxidation (combining with oxygen) and solution (dissolving in a liquid). *See* ABRASION; ATTRITION; CORRASION; CORROSION; DENUDATION; EROSION.

Wedge of high pressure A narrow ridge of high atmospheric pressure, bringing a brief spell of fine weather. *See* RIDGE OF HIGH PRESSURE.

Westerlies Winds that blow frequently and fairly regularly in belts that lie between the Horse Latitudes and the polar regions in both hemispheres. In the Northern Hemisphere they blow from the SW; in the Southern Hemisphere they blow from the NW and are known as the *Roaring Forties*. *See* HORSE LATITUDES; ROARING FORTIES.

Whirlpool A strong eddy or circular current in a body of water. It is usually caused by the meeting of two currents, but sometimes a whirlpool results when a fast current enters an obstructed channel.

Whirlwind A small, violent current of air rotating rapidly round an area of low atmospheric pressure. *See* CYCLONE; TORNADO; WATERSPOUT.

Wind Air currents moving across the surface of the earth. Wind speeds are usually classified according to the Beaufort Wind Scale. Wind direction is the direction from which the wind blows; for example, an east wind is one blowing from the E.

Winter One of the four seasons of the year. It is the coldest season. Winter is the period between December 22 and March 21 in the Northern Hemisphere; and June 21 and September 22 in the Southern Hemisphere. *See* AUTUMN; SEASON; SPRING; SUMMER.

Wold Open, uncultivated, rolling, chalk upland.

Year The time taken by the earth to make a complete revolution round the sun. There are various ways of measuring this time, but for practical purposes a *calendar year* equals 365 days—beginning on 1 January and ending on 31 December. To make the calendar more accurate, a *leap year* of 366 days is introduced every fourth year. *See* SOLAR YEAR.

Zenith The point in the heavens that is directly overhead from the observer's point of view. *See* NADIR.

Zenithal projections *See* AZIMUTHAL PROJECTIONS.

Zodiac An imaginary belt in the heavens in which the sun and all planets except Venus and Pluto appear to travel. The zone extends 8° on each side of the sun's seeming path across the sky, called the *ecliptic*. In astrology, the zodiac is divided into 12 divisions or signs each marked by a constellation (group of stars).

Zone (i) Any belt or region with its own distinct characteristics. (ii) One of the five climatic regions of the earth; the northern and southern *frigid* zones; the northern and southern *temperate* zones; and the *torrid* zone. *See* FRIGID ZONE; TEMPERATE ZONE; TORRID ZONE.

Zoography The scientific study of the distribution of the earth's animals in relation to their physical environment.

Waterfall. Several of the world's highest waterfalls spill in filmy veils down the sides of Merced Canyon, in Yosemite National Park, California. One, the Yosemite Falls, is 2,425 ft (740 m) high. The Park lies in the Sierra Nevada mountains.

THE SEVEN CONTINENTS

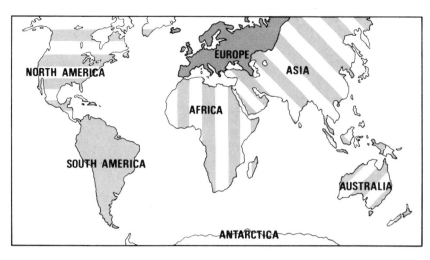

The continents are the world's great land masses. Together, they cover about one quarter of the Earth's surface. Europe is usually described as a separate continent but—geographically—it is really a peninsula of Eurasia, a land mass made up of Europe and Asia.

EUROPE

The River Rhine at Kaub, in Germany. An ancient toll-collecting castle stands on an island in the river.

Area: 4,036,000 sq miles; 10,453,000 sq km.
Greatest Distances: N to S, 4,000 miles (6,400 km); W to E, 3,000 miles (4,800 km).
Population: 692 million.
Europe is the smallest of the continents, but has influenced world history in a way that is out of all proportion to its size. It is bounded on the N by the Arctic Ocean, on the S by the Black Sea and the Mediterranean Sea, and on the W by the Atlantic Ocean. On the E, the Ural Mountains separate it from the continent of Asia.

ASIA

A floating market in Bangkok, the capital of Thailand. The city has a network of busy canals.

Area: 17,336,000 sq miles; 44,900,000 sq km.
Greatest Distances: N to S, 5,400 miles (8,700 km); W to E, 6,000 miles (9,700 km).
Population: 2,221 million.
Asia is the largest of the continents. It also has the largest population, and is the richest in natural resources. It is bounded on the N by the Arctic Ocean, on the E by the Pacific Ocean, and on the S by the Indian Ocean. On the W, the Ural Mountains form its boundary with Europe.

AFRICA

A village in the highlands north of Nairobi, in Kenya. It belongs to the large Kikuyu tribe.

Area: 11,700,000 sq miles; 30,300,000 sq km.
Greatest Distances: N to S, 5,000 miles (8,000 km); W to E, 4,700 miles (7,600 km).
Population: 380 million.
Much of Africa was not fully explored until the 1900s because deserts, jungles, and unnavigable rivers made travel extremely difficult. The continent is bounded on the N by the Mediterranean Sea, on the E by the Red Sea and the Indian Ocean, and on the W by the Atlantic Ocean.

NORTH AMERICA

Contour farming in the United States. By following the land's natural contours, soil erosion is prevented.

Area: 9,600,000 sq miles; 24,900,000 sq km.
Greatest Distances: N to S, 4,500 miles (7,200 km); W to E, 4,000 miles (6,400 km).
Population: 330 million.
Although much of it was almost uninhabited until Europeans 'discovered' it 500 years ago, North America is today the richest and most powerful of the continents. It is bounded on the N by the Arctic Ocean, on the E by the Atlantic Ocean, and on the W by the Pacific Ocean. A narrow 'bridge' joins it to South America.

SOUTH AMERICA

Banana and coffee plantations in Bolivia. The coffee berries are harvested laboriously by hand.

Area: 6,772,000 sq miles; 17,540,000 sq km.
Greatest Distances: N to S, 4,750 miles (7,640 km); W to E, 3,200 miles (5,150 km).
Population: 190 million.
There are places in the almost impenetrable forests of South America where no human being has ever walked. The continent still challenges explorers. It is bounded on the N by the Caribbean Sea and the Atlantic Ocean, on the E by the Atlantic, and on the W by the Pacific Ocean. Its southern tip approaches Antarctica.

AUSTRALIA

Herding sheep on a ranch in Australia. More than a quarter of the world's wool comes from Australia.

Area: 2,968,000 sq miles; 7,687,000 sq km.
Greatest Distances: N to S, 2,000 miles (3,200 km); W to E, 2,400 miles (3,900 km).
Population: 13,900,000.
Australia is the only continent that is also a single country. It is bounded on the N by the Timor Sea and the Arafura Sea, on the E by the Pacific Ocean, and on the W by the Indian Ocean. The term *Australasia* refers to a region comprising Australia, New Guinea, New Zealand, and other smaller islands nearby.

ANTARCTICA

Penguins in Antarctica. Their feathers and a layer of fat under the skin protect them from the cold.

Area: 5,000,000 sq miles; 13,000,000 sq km.
Greatest Distance: Wilhelm II Coast to Eights Coast, 2,700 miles (4,350 km).
Most of Antarctica, the freezing, desolate continent surrounding the South Pole, lies beneath a mass of ice and snow that is a mile thick in places. The lowest known temperatures on Earth have been recorded in Antarctica. No animals live in the inland areas of the continent. Animals can find the means of existence only near the coasts or in surrounding waters. Plants exist mainly around the edges of Antarctica.

EUROPE

Europe is part of the same land mass as Asia, but is always considered a separate continent. It is the second smallest continent and its many countries have often been devastated by war between themselves. But for thousands of years Europe has had greater influence, culturally and politically, than any other continent.

Europe is bounded on the N by the Arctic Ocean, and on the W by the Atlantic Ocean. The Mediterranean and Black Seas wash its southern shores. On the E it is linked to Asia, with the Ural Mountains and the Caspian Sea marking the boundary between the two continents. In the SE, from the Black Sea to the Caspian, stretch the Caucasus Mountains. On the other side of the Mediterranean Sea lies Africa. At the Strait of Gibraltar, where the Mediterranean is joined to the Atlantic, Europe and Africa are only 9 miles (14 km) apart.

The area of Europe is 4,036,000 sq miles (10,453,000 sq km). The greatest distance N to S is about 4,000 miles (6,400 km), and W to E 3,000 miles (4,800 km). The continent's coastline, including the coastlines of its islands, measures nearly 38,000 miles (61,000 km). The population in the mid-

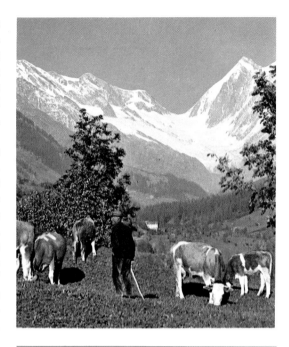

Snow-Covered Peaks. The Alps stretch in a great arc for nearly 700 miles (1,100 km) across southern Europe. Farming is confined to the valley floors and to the high summer pastures—called alps—above the tree line.

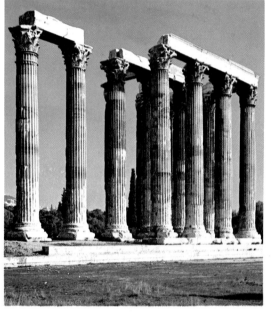

The Legacy of Greece. The temple of Olympian Zeus, near the Acropolis in Athens. Parts of it date from B.C. 515, but it was completed by the Roman Emperor Hadrian in 129 A.D. Only 15 of its 104 Corinthian columns still stand.

COUNTRIES OF EUROPE
POPULATIONS AND MONETARY UNITS

Country	Population	Monetary Unit
Albania	2,400,000	Lek
Andorra	25,000	Franc, Peseta
Austria	7,500,000	Schilling
Belgium	9,800,000	Franc
Bulgaria	8,650,000	Lev
Czechoslovakia	14,680,000	Koruna
Denmark	5,020,000	Krone
Finland	4,750,000	Markka
France	52,350,000	Franc
German Democratic Rep.	17,400,000	Mark
Germany, Federal Rep. of	62,400,000	Deutsche Mark
Gibraltar	29,000	Pound
Greece	9,100,000	Drachma
Hungary	10,420,000	Forint
Iceland	220,000	Króna
Ireland, Rep. of	3,050,000	Pound
Italy	54,900,000	Lira
Liechtenstein	22,000	Franc
Luxembourg	348,000	Franc
Malta	322,000	Pound
Monaco	24,000	Franc
Netherlands	13,490,000	Florin (Guilder)
Norway	3,970,000	Krone
Poland	33,700,000	Zloty
Portugal	9,800,000	Escudo
Romania	20,900,000	Leu
San Marino	20,000	Lira
Spain	34,200,000	Peseta
Sweden	8,240,000	Krona
Switzerland	6,460,000	Franc
Turkey (Europe)	3,200,000	Lira
U.S.S.R. (Europe)	198,200,000	Rouble
United Kingdom	56,500,000	Pound
Vatican City	1,000	Lira
Yugoslavia	21,500,000	Dinar

Mediterranean Playground. Saint-Tropez, on the Côte d'Azur in southern France. Yachts and fishing-boats share the crowded harbour. Once a quiet Provençal fishing village, Saint-Tropez became known as a haunt of painters and was soon a fashionable tourist resort.

Rugged Cliffs. 'The Mouth of Hell', a cave formation on the storm-swept Atlantic coast near Cascais in Portugal.

Peaceful Countryside. Farmland in England, carefully tended for hundreds of years.

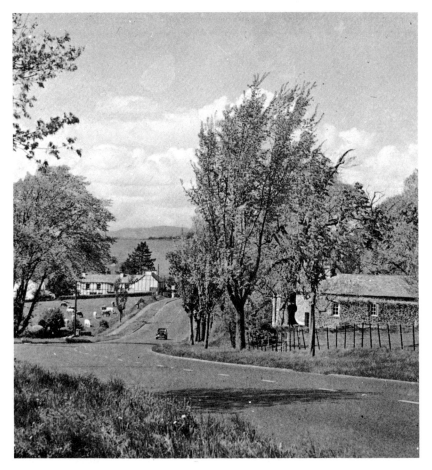

1970s was estimated at about 692 million.

The continent includes thousands of islands. The largest and most important of these are the British Isles.

Physical Features and Climate

Europe consists of three main regions: the northern mountains, the central plains, and the southern mountain chains. This pattern of plains and mountains extends from the mainland through the offshore islands.

The Northern Mountains lie in Scotland and northern England, and in Norway and Sweden. The Ural Mountains in Russia are also part of the region. Finland is not mountainous, but it is formed from the same rock system. During the Ice Age much of this region was worn away by the action of glaciers. Among the results of this glacial action are the fiords of W Norway—deep, steep-sided sea inlets. Finland's thousands of lakes were also formed by glacial action. The mountains of this northern region are covered by thin soil and are very steep. Consequently, they are difficult to farm.

The Central Plains cover by far the largest part of Europe. They stretch from Ireland in the W as far as the Ural Mountains. The greater part of the central plains is less than 500 ft (150 m) above sea-level. The eastern section of the plains is in Russia, and geologists often call it the *Russian Platform*. The highest point of the Russian Platform is just over 1,000 ft (300 m) above sea-level.

The Southern Mountain Chains are a complex series of mountain ranges. The chief mountains are the Alps, which curve in a high, rugged arc that cuts off the Italian peninsula from the rest of the continent. The highest point of the Alps is Mont Blanc, whose peak is 15,781 ft (4,810 m) above sea-level. This and the other high peaks of the Alps are always covered in snow above the 8,000 ft (2,500 m) level. The Alps extend eastwards and are continued in the Balkan Peninsula as the Balkan Mountains. The Caucasus Mountains, between the Black and Caspian Seas, are also a part of the same mountain system. To the W, the Pyrenees, lying across the northern neck of the Iberian Peninsula (Spain and Portugal), are another extension of the Alpine chain.

These mountain chains form barriers between the coastlands bordering the Mediterranean Sea and the plains of N and central Europe. But there are several breaks in the chains, such as the valley of the River Rhône in France, and these serve as major routes from N to S.

Other mountain systems are found in Spain (the Meseta), central France, southern Germany, and Czechoslovakia. They are not so high as the Alpine mountains. The upper slopes of many of them are thickly forested.

Europe has several inland seas. The North Sea, between Great Britain and the continental mainland, is a wide arm of the Atlantic. In northern Europe lies the Baltic Sea, with its two arms, the Gulf of Bothnia and the Gulf of Finland. It separates Norway, Sweden, and Finland from the central part of the continent. The Mediterranean Sea, to the S, is almost entirely landlocked. But it has a narrow link with the Atlantic in the W through the Strait of Gibraltar. The Tyrrhenian, Adriatic, Ionian, and Aegean Seas are arms or sections of the Mediterranean. Two narrow straits, the Dardanelles and the Bosporus, with the little Sea of

Marmara between them, link the Mediterranean with the Black Sea. The Sea of Azov is an almost entirely landlocked arm of the Black Sea. The Caspian Sea, although it is salt, is really a huge lake.

Europe's main rivers are the Danube, flowing eastwards through central Europe to the Black Sea, the Rhine, flowing northwards to the North Sea, and the Volga, flowing southwards through Russia to the Caspian Sea.

Most of Europe has a relatively mild climate. Northern, central, and eastern parts have warm summers and cold winters. Western parts, including the British Isles, have winters that are less cold. The countries around the Mediterranean have hot, dry summers and mild, wet winters. The extreme N lies within the Arctic Circle, but even in these extreme latitudes the warming action of the Gulf Stream can be felt.

Resources and Industry Many parts of Europe are rich in mineral resources. The continent produces about 40 per cent. of the world's coal, and half of its iron ore. Petroleum deposits lie in SE Europe, and also in the W near the Urals. Further large deposits of both petroleum and natural gas lie under the North Sea.

Other minerals produced in large amounts include bauxite, copper, lead, mercury, platinum, potash, salt, sulphur, and zinc. There are smaller deposits of a large number of minerals.

An Ancient Landscape. The low hills and rolling uplands of Sardinia. Many of the people of this Mediterranean island pursue an age-old way of life, jealously preserving traditional customs.

Village Life. A village in the beautiful valley of the River Möll in Carinthia, southern Austria. In the background is Grossglockner, Austria's highest mountain.

The world's greatest manufacturing region has developed around the coalfields of Belgium, Britain, northern France, and Germany. Its industries produce iron, steel, and all kinds of metal goods, as well as textiles —silk, cotton, wool, linen, and artificial fabrics—chemicals, pottery, and glass. In Russia the main manufacturing districts are in the neighbourhood of Moscow, in the basin of the River Don, and in the foothills of the southern Urals.

The motor-car factories of Britain, France, West Germany, and Italy make almost half of the world's vehicles. Machine-tools are produced in Britain and West Germany, and watches and similar precision machinery are a speciality of Switzerland. Heavy engineering industries are highly developed in West Germany and the U.S.S.R.

Although the first industries grew up around coalfields, and close to iron deposits, many modern industrial centres are based on the availability of hydro-electric power, particularly in northern Italy, Switzerland, and Norway and Sweden.

Much of Europe's prosperity depends on trade. Because of its long coastline, no part of western Europe is far from a seaport. Ships carry products from the factories of

MINERAL DEPOSITS						MAIN INDUSTRIAL AREAS		
● Coal	⚱ Crude oil	● Pyrite	✜ Mercury	⬓ Lead	○ Phosphates		◗ Heavy industry	⬛ Woodworking industries
◉ Iron ore	▢ Bitumen	★ Platinum	◆ Sulphur	▼ Tin	○ Potassium		◗ Light industry	～ Principal inland waterways
◎ Lignite (soft coal)	◖ Bauxite	★ Copper	■ Zinc				◗ Textile industries	

(The symbols indicate only those items which are at least 1% of world production or 10% of the consumption of the country.)

THE EUROPEAN COMMUNITY

The European Community consists of nine associated countries of Western Europe: Belgium, Denmark, France, the Federal Republic of Germany (West Germany), the Republic of Ireland, Italy, Luxembourg, the Netherlands, and the United Kingdom. It has about 80 per cent. of Western Europe's population, and only the United States has greater economic strength.

The aim of the member countries is to create a unified economy: to combine their economic resources—finance, industry, and agriculture— for the benefit of all. In time, there will be free movement of workers, money, and goods from one country to another. This economic union, it is hoped, may pave the way for closer political unity.

The Working of the Community

The European Community has its own civil service and courts, as well as a parliament and ministers. Its chief organs are the Council of Ministers, the Commission, the Assembly, and the Court of Justice.

The Council of Ministers is the lawmaking body of the Community, but it acts only on proposals put to it by the Commission. It consists of one representative of each of the nine member countries, and meets in Brussels or Luxembourg.

The Commission proposes new laws or policies to the Council, which then has to decide. It consists of nine Commissioners, appointed for 4-year terms of office. Each Commissioner has particular responsibility for one aspect of the Community's work. The Commission meets in Brussels.

The Assembly is the parliament of the Community. It controls the budget, and has power to dismiss the Commissioners. But in other matters, it is a purely advisory body. It debates the proposals of the Commission for new laws, and offers its opinion to the Commission and to the Council of Ministers. The Assembly has 142 members appointed by the parliaments of the member countries. It meets in Strasbourg or Luxembourg.

The Court of Justice consists of seven judges who have the task of settling legal disputes on Community matters. It meets in Luxembourg.

History

The idea of European unity is an old one, but it was given new life by the disastrous results of disunity in the 20th century: the two World Wars. In 1945 the French statesman Jean Monnet suggested a gradually-developing economic union. The United States aided European union with the 'Marshall Plan' and its contribution to the *Organization for European Economic Co-operation (OEEC)*. Other co-operative bodies were formed, too, including the *Council of Europe* and the *Brussels Treaty Organization*. But in practical terms, the European Community derives from three specialized bodies: the *European Coal and Steel Community (ECSC)*, the *European Economic Community (EEC),* and the *European Atomic Energy Commission (Euratom)*.

The ECSC was established by the Treaty of Paris in 1951; the EEC and Euratom by the Treaties of Rome in 1957. At first, the Community had six members. In 1973, they were joined by three more: Denmark, Ireland, and the United Kingdom. Norway decided, after a referendum, not to apply for membership.

Europe to all parts of the world. A large proportion of the world's financial business, including international banking transactions and insurance, is carried on in the continent's major cities, particularly in London and Zurich. But Europe is not only a large exporter. It has to import huge amounts of raw materials, particularly petroleum and other minerals of which it has insufficient quantities for its industrial needs.

Europe's major natural resources include the fertile soils of the central plains, where much of the continent's food is grown. Over the centuries, forests have largely been cut down, but wide tracts of Germany and other countries of central Europe are still heavily wooded. The pine forests of Scandinavia provide timber for building and furniture, and wood pulp for making paper.

The wild life of Europe has been greatly reduced by man's activities—by hunting and by the destruction of wild woodland —but a number of large animals are still found in the more remote areas. They include bears, wild boar, and wolves. Herds of reindeer roam the Arctic regions of northern Europe—an area known as *Lapland* that includes parts of Norway, Finland, and Russia. The European bison, also called the wisent, survives in parks and nature reserves. Some animal species, such as the European beaver, are increasing in number again after having been almost wiped out. The continent is rich in birds, many species of which migrate S to Africa for the winter months.

Communications in Europe are more highly developed than anywhere else in the world. The continent is criss-crossed with a dense network of railways and roads, and a busy system of waterways.

Railways run over many thousands of miles, connecting up all parts of the continent except N Scandinavia, Russia, and areas of SE Europe. Some routes serve only local needs, but there are six main trunk routes, all radiating from Paris.

The first trunk route links Paris to Calais, and then—by way of ferries to Dover—to London and other British cities. The second trunk route is that of the *Sud Express*, which runs from Paris around the western end of the Pyrenees to Madrid and Lisbon.

The third route runs S, via Dijon to Marseilles, on the Mediterranean coast. The fourth route branches off at Dijon and goes down to Italy through long tunnels under the Alps. The fifth, the *Orient Express* route, links Paris and Istanbul; the journey is via Strasbourg, Munich, Vienna, Budapest, Bel-

COMECON

The Council for Mutual Economic Assistance was formed in 1949 by a number of countries of central and Eastern Europe to co-ordinate their economic policies. All the member countries have communist forms of government; the leading member is the U.S.S.R. Comecon provides a counter-balance to the European Economic Community and such other associations as EFTA and Benelux in the West, but it has never been as active as they have. However, in the 1950s and 1960s it embarked on a number of wide-reaching schemes, which had varying success. The organization aims at common economic planning, and attempts to rationalize industrial production in the Eastern bloc countries. It also tries to make the best use of the available pool of labour, and it has proved useful in arranging for the exchange of scientific knowledge and the passing on of technical know-how. Nearly 400 million people live in the Comecon countries. The members are the U.S.S.R., Bulgaria, the German Democratic Republic, Czechoslovakia, Romania, Hungary, Poland, and Yugoslavia. Two non-European countries, Mongolia and Cuba, are also members.

EUROPE : BASES OF INDUSTRIAL PRODUCTION

● Coal	● Steel	● Electric power
●● 100 million tons		● 100 billion kWh
● ● 30 million tons		● 30 billion kWh
● ● 20 million tons		● 20 billion kWh
· · 10 million tons		· 10 billion kWh
· · 1 million tons		· 1 billion kWh

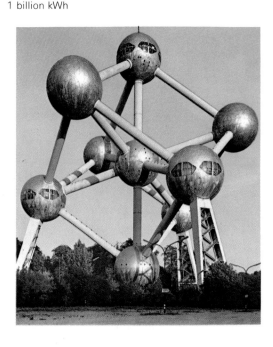

Symbol of the Atomic Age. The Atomium, a steel and aluminium model of a metal molecule, in the park of Laeken in Brussels. A relic of the World Fair of 1958, it is 335 ft (102 m) high and houses a scientific exhibition.

grade, and Sofia. The sixth line is via Berlin and Warsaw to Moscow. Though Russian railways run on broad gauge rails, some Russian carriages can be adapted to run on the standard gauge rails of other European countries. From Moscow, the Trans-Siberian Railway crosses Asia to Vladivostok.

Roads. The tremendous increase in the number of road vehicles in recent years has been paralleled by improvements in roads and highways. Today, fast motorways connect all the main centres of population. The first of these major highways were the German *Autobahnen* constructed in the 1930s. Much heavy freight that once would have been carried by rail is now transported on the road in huge lorries popularly called 'juggernauts'. Increasingly, too, passengers—particularly tourists—travel by long-distance bus instead of by train.

Air Services link the major cities of Europe with one another and with the rest of the world. But there are still relatively few local airports.

Rivers and Canals provide cheap and efficient transport in the central and western low-lying regions. The main waterways are the Rhine and the Danube with their tributaries and linking canals in western Europe; and the Volga, Don, and Dnieper in Russia.

Railways, roads, and waterways all serve the continent's ports, most of which remain ice-free all the year round. The only ports that are closed in winter are Archangel on the White Sea, those along the Baltic coasts, and some Black Sea ports—which, however, can be kept open by icebreakers.

Other forms of communication are also plentiful. Networks of pipelines carry water, petroleum, and gas. Telephones link nearly all populated areas. Europe's newspapers have some of the world's largest circulations. The Russian newspaper *Pravda* sells more than 6 million copies a day, and several British newspapers have circulations of 2 million or more.

Agriculture, Fisheries, and Land Use

Because of Europe's temperate climate and good rainfall, most of its farms are very productive. Even so, it has to import about one fifth of its food from other continents. Some heavily-populated countries, such as the United Kingdom, produce much less food than they need. Others, such as the Netherlands, produce more than they need of certain kinds of foodstuffs, and their food exports make a valuable contribution to their budgets.

The continent has three important farming regions: the lands around the Mediterra-

Castles and Garages. The ancient town of Bellinzona in Switzerland. Few towns and villages are without their petrol and service stations, successors to the blacksmiths and stables of the past.

Europe's Largest Port. Rotterdam, in the Netherlands, is 15 miles (24 km) from the North Sea, but a canal system enables it to be used by even the biggest ships. Because of its position on an arm of the River Rhine, it is also a great inland port.

Fishery Resources. Tuna fishing, a major item in the economies of Sicily, Sardinia, and other Mediterranean islands. Deep sea fishing is an important industry in several countries on the Atlantic seaboard.

Inland Waterways. A heavily-laden barge on the Rhine, at Bacharach in West Germany. The Rhine, one of the great trade arteries of Europe, provides an outlet to the sea at the port of Rotterdam in the Netherlands. It passes through several countries, and is linked by canal to many other rivers. The huge Rhine barges carry mineral ores, oil, and other heavy cargoes. Often a barge is 'home' to its crew, which may be a family. The family car is sometimes carried on deck for use in port.

nean Sea; the western part of the plains; and the eastern lowlands of Russia.

The Mediterranean lands have dry, sunny summers. There, farmers can grow large crops of such fruits as almonds, apricots, lemons, olives, oranges, peaches, and pomegranates. Wheat, barley, maize, and rice are grown too. There are few good pasture lands for cattle, but goats and sheep graze on the scanty vegetation of the mountainsides.

The western plains have fertile soil, and the most efficient and intensive farming takes place in this region. Farmers practise *rotation farming*, growing different crops each year on their land over a cycle of three or four years. By using modern machinery and chemical fertilizers, they obtain higher crop yields than is achieved almost anywhere else in the world. Wheat, oats, barley, and rye are the chief cereal crops, and root crops include sugar beet and potatoes. Beet pulp is used as cattle feed. Many farmers in the north-western areas have mixed farms —that is, they grow some crops and also raise some animals. One very important crop on such farms is lucerne (also called *alfalfa*). It provides feed for cattle and is also a good fertilizer when ploughed into the soil.

In this region the commonest farm animals are dairy cattle. Some European cattle breeds, such as Jersey, Hereford, Friesian, and Brown Swiss, have become famous as stock all over the world. Belgium, Denmark, and the Netherlands are the chief exporters of dairy produce—milk, butter, and cheese. The most intensive sheep farming is in the British Isles, where there are, on average, more animals to each sq mile than anywhere else.

The eastern lowlands of Russia also have good soil, but the farming methods used are generally less efficient than in the W. As a result, the yields of crops are much lower. The main crops are rye and flax in the central areas, and wheat on the rich black earth of the steppe-land farther S.

In Alpine regions, *transhumation* is frequent—that is, farmers drive their cattle, sheep, and goats up to the high pastures during the warm summer months. Then they bring them down again to the lowlands in the cold weather when the highlands are covered with snow.

Fisheries are another important part of Europe's food industry. The leading fishing country is Norway, which has about 60,000 fishermen. Other leading fishing countries are Britain and Iceland. Some of the most

COUNTRIES OF EUROPE AREAS		
Country	Sq Miles	Sq Kilometres
Albania	10,630	27,532
Andorra	191	495
Austria	32,373	83,846
Belgium	11,774	30,495
Bulgaria	42,858	111,002
Czechoslovakia	49,377	127,886
Denmark	16,620	43,045
Finland	130,165	337,126
France	211,200	547,000
German Democratic Rep.	41,660	107,900
Germany, Federal Rep. of	95,980	248,590
Gibraltar	2.25	5.8
Greece	50,944	131,944
Hungary	35,919	93,030
Iceland	39,709	102,846
Ireland, Rep. of	27,135	70,279
Italy	116,303	301,223
Liechtenstein	62	161
Luxembourg	999	2,587
Malta	122	316
Monaco	1	1.8
Netherlands	13,500	34,965
Norway	125,181	324,217
Poland	120,360	311,730
Portugal	34,240	88,680
Romania	91,700	237,500
San Marino	24	62
Spain	194,885	504,750
Sweden	173,400	449,110
Switzerland	15,940	41,284
Turkey (Europe)	9,121	23,623
U.S.S.R. (Europe)	2,151,000	5,571,000
United Kingdom	94,216	244,018
Vatican City	108 acres	44 hectares
Yugoslavia	98,750	255,760

Terrace Farming. A village in central Italy, built on a mountain slope. Over many laborious years, the villagers have constructed terraces on the rocky slopes to preserve the thin soil and to hold rainfall.

Conquest from the Sea. The windmills of Kinderdijk in the Netherlands in a region of 'polder' land, mostly below sea-level, that has been reclaimed from the sea by the building of dikes. Many of the famous Dutch windmills were pumps for draining the land; others were used for grinding corn. Today, the windmills of Kinderdijk are no longer in use, but are preserved as relics of a day that has gone.

important fishing grounds are in the North Sea, but many fishing vessels go out into the North Atlantic in search of herring and cod. Russian fishermen catch sturgeon in the Black and Caspian Seas. Sturgeon roes are the delicacy called *caviar.* Small quantities of fish are caught in the rivers and fresh-water lakes of the continent, and some are greatly prized as food.

Four countries of Europe are much more densely populated than any others. They are Belgium, England, West Germany, and the Netherlands, which have an average of about 500 inhabitants to each sq mile (more than 190 per sq km). Most of the rest of Europe has a population of between 33 and 256 per sq mile (12 and 98 per sq km). Most of the areas of high population lie around the regions of the coalfields, where heavy industries grew up and many of the biggest cities are.

Other major areas of high population lie in N Italy; around Oporto and Lisbon in Portugal; around Madrid and Barcelona in Spain; and around Moscow. The valleys of the Rhine and Rhône rivers are also heavily populated, partly because of the high level of agriculture in these regions.

FISHING
Sardines
Herrings
Cod
Tuna
Sponges

More than ⅓ of workers in industry; others mainly in agriculture
High population density in rural areas (more than 50 per sq. km.)
Modest population density in rural areas (25–50 per sq. km.)
Low population density in rural areas (less than 25 per sq. km.)
Economically unusable areas (mountains, desert and tundra)

Northern limit of vine growing — — — Northern limit of olive groves
Forest Maize Sugarbeet Rye Sugar cane
Wheat Rice Silkworms Hemp Fruit Vegetables
Potatoes Flax Tobacco Cotton Dates

STOCKBREEDING: ○ Cattle ◎ Sheep

People and Ways of Life

Nobody knows quite how long people have lived in Europe, but fossilized skeletons suggest that the continent has been inhabited for at least 300,000 years. Cave paintings that still exist in France and Spain were made at least 30,000 years ago.

The present population of Europe consists mainly of members of the Caucasian race—commonly called 'white people'. Their ancestors moved into Europe a long time ago from the plains of central Asia. There are three main groups of European peoples, and though they have intermarried over the years each one of these types is still more common in certain areas than the others are.

Nordic people have long skulls, fair hair, fair complexions, and blue eyes. They are generally tall. They are still found in northwestern Europe—in Scandinavia, northern Germany, and in smaller numbers the Netherlands.

Mediterranean people have long skulls too, but they have dark hair and complexions, and brown eyes. They are shorter than Nordic people. They are found chiefly in countries bordering the Mediterranean Sea.

EFTA

The European Free Trade Association includes a number of nations of Western Europe that do not belong to the European Community. Its purpose is to promote trade among its members, and to eliminate tariffs and quota restrictions. EFTA has a working relationship with the Community. It was formed in 1959, and its original members, the *Outer Seven*, were Austria, Denmark, Norway, Portugal, Sweden, Switzerland, and the United Kingdom. Later Iceland became a member, and Finland an associate member. Denmark and the United Kingdom withdrew from EFTA in 1973 on joining the European Community.

BENELUX

Benelux, one of the most successful of international groupings, is an economic union of Belgium, the Netherlands, and Luxembourg. It was formed in 1948. Its name is made up of the first letters of the names of its members. It was the earliest of the European 'common markets', and its aims were to standardize taxes, wages, prices, and social security benefits. It also gradually abolished trade restrictions from one country to another. Much of the success of Benelux derived from the community of interest of its members, and the fact that their recent histories had followed the same economic and political course.

69

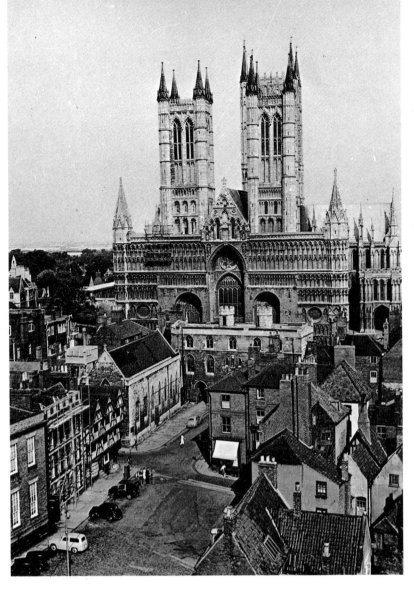

Past and Present. Above: Lincoln Cathedral in England, a splendid example of Early English Gothic, stands as testimony to beliefs and aspirations that have played a large part in shaping European life. Below: Old houses in a market place in Bergen, Norway, provide a background to shoppers and sightseers.

Alpine people have broader skulls than the other two types, have hair and complexions that are medium to dark, and have a sturdy build. They are found in the mountainous regions of central and eastern Europe.

Two other major groups are the *Dinaric* people, who have round heads and are found in the Balkans, and the *East Baltics*, who have broad, round heads, and are found in eastern Europe, including parts of Russia.

Today the peoples of Europe are grouped in different 'nationalities'. The continent is divided politically into 35 countries, and the people of those countries speak about 60 different languages. Most of the languages belong to three main groups.

The *Germanic* languages include Danish, Dutch, English, German, Norwegian, and Swedish. The *Romance* or *Latin* languages include French, Italian, Portuguese, Romanian, and Spanish. The *Slavic* languages include Bulgarian, Czech, Serbo-Croatian, Polish, Russian, and Ukrainian. Other language groups include the Celtic tongues —Gaelic, Welsh, and Breton—and Greek and Albanian groups. All these different languages belong to the major language family known as *Indo-European*, which also includes Hindi and Persian.

The people of Finland and Hungary speak languages akin to those of central Asia. The Basques, a people who live on the borders of France and Spain, have a language whose connexions are not clear.

All these variations of language have helped to keep the nations of Europe divided. For hundreds of years Latin—the language of Ancient Rome—was used as a universal language by scholars, and was in daily use as an international language by the Catholic Church until the 1970s. Today, English, French, German, and Russian are the most useful languages for a European. The most widely-spoken language in Europe is German.

Since the end of World War II many peoples have emigrated to Europe from other continents. The United Kingdom, for example, now has a large number of Negro, Indian, and Pakistani citizens, many people from Asian Turkey are found in West Germany, and many North Africans in France.

Today, the nations of Europe often tend to group themselves into the non-communist 'Western' countries and the communist 'Eastern' countries. A few, such as Austria and Switzerland, remain 'neutral'. Nine countries of western Europe form the *European Community*, an economic alliance, still in its formative stages, which many hope will become a political union.

Relaxation. The village of Portofino, on the Gulf of Genoa in northern Italy. Tourism has become a major local industry in many parts of Europe. Holidaymakers find in such spots as Portofino their dreamed-of holiday combination: the sun, the sea, and the feeling of being 'away from it all'.

IMPORTANT DATES IN THE HISTORY OF EUROPE

3000-4000 B.C. The Minoan civilization developed on Crete, in the Aegean Sea.

1194-1184 B.C. According to tradition, the Helladic peoples (the Achaens) fought against Troy, the war celebrated in Homer's *Iliad*.

490 B.C. The Greeks defeated the Persians at the Battle of Marathon.

461-431 B.C. The Golden Age of Athens, Greek civilization reaching its peak.

146 B.C. Greece became a Roman province.

55-54 B.C. Julius Caesar invaded Britain.

27 B.C. Augustus became the first Roman emperor.

100s A.D. Rome attained its greatest power.

293 The emperor Diocletian divided the Empire.

330 Constantine, the first Christian emperor, moved the capital to Byzantium.

476 The 'barbarians' under Odoacer deposed the last Western Roman emperor, Romulus Augustulus.

800 Charlemagne, king of the Franks, crowned emperor of the Romans by the Pope.

962 Otto of Germany founded the Holy Roman Empire.

1096-1099 The First Crusade sought to take the Holy Land from the *Saracens*, the Muslims; it established a Christian kingdom of Jerusalem.

1147-1291 Several later Crusades ended in failure and dissension among the Crusaders, who, during the Fourth Crusade (1202-04), sacked Constantinople and deposed the Eastern Roman emperor.

1337-1453 The Hundred Years' War between England and France. At its end only Calais was left in English hands.

1300s-1500s The Renaissance followed the medieval period, and with it came a new 'humanist' attitude to art and knowledge and a new interest in Europe's past.

1453 The Turks captured Constantinople.

1492 The last Muslim kingdom in Spain destroyed by the Spaniards.

1492 Christopher Columbus 'discovered' the New World.

1517 The Protestant Reformation began with Martin Luther's protest at the sale of indulgences.

1545-1563 The Council of Trent reformed the discipline of the Catholic Church, and began the Counter-Reformation.

1555 The Peace of Augsburg left the German lands half Catholic and half Protestant.

1571 The 'Turkish threat' ended by the defeat of the Turkish fleet by Don Juan of Austria at Lepanto.

1683 On land, the Turkish advance was ended by the victory of John Sobieski, king of Poland, at Vienna.

1618-1648 The Thirty Years' War limited the power of the German emperor and increased the influence of France.

1700s-1800s The Industrial Revolution began in England and led to the growth of industrial towns.

1783 End of the American War of Independence. England lost the American colonies.

1789-1799 The French Revolution began. The rise of Napoleon, and the first of the victories that carried him through Europe.

1815 Final defeat of Napoleon by the British and Prussians at Waterloo.

1848 The Year of Revolutions.

1870-1871 The Franco-Prussian War ended in the defeat of France and the creation of a new German Empire.

1914-1918 The First World War. Defeat of the Central Powers by the Allies, destruction of Austria-Hungary and abdication of the German emperor.

1917 The October Revolution in Russia led eventually to the victory of the Bolsheviks.

1933 Adolf Hitler and the National Socialists in power in Germany.

1939-1945 The Second World War. Defeat of Germany, Japan, and supporters by the Allies. Europe left in ruins.

1959 Treaties of Rome establish the European Economic Community.

THE COUNTRIES OF EUROPE

ANDORRA

Tiny republic in the E part of the Pyrenees, between France and Spain. The President of France and the bishop of Urgel in Spain are joint suzerains. The country consists of mountains and high valleys. Its river, the Valira, has a number of branches. Cattle and sheep graze on the hills, and crops are grown in the valleys. Tourism is important. The people speak Catalan, French, and Spanish. Andorra la Vella, a small town, is the cap., and there are several villages. Area 191 sq miles (495 sq km). For population, see Table on page 61.

AUSTRIA

Republic of central Europe, until 1918 the most important part of the Austro-Hungarian Empire. Austria has no sea-coast, and is one of the continent's most mountainous countries. The Alps, in the W and S, extend across two-thirds of the country. The highest peaks are Grossglockner (12,461 ft; 3,798 m) and Wildspitze (12,383 ft; 3,774 m). N of the Alps is the plain of the River Danube, with the wide Vienna plain at its E end. Beyond the Danube, the land rises northwards towards the mountains. The Danube, one of Europe's largest rivers, enters Austria from West Germany and flows E across the country to Czechoslovakia. Its tributaries include the Inn, Traun, Enns, and Ybbs, in the centre, and the Drau and Mur in the S. Austria has several large lakes, the chief of which are the Neusiedler Lake on the border with Hungary, and the Atter and Traun lakes in the NW. They are famous for their beauty. The Danubian plain has cold winters and warm summers. In Alpine areas the weather is often very cold and wet. In the valleys, most people live by agriculture. They raise cattle, sheep, and goats and, where the land is suitable, grow fruits and cereals, including rye and wheat. Manufacturing industries are sited chiefly in the low-lying areas of the N and E. There, factories produce iron and steel goods, textiles, leather goods, chemicals, paper, and processed foods. The main source of power is hydro-electricity. The country has valuable mineral resources, including iron, copper, coal, lignite, petroleum, and graphite. Tourism is an important source of income. Millions of summer visitors are attracted by the mountain scenery of the Tyrol and other areas, and by the

Austria. The castle of Werfen in the valley of the Salzach, south of Salzburg. Many of the Alpine valleys have similar medieval fortresses built in commanding positions.

ALBANIA

Shqiperia, 'the Land of the Eagle'. Republic in the Balkan Peninsula, on the E shore of the Adriatic Sea. The Albanian Alps cover most of the country, but there are lowlands in the W. Several rivers, including the Drin and the Semeni, flow W to the sea. Lake Scutari is on the N border, lakes Ohrid and Prespa on the E. The climate is Mediterranean in the W, severe in the mountains. There is little industry, but there are deposits of petroleum, coal, oil, and copper. The Albanian language has two dialect groupings—*Gheg* in the N, *Tosk* in the S. Chief towns: Tiranë (the cap.), Shkoder (Scutari), Korcë, Durrës (Durazzo). Area 10,630 sq miles (27,532 sq km). For population, see Table on page 61.

baroque towns and picturesque villages. In winter, some Alpine resorts are thronged with people who have come to enjoy ski-ing and other winter sports. Music lovers from all parts of the world visit the music festivals in Salzburg. Vienna is the cap. and largest city. Other major cities are Graz, Linz, Salzburg, and Innsbruck. Even the mountain areas have good road and rail communications. The Danube is an important international trade route, and can be navigated by heavy barges. By far the heaviest concentration of population is in the low-lying N and E. About a quarter of the people live in Vienna. Almost the whole population speaks German, except in some border districts. Area 32,373 sq. miles (83,846 sq km). For population, see Table on page 61.

Belgique (French); *Belgie* (Flemish). Country of W Europe, a constitutional monarchy. Its capital is Brussels. With Luxembourg and the Netherlands, it forms the customs union *Benelux*. In the W Belgium has a short coastline on the North Sea. The N and central half of the country is generally flat and rather featureless, and is crossed by the River Schelde and its tributaries. In the NW is Flanders. The flat land in the N, on the border with the Netherlands, is known as the Campine (Kampen). S of the plain are the valleys of the River Meuse and its tributaries the Sambre and Ourthe. Farther S and to the E are the beautiful wooded hills of the Ardennes. Signal de Botrange (2,277 ft; 694 m) in the Ardennes is the highest point in Belgium. Belgium has a mild climate, with warm summers and damp winters. Most of the country has less than 30 in

BELGIUM

Belgium. Brugge (Bruges) in Flanders, known for its romantic buildings and bridges and its quiet canals, was once a great commercial city.

(762 mm) of rain each year. Flanders and the central plain are regions of productive farmland. In the Campine the land is naturally poor, but some of it has been improved and put to agricultural use. Most Belgian farms are small. The chief crops include barley, oats, wheat, and sugar-beet. The only large-scale mineral resource is coal, but Belgium has nevertheless developed into one of the most highly-industrialized countries of Europe. Large amounts of raw materials are imported to feed its manufacturing industries. The chief industries are located in the N and in the valley of the Meuse and the Sambre. Liège, on the Meuse, is the country's chief steel city, and has many other industries too. Namur, Charleroi, and Mons, in the same area, are also centres of heavy industry. Several cities have important textile industries. Other major industries include brewing, chemicals, and glass. Antwerp is the largest city and chief port. Other large cities are Brussels, Ghent, Liège, Mechelen (Malines), and Schaarbeek. The road and railway systems carry much international traffic. Barges on the canals carry raw materials from Antwerp to the manufacturing centres, and take goods for export back to the port. The country is thickly populated, the heaviest concentrations being in the N and centre. There are two distinct groups of people: the Flemish-speaking northerners and the French-speaking Walloons of the S. Flemish is a Dutch dialect. Area 11,774 sq miles (30,495 sq km). For population, see Table on p. 61.

Belgium. The Grand-Place in Brussels, one of the finest squares in Europe. Its superb flamboyant buildings are intricately carved and gilded.

The People's Republic of Bulgaria lies in the E half of the Balkan Peninsula, and has a coastline on the Black Sea. The centre, W, and S of the country are mountainous. The Balkan Mountains run E-W across the centre, and the wild and rugged Rhodope Mountains across the SW part. Between them lies the broad valley of the River Maritsa. To the N, the great River Danube forms most of the boundary between Bulgaria and Romania. The flat plateau which lies between the Danube and the mountains is edged in places with sharp bluffs that drop as much as 600 ft (180 m) to the river. The climate varies considerably. The plains in the N are bitterly cold in winter and hot in summer. In the mountains, the climate is also severe, but it is generally mild in the Maritsa valley. Bulgaria is primarily an agricultural country. On the Danubian plain, farmers grow barley, maize, rye, and wheat, and raise sheep and other livestock. In the Maritsa valley, with its milder climate, tobacco and fruits are also grown. Plums are a valuable crop and are used for making slivovitz, a plum brandy. Sunflowers are cultivated for making cooking oil, and damask roses for the manufacture of attar of roses. Many manufacturing industries are related to agriculture, but there are also heavy industries producing chemicals, metals, and machinery. The country has sizeable deposits of coal, iron, lead, zinc, copper, manganese, and oil. Tourism is important. Visitors are attracted by the spectacular mountain scenery, and the beautiful beaches of the Black Sea coast. The people of Bulgaria come from two main stocks: Bulgar and Slav. Their language, Bulgarian, is Slavic. The chief cities are Sofia (the cap.), Plovdiv, Varna, Burgas, Dimitrovo (Pernik), and Ruse. Area 42,858 sq miles (111,002 sq km). For population, see Table on page 61.

Bulgaria. The Bay of Nesebur on the shore of the Black Sea. It lies between Varna and Burgas.

Czechoslovakia. The Tatra Mountains in the eastern half of the country are part of the Carpathian Range. There are numerous lakes in the valleys. Ski-ing and other winter sports attract many visitors.

Country in central Europe formed after World War I from territories that were formerly part of Austria-Hungary. It is bordered to the W by West Germany, to the N by East Germany and Poland, to the E by the U.S.S.R., and to the S by Hungary and Austria. Czechoslovakia has a communist form of government, although it retains many links with the countries of W Europe. Some powers are left to regional assemblies representing the main national groups. The country has three natural regions, which also correspond to national groupings. The western, Czech, part of the country is Bohemia. It is a basin-shaped plateau, fringed by mountains. In the low-lying centre is the city of Prague, on the Vltava River. To the NW is the mountain range of the Erzgebirge (Krušné Hory in Czech), and in the SW are the wooded hills of the Bohemian Forest. Moravia, the central section of the country, is a fertile region of low hills. It is drained by the Morava River. The eastern part of Czechoslovakia is Slovakia. It is largely occupied by ranges of the Carpathian system. The High Tatra, famous for its beautiful scenery, includes Czechoslovakia's highest mountain. This is the Gerlachovka, which rises to 8,737 ft (2,663 m). The southern part of Slovakia is extremely fertile. It lies in the Danube basin. The Danube, called the *Duna*, forms part of the country's southern border. On it is Bratislava, the chief city of Slovakia and an important river port. Many tributaries flow to the Danube from the hills of Moravia and Slovakia. From Bohemia, the Vltava (Moldau) and Labe (Elbe) rivers flow northwards into Germany. There are many small lakes in the mountains in all parts of the country. Czechoslovakia has, in general, hot summers and cold winters. There is heavy rainfall in most of the mountainous regions. Deposits of coal, iron, and uranium are the principal mineral resources. The country also has china clay, gold, graphite, lead, manganese, mercury, silver, zinc, and a silica sand suitable for making glass. But it has to import much of its raw materials. The main industries are metal refining, heavy engineering, glass-making, and textile manufacture. Farms in the lowland areas produce rich crops of cereals, potatoes, and sugar-beet. The chief languages are Czech, spoken in the W, and Slovak, spoken in the E. A few people speak German or Magyar. Chief towns: Prague (the cap.), Brno, Bratislava, Ostrava, Kosiče, and Plzen. Area 49,377 sq miles (127,886 sq km). For population, see Table on page 61.

The small country of Denmark in NW Europe is a constitutional monarchy. It is one of the oldest independent countries in the world. Denmark consists of a large number of islands, and an irregular peninsula that extends northwards from the N coast of West Germany. The peninsula, Jutland, is separated from the Scandinavian Peninsula by the Skagerrak on the NW and the Kattegat on the NE. These channels form part of the waters linking the North Sea with the Baltic Sea. The largest of Denmark's islands is Zealand (Sjaelland), which is separated from Sweden by a narrow passage called the *Sound* (Öresund).

Czechoslovakia. The 14th-century Charles Bridge in Prague, leading from Hradcany Castle to the Old Town on the east bank of the Vltava (Moldau) River. It is named in honour of the Holy Roman Emperor Charles VI (1316-78).

DENMARK

Denmark's territory includes the Faeroe Islands, lying between Iceland and Scotland, and the great barren land of Greenland, the largest island in the world. Most of Greenland—which is only a few miles from the coast of Canada, lies within the Arctic Circle. Despite its size, 840,000 sq. miles (2.176,000 sq. km), only about 37,000 people live there.

Denmark. The city hall in Copenhagen. Its clock tower is a well-known landmark.

Copenhagen, the country's capital, is on Zealand. Between Zealand and Jutland is another large island, Fünen (Fyn). It is separated from Zealand by a passage called the *Great Belt*, and from Jutland by the *Little Belt*. The next largest island is Lolland, to the S of Zealand and Fünen. Altogether, Denmark has about 500 islands, of which 100 or so are inhabited. Jutland is mostly low-lying. There are low hills in the centre, but they rise only some 500 ft (150 m) above sea-level. The W coast, on the North Sea, has sandy beaches and lagoons. The country inland from them is largely unproductive. On the E coast of Jutland there are several long, narrow fiords. The largest of them, the Lim Fiord, stretches most of the way across the country, almost cutting off the northern tip. A short canal, the Tyboron Canal, has been constructed to provide a passage through to the North Sea. Denmark has a considerable number of lakes, none of them large. It also has many rivers. The climate is temperate, and the rainfall is low, but because of its position Denmark is subject to gales sweeping in from the Atlantic. The country has almost no mineral resources. But it is one of Europe's leading dairy-farming countries, renowned for its butter and cheese. It is also a leading producer of high-quality bacon, and exports eggs and poultry to many countries. Much of its arable land is given over to the production of feeding-stuffs for livestock. In places, beet and potatoes are cultivated. The fishing industry is an important source of revenue. There are some manufacturing industries, chiefly based on engineering. They rely on imported raw materials. The people speak Danish, a language related to Swedish and Norwegian. The chief towns are Copenhagen (the cap.), Aarhus, Odense, and Aalborg. Area 16,620 sq miles (43,045 sq km). For population, see Table on page 61.

Finland, situated between Scandinavia and the U.S.S.R., is a land of lakes and forests. Its name in Finnish, *Suomi*, means 'the Land of the Lakes'. In the W it has a long coastline on the Gulf of Bothnia. Nearly all of the country is low-lying, but there are some mountains in the N. The N regions form part of Lapland. About one-tenth of Finland consists of lakes, the greatest concentration being in the S. Many of them form lake-systems—winding, irregular sheets of water, containing numerous islands. The largest lakes are Saimaa (680 sq miles; 1,761 sq km) and Päijänne (608 sq miles; 1,575 sq km). There are many short rivers, some with cataracts. The waters off the S coast are dotted with islands, including a major island group, the Åland Archipelago. About a third of the country is within the Arctic Circle, and has a severe climate. Elsewhere, the summers are warm though the winters are long and cold. Some two-thirds of Finland is forested, and timber is the basis of the chief industries, which include the manufacture of paper, chemicals, and furniture. There are also textile,

metal, and electrical industries. Though only a tenth of the country is farmland, cattle are raised and some cereals are grown. Many of the people of Finland are fair-haired and fair-skinned. Their language, Finnish, belongs to the same group as the languages of Hungary and Estonia. Swedish is recognized as a second official language. The chief cities are the cap. Helsinki (Helsingfors), Tampere (Tammerfors), Turku (Åbo), and Lahti. Area 130,165 sq miles (337,126 sq km). For population, see Table on page 61.

The French Republic is the largest country in western Europe, and one of the world's chief industrial nations. Although France suffered immense material damage in both World Wars, its economy today is stronger than ever before. At the same time, it has retained its leading position in relation to many aspects of 'good living'—in food and wines, in fashion, and in the production of such luxury goods as perfumes. The capital of France is Paris, on the River Seine in the NW of the country. Area 211,200 sq miles (547,000 sq km). For population, see Table on page 61.

Land and Climate In shape, France is roughly hexagonal. On the NE it has borders with Belgium and Luxembourg, and on the E with West Germany, Switzerland, and Italy. In the SE it has a coastline on the Mediterranean Sea. The independent principality of Monaco forms a tiny enclave on this coast. France's southern boundary consists of the high, wild Pyrenees Mountains, which separate it from Spain. In the E of the Pyrenees there is another tiny state, Andorra. And on the W and NW France has a long, irregular coastline on the Atlantic. The large island of Corsica in the Mediterranean is part of France.

Generally speaking, France is low-lying in the N and W, and high in the S and E. The various parts of the lowlands have their distinctive features. Gascony, in the SW, has marsh and grasslands. To the N of it is the rich region watered by the Garonne and other great rivers. Farther N are the broad plains of NW France, divided into two principal regions: the basin of the River Loire; and the *Paris basin*, whose heart is the Île de France.

The highlands are even more varied. In the NW, the hills of the Armorican Massif occupy Brittany and part of Normandy. The Massif Central in the centre and SE is a vast plateau that occupies about one-sixth of the total area of the country. To the NE are the

FRANCE: MINERALS, SOURCES OF ENERGY, INDUSTRY

▢ Coalfields
The size of the square is proportionate to production

⬭ Deposits of oil and natural gas and localities of extraction

⬭ Iron deposits

▨ Hydro-electric stations
▨ Other power stations } The size of the symbol is proportionate to production

◎ Chief industrial cities

◎ Other cities with important industries
The size of the circle is proportionate to population

🏭 Oil refineries   ◊ Natural gas

wooded hills of the Jura, the Vosges, and the Ardennes. The Alps rise along the borders of Switzerland and Italy in the SE. The highest Alpine peak in France is Mont Blanc (15,781 ft; 4,810 m). In the extreme S, the Pyrenees form a high, relatively narrow barrier to the Iberian Peninsula.

France's longest river is the Loire, which rises in the Massif Central and flows W for 650 miles (1,050 km) to the Bay of Biscay. The Garonne also empties into the Bay of Biscay. It rises in Spain, and flows NW to join the River Dordogne near Bordeaux. Together they form the wide estuary called the *Gironde.* Another great river, the Rhône, enters France from Switzerland, and flows W then S to the Mediterranean. In the N the chief rivers are the Seine and its tributaries, which empty into the English Channel.

Western and northern France has cool summers and mild winters. In the Massif Central winters are often very cold. Southern regions have a Mediterranean climate, with hot, dry summers.

Agriculture and Industry France is Europe's most productive agricultural country. The best farmlands are in the plains. Cereals are grown in most regions, but principally on the relatively large farms of the Paris basin and the areas running up to the Belgian border. The chief cereal crop is wheat, but barley, maize, oats, and rye are also grown extensively. Rice is cultivated in the marshy Rhône delta, called the *Camargue.*

One of the most important crops is sugarbeet. There are also high yields of potatoes and fodder crops. Market-gardening is common, particularly in the S. Normandy, Brittany, and Maine grow apples and pears, part of the crop being made into cider.

All parts of the country, except the N, have vineyards. The great vineyards of Bordeaux, the Loire, Champagne, Burgundy, Beaujolais, the Rhône, and Alsace give France its pre-eminent position among the world's wine-producing countries.

France has a historic textile industry, though other industries have overtaken it in importance. Chief among these are the production of iron and steel; and industries based on engineering and chemical technologies, such as the manufacture of motor vehicles, aircraft, precision instruments, electrical goods, plastics, and drugs. There are major rubber and oil industries. The chief mineral resource is coal. Other minerals include iron, aluminium, zinc, and salt.

Cities and Communications France's largest and most famous city is Paris, the

France. The château of Langeais on the Loire, built in about 1465 by one of the ministers of Louis XI, still has the menacing air of a feudal fortress. It contains fine tapestries and paintings.

The German Democratic Republic (East Germany) is the part of Germany that formed the Russian occupation zone after World War II. Apart from the U.S.S.R. it is the richest of the communist countries of Europe. East Germany is a NE segment of pre-war Germany, and includes Brandenburg, Saxony, Mecklenburg, and Thuringia. Although the city of Berlin is within its territory, only part of the city, *East Berlin*, belongs to East Germany. It is the country's capital. Area 41,660 sq miles (107,900 sq km). For population, see Table on page 61.

capital. Other large cities include Marseilles, Lyons, Toulouse, Nice, Bordeaux, Nantes, Strasbourg, St-Étienne, Le Havre, and Lille. All major cities are linked by road and rail, and there are some 5,000 miles (8,000 km) of waterways. There are several international airports.

The People The people of France vary greatly in type and ancestry. The people of Brittany, for example, are largely of Celtic origin and many still speak Breton, a Celtic language. Their neighbours, the Normans, are descended from Norse invaders. In the E many people are of Germanic origin, and those in Mediterranean areas are frequently of Italian descent. The Basques—who live in the Pyrenees and are found also in Spain—have their own language; its origin is unknown.

France. The Renaissance château of Chenonceaux in the Loire valley, spanning the River Cher. Its great gallery is nearly 200 ft (60 m) long. The château once belonged to Catherine de Medici.

Germany. The town of Braunfels on the River Lahn, near Wetzlar. The castle overlooking the town, formerly the seat of the princes of Solms-Braunfels, dates partly from the 1300s.

Land and Climate On the N, East Germany has a coastline on the Baltic Sea. Its E boundary follows the line of the rivers Oder and Neisse; beyond them is Poland. On the SE, the border with Czechoslovakia lies along the mountains of the Erzgebirge. To the SW and W is West Germany. The greater part of East Germany is low-lying. Its northern two-thirds is part of the great N European plain, which slopes generally downwards towards the N and NW. Some of it is marshy. The most northerly part of this region, the Mecklenburg plateau, has a great number of lakes and wooded hills.

The southern part of the country is mountainous. It includes, in the SW, the thickly-wooded hills of the Thuringian Forest, and in the SE the 'Ore Mountains', the Erzgebirge. The highest peak in the Erzgebirge is Keilberg (4,080 ft; 1,245 m). On the W, on the border with West Germany, are the lower Harz Mountains, the highest of which, Brocken (3,747 ft; 1,142 m), is famous in German legend.

Except in the S, East Germany has many lakes. Its chief river is the Elbe, which flows SE—NW across the country. Most of East Germany has a mild climate. But winters in the E are severe.

Agriculture and Industry Almost all agricultural land has been allocated to state farms or to collectives. Consequently, the farms are large by the standards of most European countries. In the central areas, where the soils are relatively poor, there is mixed farming. The chief crops are rye and potatoes, and great numbers of pigs are reared. Cattle, sheep, and poultry are also kept. The richest land in the country is in a belt running SE from the Harz Mountains. There, farmers produce large harvests of sugar-beet and wheat. Wheat is grown, too, in the N. In these Baltic lands, dairy farming is important, and consequently fodder crops are cultivated extensively.

East Germany is a major industrial country. Its principal industries are concerned with engineering, such as the manufacture of machinery and machine tools, and ship-building. Some of the cities of the S have long-established reputations for precision engineering and the manufacture of optical equipment and electrical goods. There are also highly-developed chemical and textile industries. Mineral deposits are few.

Cities and Communications The largest city is the capital, East Berlin, which has a population of more than one million. Other important cities are Leipzig, Dresden, Karl-Marx-Stadt (Chemnitz), Magdeburg, Halle, Erfurt, and Rostock. Although there are extensive networks of road and rail communications, many roads are in a bad state of repair.

The People The population of East Germany was swelled after World War II by the influx of refugees from the former German lands E of the Oder-Neisse line. But it later dropped as millions of people left to seek a better life in West Germany. The closing of the frontier halted this decline. But East Germany still suffers from a labour shortage. In recent years, there has been a growing sense of pride among the people because of their achievements in rebuilding their country with little outside aid.

The Federal Republic of Germany (West Germany) is one of the most powerful industrial countries in the world. It is composed of the areas of Germany that formed the British, American, and French zones of occupation after World War II, and is more than twice as large as East Germany. The federation consists of 11 *Länder* (states). One *Land* is West Berlin, which is within the territory of East Germany. But since West Berlin is still under military occupation, federal laws do not automatically apply there. The capital of West Germany is Bonn. Area 95,980 sq miles (248,590 sq km). For population, see Table on page 61.

Land and Climate On the N, West Germany has coastlines on the North Sea and the Baltic Sea, as well as a short border with Denmark in the Jutland Peninsula. In the E, it borders on East Germany and Czechoslovakia, and in the S, on Austria and Switzerland. On the W, it has frontiers with France, Luxembourg, Belgium, and the Netherlands.

The northern lowlands are part of the vast N European plain, which stretches right across the continent. Its most fertile areas are those to the S. The coastal lands bordering the North Sea are marshy in many places, but some bogland has been reclaimed by the building of dikes. Off-shore, the waters are shallow, and there are sandy islands. The Baltic coast has a quite different character, with many long inlets.

The central uplands form a much broken up region of hills, valleys, and plateaux. This region has some of the most varied scenery in Europe. In the W are the small hills and round, crater lakes of the Eifel.

GERMANY, FEDERAL REPUBLIC OF

Germany. An old street in Bad Wimpfen, on the River Neckar in Baden-Württemberg. The Blue Tower in the background is part of the ruined medieval palace of the Hohenstaufens.

Farther S and E are the most beautiful sections of the Rhine and Mosel (Moselle) valleys, with their steep, vine-covered and castle-crowned sides.

The several mountain ranges of the southern highlands include the Schwarzwald (Black Forest) in the SW, the highest peak

Germany. Below: The ornate Baroque church of the former Benedictine abbey of Zwiefalten, in the Swabian Jura. It dates from the mid-1700s, and incorporates a much older building.

Map legend:

1·5 sq mm = 2 million tons/km

☐ Trade upstream
▨ Trade downstream
····· Canals

Sources of energy in the Rhine basin

⬮ Coal deposits
☐ Chief centres of hydro-electric power
▨ Other important electric power installations
--- High tension power lines

🏭 Chief oil refineries

····· Oil pipelines

Cities

⊙ More than 500,000 inhabitants
○ 100,000 to 500,000 inhabitants

Danube flows SE into Austria, eventually reaching the Black Sea.

The climate of West Germany is generally mild. But the Alpine areas often have severe winters.

Agriculture and Industry Most German farms are small, generally run as family holdings. This is particularly true of the central uplands where the soil is poor in many sections and mixed farming is common. The lands just N of the uplands are extremely fertile, however, and produce large crops of wheat and barley, as well as sugar-beet and vegetables. Rye is another important crop.

In the NW, large areas of grassland are given over to stock raising. The land in this part of the country is not very suitable for cultivation. Livestock raising is also of major importance in the southern highlands, where the dairy industry has a high reputation. The valleys of the Rhine and its tributaries are known for their vineyards.

German industry is famous for its high productivity and its inventiveness. The chief industrial area is the *Ruhrgebiet*, a complex of industrial cities in the rich mining area of the Ruhr valley; the Ruhr is a NW tributary of the Rhine. West Germany's industries include almost all kinds of processing and manufacturing: steel-making; the manufacture of motor vehicles, engines, machine tools, electrical goods, precision instruments, drugs, fertilizers, dyes, textiles, plastics, optical instruments, furniture, toys, and ceramics; brewing and food processing; and printing.

Some of Germany's industry is based on its mineral resources. These include coal, potash, iron, oil, silver, copper, and lead.

Cities and Communications West Germany's largest cities are West Berlin, Hamburg, Munich, Köln (Cologne), Essen, Frankfurt am Main, Dortmund, Düsseldorf, and Stuttgart. The country's transport and communication systems are among the most highly-developed in the world. *Autobahnen* (fast motor roads) link the major cities, and are complemented by efficient railways. A complex waterway system, with the Rhine as one of its major components, plays a large part in industrial transport. There are several important ports in the N, and a number of international airports.

The People The largest concentrations of population are on the northern seaboard and in the Rhine and Ruhr valleys. The people are about 50 per cent. Protestant and 45 per cent. Roman Catholic.

of which is the Feldberg (4,695 ft; 1,430 m). The hills of the Schwarzwald, covered in thick, dark woods, figure in many German legends. To the extreme S are the Alps. Although the highest peak of the Bavarian Alps (Zugspitze) rises only to 9,720 ft (2,960 m), the scenery is among the most spectacular in Europe.

West Germany is a land of large rivers. The chief one is the Rhine, Europe's principal waterway. Other rivers that flow to the North Sea include the Ems, Weser, and Elbe. In the NE of the country, the Oder empties into the Baltic. And in the S, the

Germany. Bernkastel-Kues, a famous wine village in the Mosel (Moselle) valley. Among its buildings is the Cusanusstift, a refuge and hospital founded by the humanist Cardinal Nicholas Cusanus in the 1400s; its work still continues today.

GIBRALTAR

Gibraltar is a British territory at the southern end of the Iberian Peninsula. It is a small, rocky peninsula jutting out from the Spanish mainland, with which it is connected by a low isthmus. Morocco, in Africa, is about 14 miles (22 km) distant across the Strait of Gibraltar. The chief feature of Gibraltar is "the Rock"—the limestone promontory that rises 1,408 ft (429 m) above the waters of the Mediterranean. The rock is famous for the Barbary apes that live on it. On the W slopes of the rock is the town of Gibraltar. Gibraltar has been a British fortress and naval base since the early 1700s, but the Spanish government claims it as part of the territory of Spain. To the ancient Greeks it was one of the *Pillars of Hercules*. For population, see Table on page 61.

Greece is a mountainous republic in the eastern Mediterranean. Its people call it *Hellas*. The largest part of its territory is in the Balkan Peninsula, but it also includes a great number of islands. Greece is bordered in the N by Albania, Yugoslavia, and Bulgaria, and in the NE by European Turkey. The Ionian and Aegean Seas, arms of the Mediterranean, lie respectively W and E of the country. The S part of mainland Greece, Peloponnese, is almost an island. But it is linked to the rest by the narrow Isthmus of Corinth, through which runs a canal. Peloponnese is often called *Morea* because it resembles a mulberry leaf in shape. Crete, the largest island of Greece, lies SE of Peloponnese. Euboea, the second largest, is in the Aegean Sea, just off the E coast of the peninsula. Two famous groups of small islands also lie in the Aegean. They are the Cyclades and the Dodecanese. One of the Dodecanese is Rhodes. Altogether Greece has about 400 islands, most of them rocky and many of them vine-covered on their S-facing slopes. The low-lying coastland of NE Greece is the most densely-populated part. The N of the peninsula is mountainous. The Pindus Mountains, extending N-S, separate Epirus from Thessaly. They have peaks rising to more than 7,000 ft (2,130 m). Mount Olympus, N of Thessaly, rises to 9,558 ft (2,914 m) above sea-level. The south-eastern part of the mainland has broken mountains and lowlands. Peloponnese is

GREECE

Greece. A village in the Peloponnese, the southern peninsula of the Greek mainland. The villages of this part of Greece have a simple beauty, but life is hard for those who have to wrest a living from the sparse areas of arable land.

again mountainous. Towns and farmlands lie in valleys between the hills. Rivers flow fast in winter, but tend to dry up in summer. Greece has hot summers. In winter the coastlands have mild weather but in the mountains it is very cold. Greece has some mineral resources. There are large deposits of soft coal, and small amounts of various metals. Industry is increasing, but the country is predominantly agricultural. Farmers keep goats and sheep, and grow olives, grapes, currants, cotton, and tobacco. The language of nearly all the people is Greek. The chief towns are Athens (the cap.), Salonika, Patras, Piraeus, Vólos, Larissa, and Iráklion (Crete). Area 50,944 sq miles (131,944 sq km). For population, see Table on page 61.

HUNGARY

Hungary. Budapest, on the Danube, has two distinct sections: Buda on the hilly west bank of the river, and Pest on the flat east bank. The Margaret Bridge, in the background of the picture, connects the Margitsziget (Margaret Island) to the rest of the city. Budapest is the centre of Magyar culture.

Republic in central Europe that formed part of Austria-Hungary until the end of World War I. Today it has a communist system of government. Hungary is bordered on the N by Czechoslovakia and the U.S.S.R., on the E by Romania, on the S by Yugoslavia, and on the W by Austria. Most of the country is a lowland, consisting mainly of two plains —the Little Alföld in the NW and the Great Alföld stretching across the centre and E. The Great Alföld is split into three parts by two large rivers. The River Duna (Danube) forms part of the country's northern border, then swings southwards and flows to the southern border. Farther E, the River Tisza, a tributary, follows an extremely winding course, but one that is generally parallel to that of the Duna. In the W is central Europe's biggest lake, Lake Balaton, which covers an area of 245 sq miles (635 sq km). To the NW the lake is bordered by the wooded hills of the Bakony Forest. The Mecsek Mountains in the SW rise to more than 2,200 ft (670 m). In the N there are further ranges of hills, including the Matra and the Bükk Mountains. The Matra has Hungary's highest peak, Mt Kekes (3,330 ft;

1,015 m). The climate is cold in winter and hot and dry in summer. Most rain comes in the autumn. Hungary's mineral resources include bauxite, coal, iron, manganese, natural gas, petroleum, and uranium. The soil of the Alföld is fertile in the basin of the Duna. The principal crops include maize, wheat, sugar-beet, potatoes, barley, and rye. In the western part of the plain there are fine grazing lands. Fruit is an important cash crop in Hungary, and vineyards produce high-quality wines, one of which is the famous Tokay. Food processing is a major industry, but almost as important are the manufacture of road vehicles and rolling stock and the making of textiles. Hungary has to import the larger part of the raw materials needed by its heavy industry. Most of its trade is with other communist countries. The Duna provides a valuable trade route S and E to the Black Sea, and W to Germany. The country's main language is Magyar, a language related to Finnish. The chief cities are Budapest (the cap.) on the Duna, Miskolc, Debrecen, Pécs, and Szeged. Area 35,919 sq miles (93,030 sq km). For population, see Table on page 61.

Island republic in the North Atlantic, its N tip being just inside the Arctic Circle. Much of the island is a high, ridged plateau, with more than 200 volcanoes. The best known is the active volcano Mt Hekla (4,747 ft; 1,447 m) in the S. Clouds of steam drift from hot springs and geysers: the word *geyser* comes from the Great Geysir in the SW. Ice sheets cover one-eighth of the land. Elsewhere there is desert, scrubby moorland, and bog. Rivers carry away the water from melting glaciers, passing over waterfalls on their way to the sea. Because of the effect of the

ICELAND

Gulf Stream, Iceland's climate is less severe than might be expected. The chief industries are fishing and the processing of fish. In the limited areas suitable for farming, cattle, sheep, and horses are raised, and hay and vegetables are grown. The people of Iceland are descended from Viking settlers. Despite the rigours of life on the island, their standard of living is high. Half of them live in Reykjavík, the cap. and the only sizeable town, but there are a few other towns, too. Area 39,709 sq miles (102,846 sq km). For population, see Table on page 61.

IRELAND, REPUBLIC OF

The Republic of Ireland comprises five-sixths of Ireland, the second largest island of the British Isles. In Gaelic, its name is *Eire*. The rest of Ireland, the six NE counties of Ulster, form *Northern Ireland*, a part of the United Kingdom. But some of Ulster, including the most northerly part, is in the Republic. On the E, the Irish Sea and St. George's Channel separate the country from Great Britain. On the W and N is the Atlantic Ocean. The Republic of Ireland has a wide central lowland plain, consisting of good pasture and arable land and peat bogs—called turf bogs. Many low mountain ranges fringe the plain. Those in the W and S are particularly beautiful. MacGillycuddy's Reeks in the SW include Carrantuohill (3,414 ft; 1,040 m), the highest mountain in Ireland. The W coast is irregular, with many bays and long, narrow inlets. The River Shannon flows for 215 miles (346 km) from its source in the NW to its estuary on the Atlantic in the SW. It is the longest river in the British Isles. It passes through several lakes, including Lough Ree and Lough Derg. Among other lakes are the famous Lakes of Killarney in the SW and the remote, dark loughs of Connemara in the W. In the E the River Liffey flows into the Irish Sea at Dublin. S of Dublin are the Wicklow Hills, popular with climbers and walkers. Ireland's climate consists of mild winters and warm summers. There are some mineral deposits: copper, lead, zinc, and coal. Turf (peat) is in widespread use as fuel, either in its natural or in processed form. But Ireland's most important natural resource is its rich farmland. Agriculture is the chief industry, the raising of cattle and the production of milk, butter, and cheese taking

the leading place. Pig-breeding is also a major farming activity, and the country is celebrated for its horses. The principal crops include cereals and vegetables. The official languages of the country are Gaelic and English. The chief cities are Dublin (the cap.), Cork, Limerick, Dun Laoghaire, Waterford, and Galway. Area 27,135 sq miles (70,279 sq km). For population, see Table on page 61.

Ireland. The deserted beauty of the western seaboard, with its dark hills, lakes, moors, and boglands. Everywhere, ruined castles and monasteries tell of a turbulent past.

Italy. The Rialto Bridge in Venice, built in the 1500s, the oldest of the bridges across the Grand Canal. The bridge, a single marble span, has arcades lined with shops. It is named after the Rialto Islands on which part of the city is built.

The republic of Italy is a long, slender boot-shaped peninsula in the central Mediterranean. At the toe of the 'boot' is the large island of Sicily; and Italy also includes the island of Sardinia to the W. Italy is a country with a long history of civilization. Millions of tourists from other countries visit it each year because of its sunny climate, its beautiful scenery, its famous buildings, and its art treasures. Its capital is Rome, built on the River Tiber in central

ITALY

Italy, near the W coast. Area 116,303 sq miles (301,223 sq km). For population, see Table on page 61.

Land and Climate Right across the N of Italy is the great barrier of the Alps. To the E of the country is the Adriatic Sea, lying between the peninsula and Yugoslavia and Albania. To the S is the Ionian Sea, with its northern inlet, the Gulf of Taranto, forming the instep of the Italian 'boot'. To the W

83

Italy. The present and the heritage of the past. Above: Modern buildings in Milan seen across the pinnacles of the vast Gothic cathedral. Below: A legacy of classical times at Tivoli, near Rome. It was here that the Emperor Hadrian built his luxurious villa, said to have covered 7 sq miles (18 sq km).

Italy has a long coastline on the Tyrrhenian Sea, and in the NW on the Ligurian Sea.

Peninsular Italy—that is, the mainland of the country—has several distinct natural regions:

The Alpine Region includes a series of spectacular mountain ranges, with many peaks rising above 13,000 ft (3,900 m). The lower Alpine slopes are terraced for growing fruit, vines, and vegetables; and on the higher slopes there are Alpine meadows in many areas.

The Valley of the Po is a broad triangular lowland between the Alps and the Apennines. It is the most densely populated region of Italy, has many of the largest cities, and has much of the country's industry. It is also the principal agricultural region.

The North Adriatic Lowland extends NE from the Po valley. It includes part of the stony *Karst*, most of which is in Yugoslavia.

The Apennines, Italy's backbone, are more than 700 miles (1,100 km) long. They tower above the Ligurian coast and then, crossing the peninsula, follow the line of the Adriatic coast. On this coast, their foothills are much serrated. The average height is about 4,000 ft (1,200 m), but there are higher peaks, and Monte Corno in the Gran Sasso d'Italia rises to 9,560 ft (2,914 m).

The Western Coastal Plain lies along the Tyrrhenian Sea from La Spezia in the N to Naples in the S. It includes the hilly land of Tuscany. The Pontine Marshes and other unhealthy marshy areas near Rome have

Italy. The art and architecture of the Baroque. The Piazza Navona in Rome, with Bernini's superb 'Fountain of the Rivers'.

been reclaimed and turned into good agricultural land.

Rivers and Lakes. Most of Italy's rivers have their sources in the Apennines, but the two great rivers of the N, the Po and the Adige, flow from the Alps. The two other chief rivers are the Arno, on which Florence stands, and the Tiber. Alpine Italy has several lakes celebrated for their scenery: Maggiore, Lugano, Como, and Garda. Central Italy has a number of crater lakes. The largest is Trasimeno.

Climate. Italy's climate is varied. The N tends to have hot summers and cold winters, with much rain. In a large part of the peninsula and the islands, the winters are warm and there is only moderate rainfall. Some places suffer from drought.

Agriculture and Industry Italy has many minerals, but few of the deposits are large. They include coal, manganese, bauxite, and sulphur. In the basin of the Po, wheat and maize are important crops, and sugar-beet is grown, particularly in the W. Wheat is also produced on a lesser scale in the centre and the S. Other major crops include olives, citrus and other fruits, and vegetables. Vineyards are widespread: Italy produces some of the world's most celebrated wines. Fodder crops are grown to support the valuable dairying industry.

Much of the country's industry is centred in the NW, in the famous Milan-Genoa-Turin triangle. But some new industry has been established in the S, sometimes supplanting older industries of the N. The most important industries are based on steel and engineering technology. Of primary importance is the manufacture of motor vehicles and components. Italy is also one of Europe's leading nations in the textile and chemical industries.

Cities and Communications The chief cities are Rome (the cap.), Milan, Naples, Turin, Genoa, Palermo (Sicily), Bologna, and Florence. Communications are highly developed. To a large extent, the railways are electrified. Italy's *autostrade* (motorways) are among the finest highways in the world. There are many airports.

The People The highest concentrations of population are in the N and W. By contrast, the barren region of Calabria in the S has relatively few inhabitants. The language of Italy, Italian, has a 'standard' spoken and written version, but many dialects are used.

LIECHTENSTEIN

The tiny country of Liechtenstein lies in the valley of the upper Rhine, between Austria and Switzerland. It is ruled by an hereditary prince, but has a postal and customs union with Switzerland and uses Swiss money. The W half of the country is a plain bordering the Rhine. Most of the people live on this plain. To the E rise ridges of the Rhaetian Alps, some of them reaching heights of more than 8,000 ft (2,440 m). The lowland farms are small. They grow mixed crops, and there are many orchards. Livestock is raised on the lower mountain slopes, and there are terraced vineyards. Postage stamps are an important source of revenue, and so is tourism. The people are mainly Austrian in origin, and speak German. There are several villages, but only one town, Vaduz, the cap. Area 62 sq miles (161 sq km). For population, see Table on page 61.

Independent grand duchy in W Europe, situated between Belgium, West Germany, and France. With Belgium and the Netherlands it forms the customs union *Benelux*. Most of the country lies on the plateau of Lorraine, but in the N there are some of the beautiful wooded hills of the Ardennes. In

LUXEMBOURG

the E, the canalized River Moselle runs along the border with West Germany. There are also several smaller rivers. The country has a mild climate, but the winters are rather cold. Although Luxembourg is one of the smallest countries in Europe, it is one of the most prosperous. It has valuable iron deposits in the S which are the basis of its important iron and steel industry. Other industries include leather tanning and the manufacture of textiles and chemicals. Beer and wine are produced. Except in the N, the farms are productive, and cattle and other animals are raised. The people of Luxembourg speak a language called *Luxembourgeois*. Many also speak French and German. The chief towns are Luxembourg (the cap.), Esch-sur-Alzette, Differdange, Dudelange, and Pétange. Area 999 sq miles (2,587 sq km). For population, see Table on page 61.

Luxembourg. The city of Luxembourg, capital of the Grand Duchy, built on steep rocky heights above the River Alzette. The Pétrus gorge, crossed by bridges, separates the old town from the new.

Malta. Valletta, on the north-east coast of the island, is one of the best anchorages in the Mediterranean, and was for long an important naval base. The city is built on a rocky peninsula between two deep harbours. It has many reminders of the Knights of St. John of Jerusalem—the Knights of Malta—who ruled the island for more than 250 years.

MALTA

Island country in the Mediterranean Sea, 58 miles (93 km) S of Sicily. It consists of the islands of Malta and Gozo, and the smaller islands of Comino, Cominotto, and Filfla. From 1814 to the 1960s it was a British colony and was an important naval base. During World War II the island was awarded the George Cross by King George VI of the United Kingdom because of its people's bravery in the face of air attack. The country is generally low and rocky, with thin soil retained by terracing. There are no rivers or lakes. The climate is hot in summer and mild in winter, with moderate rainfall. Apart from the dockyard, industry is small, though growing. Most people live by agriculture or tourism. Malta exports some fruit and vegetables. The official languages are Maltese and English. The chief towns are Valletta (the cap.), Sliema, Paola, Rabat, and Victoria (Gozo). Area 122 sq miles (316 sq km). For population, see Table on page 61.

MONACO

The tiny country of Monaco lies on the Mediterranean Sea at the NE end of the French Riviera. It is an hereditary principality, and consists of a short strip of coast behind which rise foothills of the Maritime Alps. Almost the whole area is heavily built up. The chief source of income is tourism. The principality is famous for its gambling casino at Monte Carlo, for its motor sport events—the Monaco Grand Prix and the Monte Carlo Rally—and for its yachting harbour. The people are called *Monégasques*. Area 0·7 sq miles (1·8 sq km). For population, see Table on page 61.

A kingdom in NW Europe, the Netherlands is often called *Holland*. But, properly, *Holland* is the name only of its two western provinces. The Netherlands is one of the flattest countries in Europe, and some of it consists of land reclaimed from the sea. A famous feature of the Dutch landscape is the windmill. The country is dotted with hundreds of windmills, some of which were used for grinding grain, and others of which helped to pump away surplus water. The capital of the Netherlands is Amsterdam, though The Hague (s'Gravenhage) is the seat of the government. In addition to the home country, two oversea territories are considered part of the Netherlands: Surinam (Dutch Guiana) in South America, and the Netherlands Antilles. Area 13,500 sq miles (34,965 sq km). For population, see Table on page 61.

NETHERLANDS

Land and Climate The Netherlands is bordered on the E by West Germany, on the S by Belgium, and on the W and N by the North Sea. It has the deltas of three of Europe's largest rivers, the Rhine, Maas (Meuse), and Scheldt. It is very low-lying, its highest point—in the extreme SE—being only 1,057 ft (322 m) above sea-level. Many parts of the country are as much as 20 ft (6m) below sea-level, and are protected from the sea by dikes. The Netherlands has a long, low coastline, largely consisting of sand dunes that have been reinforced artificially. Just inland from the coastal dunes are the *polders,* the huge areas of land reclaimed from the sea, from swamp, or from fresh-water lakes. The largest reclamation area is around the IJsselmeer, formerly called the *Zuider Zee.* The IJsselmeer was once a bay of the North Sea, but is now a lake. Reclamation of further areas continues. Almost half the country is polder, and two-thirds of the polders are below sea-level. The eastern part of the country is relatively high, and is a region of sandy soil

and peat bogs. This generally infertile region is being reclaimed by scientific changing of the soil composition to make good crop-growing land. There are forests in this region.

The many rivers of the Netherlands are mostly delta arms of the Rhine and the

Netherlands. The Gothic tower of the cathedral of Utrecht, generally considered the most beautiful tower in the country. It rises some 370 ft (112 m), and its topmost stage is octagonal. It was built between 1321 and 1382.

Netherlands. The Voorburgwal in Amsterdam, lined with houses of many styles and periods. In the background is the Oude Kerk ('old church') of St. Nicolas, which dates from the 1300s. Amsterdam, sometimes called 'the Venice of the North', has about 50 miles (80 km) of canals, crossed by more than 400 bridges.

Maas, linked by a network of canals that serve both for transport and for drainage. The country's largest lake is the IJsselmeer. There are many smaller lakes in the polder regions. The climate of the Netherlands consists of mild to cold winters and cool summers, with moderate rainfall.

Agriculture and Industry The Netherlands has very rich deposits of natural gas, and good deposits of coal and petroleum. There are few mineral ores, but clay for brick-making and pottery is available, and brine yields quantities of table salt. The reclaimed polder lands provide fertile soil for growing crops and for providing good grazing. The main crops are wheat, sugarbeet, and vegetables, plus fodder for livestock. In the SW, hothouses produce tomatoes, lettuce, and cucumbers for export. Nearby, the sandy soil is planted with the bulbs that produce the fields of flowers for which the Netherlands is famous—particularly the fields of tulips.

Dairy products are of primary importance. Condensed and powdered milk, butter, and cheese are processed for export, and the country is Europe's largest exporter of eggs. Dutch fisheries provide fish for the home market.

The Dutch industries are large and flourishing, even though the raw materials have to be imported. Metal refining and engineering are the principal industries, with chemical manufactures a good third. Shipbuilding is another major industry. The Dutch also

BENELUX: ECONOMY

- ⊙ Chief urban areas
- Industrial areas
- Coalfields
- Iron deposits
- ● Coal-mining centres
- □ Metallurgical centres
- ◆ Textile centres
- ○ Centres of mixed industry
- ◍ Transport and other centres
- Internal waterways
- — Railways

specialize in cutting and polishing diamonds. A further major source of wealth is provided by the large volume of trade for central Europe that passes through the Netherlands or is carried in Dutch ships. The Dutch also provide banking services for international trade. Economically, the country is closely linked with its neighbours Belgium and Luxembourg to form the *Benelux* group, and with other members of the European Economic Community.

Cities and Communications The chief cities are Amsterdam, The Hague, Rotterdam, Utrecht, Eindhoven, Haarlem, Groningen, Tilburg, and Nijmegen. Road and rail communications are highly developed. So is the waterway system: there are more than 4,000 miles (6,400 km) of canals and other waterways.

The People The Netherlands is one of the most heavily populated countries in Europe. The people are mainly of Germanic origin. Their language is Dutch.

Netherlands. The crowded 17th- and 18th-century houses of the old part of Amsterdam, with their sharply-pointed gables.

Kingdom of northern Europe, forming the western part of the Scandinavian Peninsula. It is bounded on the W by the Atlantic Ocean and the North Sea, on the N by the Arctic Ocean, and on the E by the U.S.S.R., Finland, and Sweden. To the S is the Skagerrak, separating Norway from the peninsula of Jutland in Denmark. In the Arctic Ocean, 400 miles (650 km) N of peninsular Norway, is the large archipelago of Svalbard (Spitzbergen). High mountains rise sharply from the jagged western coast, which is indented with many fiords, mostly of spectacular beauty. Tens of thousands of islands fringe the coast, but most of them are very small and are uninhabited. They help to protect coastal shipping from Atlantic storms. The chief island group, the Lofoten Islands, lies N of the Arctic Circle. The highest mountain in Norway, Galdhöpiggen (8,140 ft; 2,481 m), is in the Jotunheimen Mountains in the S. Many swift rivers rush down the mountainsides, and there are many mountain lakes, the largest being Lake Mjösa (140 sq miles; 360 sq km) N of Oslo. The climate is cold in the mountains and the N, but warmer along the W coast as a result of the Gulf Stream. Summers are short and mild. Rainfall is heavy. In winter the mountains are covered with snow. Norway's natural resources include dense for-

NORWAY

Norway. Geiranger Fiord, in the south-west, one of the innumerable long and narrow inlets along the Norwegian coasts. Some fiords penetrate far inland, and ships can sail up them for many miles.

Norway. The Lofoten Islands, off the north-west coast, in the Norwegian Sea. Though they lie within the Arctic Circle, the islands have a relatively moderate climate because of the warming effect of the North Atlantic Drift.

ests of evergreen and hardwood trees. They are exploited for their timber, but replanting is practised. The main mineral resource is iron, and there are also some other metals. Large deposits of petroleum lie off the W coast. Dairying is the main agricultural activity. Norway has a large fishing fleet. There are two languages, Nynorsk (Landsmål) and Bokmål (Riksmål). The chief towns are Oslo (the cap.), Bergen, Trondheim, Stavanger, Drammen, and Haugesund. Area 125,181 sq miles (324,217 sq km). For population, see Table on page 61.

POLAND

Poland, in central Europe, is a country with few mountains but many forests, marshes, and lakes. It lies on the N European plain, and has a long coastline in the N on the Baltic Sea. On the E it is bordered by the U.S.S.R., on the S by Czechoslovakia, and on the W by East Germany. Like all of these countries it has a communist form of government. Most of Poland is flat, but the land rises slowly towards the Carpathian and Sudeten mountains in the S. The chief river is the Vistula, which rises in the Carpathians and flows northwards for 652 miles (1,050 km) to Gdańsk on the Baltic. With its tributaries, including the Bug and the Pilica, the Vistula drains the greater part of the country. Two other large rivers, the Oder and the Neisse, form the boundary with East Germany. Most of the lakes are in the N. Large areas of the E are occupied by dark and gloomy marshlands. The Polish climate tends to be warm and wet in summer, and cold in winter. Some rivers are frozen for several months a year. Poland is rich in minerals, including coal, lead, zinc, copper, sulphur, iron, and potassium. Many of the country's farms are small family holdings. The chief crops are potatoes, rye, wheat, oats, barley, and sugar-beet. Livestock includes pigs and cattle, as well as the horses for which Poland has long been famous. Among industries, the manufacture of steel and the making of various kinds of textiles and other clothing materials are important. Engineering works produce heavy machinery; Polish shipbuilding has a high reputation. The people of Poland are mainly Slavs. The country's name comes from that of a Slavic tribe, the *Polians* or 'plain dwellers'. The chief cities are Warsaw (the cap.), Lódź, Kraków, Poznań, Lublin, and Gdańsk. Area 120,360 sq miles (311,730 sq km). For population, see Table on page 61.

Poland. The Palace of Culture in Warsaw, showing the post-war influence of the modern Russian grandiose style.

PORTUGAL

Portugal. Below: The 13th-century castle in the walled village of Obidos.

Republic in the W of the Iberian Peninsula, with a long coastline on the Atlantic. On the N and E it is bordered by Spain. The high central plateau of Spain extends into the northern half of Portugal. The country here is wild and mountainous and much broken-up by valleys. One peak, in the Serra da Estrella, rises to 6,532 ft (1,991 m). Along the Atlantic coast and in the S the land is low-lying and much of it is very fertile. The E-central part of the country is occupied by highlands. There are several large rivers. They all rise in Spain, and flow W or S across Portugal to the Atlantic. From N to S, they are the Minho, the Douro, the Tagus, and the Guadiana. Their valleys are generally narrow and rocky, but in some places they are wide and contain rich farmlands. Long stretches of Portugal's coast have fine sandy beaches popular with tourists. There are few natural harbours except at the river mouths. In the lowland regions the climate is warm and equable, but in mountain areas the winters are severe. Portugal has some mineral resources of importance, including wolframite, coal, lead, copper, and antimony. Its forests are a major source of revenue; Portuguese cork oaks produce about half of the world's cork. The country is predominantly agricultural, the main farming regions being in the central highlands and the W. Crops include rice, wheat, maize, barley, potatoes, and beans. Olive trees are found in most areas. In the S, citrus fruits, figs, and

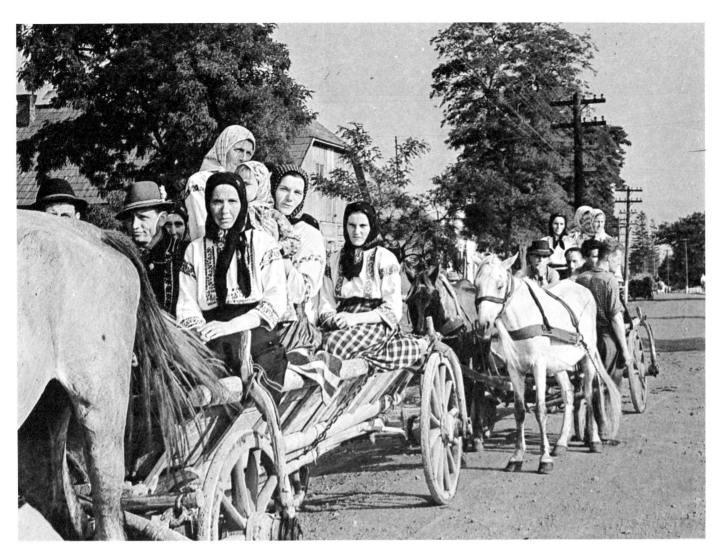

almonds are grown. Portugal is a leading wine-producing country. Its most famous wine, port, is made from grapes grown in the Douro valley and is named for the town of Oporto. Portuguese fishermen catch cod, tuna, and sardines. Sardines are a profitable export. Manufacturing industry is not highly developed, but there are factories producing textiles, steel, and ceramics. The country's language, Portuguese, is not unlike Spanish. The chief towns are Lisbon, Oporto, Setúbal, and Vila Nova a Gaia. Area 34,240 sq miles (88,680 sq km). Until the mid-1970s, Portugal's oversea possessions included the vast African territories Angola and Mozambique. The Madeira Islands and the Azores are part of Portugal. For population, see Table on page 61.

Romania. Villagers 'dressed for the occasion'. The people of the country's rich farmlands have a strong sense of tradition, and faithfully preserve the customs of their ancestors.

Republic of SE Europe with a coastline on the Black Sea. Its name means 'land of the Romans'. It has a communist form of government. Romania is bordered on the N by the U.S.S.R., on the S by Bulgaria, and on the W by Yugoslavia and Hungary. The country is extremely mountainous: the Carpathian Mountains and the Transylvanian Alps form a great arc from N to W. They

ROMANIA

enclose the plateau of Transylvania in the NW. The wide plain in the S of the country lies in the basin of the River Danube, which runs along the southern boundary. Summers are warm, except in the high mountain regions, but winters tend to be severe. Romania has rich mineral resources, particularly petroleum and natural gas. The famous oilfields of Ploeşti and Buzău are in the SE. Other valuable natural products are coal, iron, rock salt, and timber from the great forests. On the vast plains of the S, farmers grow wheat and maize. Romania also produces sugar-beet, fruit, and potatoes. And it is known for its wines, some of them of high quality. Among the products of its fishing industry is caviare—sturgeon roes. The government is trying to extend the manufacturing industries, particularly those based on metal and chemical technologies. The early inhabitants of what is now Romania were the Dacians. Today's Romanians are descended from them and from all the many peoples of S and E Europe who later arrived as invaders or settlers. The chief cities are Bucharest (the cap.), Braşov, Ploeşti, and Iaşi. Area 91,700 sq miles (237,500 sq km). For population, see Table on page 61.

SAN MARINO

San Marino, which was founded in the A.D. 300s, claims to be the oldest republic in the world. It is also the world's smallest republic. It is situated in northern Italy, on the E slopes of the Apennine Mountains, near the Italian town of Rimini. The republic's capital, a walled town also named San Marino, is built at the top of a steep, rocky, three-peaked hill, 2,420 ft (737 m) high. The chief sources of revenue are tourism and the sale of postage stamps, but there are several small industries, including the manufacture of cloth, dyes, and ceramics. Quarries provide stone for building. The people make oil from olives, and grow fruit and some cereals. There are many vineyards, in which dessert grapes are grown as well as grapes for wine-making. The citizens of San Marino, known as *Sammarinesi*, speak Italian. They have an elected parliament of 60 members. A grand council elects two regent captains to perform the functions of government. Area 24 sq miles (62 sq km). For population, see Table on page 61.

SPAIN AND PORTUGAL: INDUSTRY

PRIMARY INDUSTRIES   The size of the symbol is proportionate to the amount of production
○ Coal   ● Pyrite   ⋀⋀ Salt   ▢ Lead   ◆ Manganese   ⊙ Tungsten
○ Iron   △ Potassium   ★ Copper   ▪ Zinc   ▼ Tin   ✦ Mercury

SECONDARY INDUSTRIES   The size of the symbol indicates the amount of production, the refinery capacity or the power output
▢ Metallurgical   ◆ Textile (cotton and wool)   ▼ Food   ▫ Hydro-electric stations
◇ Chemical   ▲ Man-made fibre   🏭 Refineries   ▨ Other power stations

SPAIN

Spain occupies the greater part of the Iberian Peninsula in SW Europe. Its extreme southern tip on the Strait of Gibraltar is only 9 miles (14 km) from Morocco, in Africa. The Canary Islands in the Atlantic and the Balearic Islands in the Mediterranean are ruled as part of the country. The capital is Madrid. Area 194,885 sq miles (504,750 sq km). For population, see Table on page 61.

Land and Climate On the E, Spain has a long coastline on the Mediterranean Sea, and on the W and N an even longer coastline on the Atlantic. The Pyrenees in the NE form the boundary with France. Three-quarters of the country is a vast plateau, the *Meseta*, parts of which are as high as 6,000 ft (1,800 m) above sea-level. Round the edges of the plateau is a coastal plain, which is widest in the SW and W. Several mountain ranges lie across the country in a W to E direction. The northernmost range, the Galician and Cantabrian Mountains, are a westward extension of the Pyrenees. The chief rivers are the Douro, the Tagus, the Guadiana, the Guadalquivir, the Ebro, and the Minho. The S and E coastal areas have mild winters and warm summers. The rest of the country has greater extremes of climate.

Agriculture and Industry On the plateau, agriculture suffers from lack of water. But in most places some animals graze. In the plains, the most important crop is wheat, and some upland areas have oats and rye. Other cereal crops grown are rice,

92

maize, and barley. Olives contribute to the livelihood of many farmers. There are thousands of sq miles of vineyards. And the E and S of the country are famous for their citrus fruits, almonds, peppers, figs, tomatoes, and onions. Spain has valuable mineral deposits—mercury, coal, iron, copper, and salt. The chief manufacturing industries are in the N. They include textile, engineering, and paper industries. Tourism is important to the country's economy.

Cities and Communications Spain has many beautiful and historic cities. The largest city is Madrid, the capital. Barcelona is an important industrial city and port. Other large towns are Valencia, Seville, Zaragoza, Malaga, and Bilbao. In general, communications are not highly developed, because the country has few natural routes.

The People The Spanish people are descended from the Iberians and from the Celts, Carthaginians, Romans, Moors, and other peoples who later settled in the peninsula. Most of them speak Spanish. In the NE, around Barcelona, Catalan is spoken, and in the NW, Galician, a language similar to Portuguese, is used. In the western Pyrenees the people speak Basque.

Spain. The Alhambra at Granada, built in the 1200s and 1300s, the most splendid example of Moorish art in Europe. Its buildings formed the palace and citadel of the Moorish kings of Spain. Exquisitely-decorated halls, rooms, and alcoves surround open gardens and courts, the most famous of which is the Court of Lions, with its delicate white marble arcades.

SCANDINAVIA: ECONOMY

Legend:
- Agriculture and industry
- Agriculture
- Forested areas
- Pasturage
- Fishing

- Mining districts
- Iron
- ← Rail transport routes
- ⇨ Iron export routes
- ● Coal
- ☆ Gold
- ☆ Silver
- ★ Copper
- ⬡ Lead
- ▪ Zinc
- ◆ Manganese
- ○ Tungsten
- ▽ Nickel
- ◇ Molybdenum

SWEDEN

The people of Sweden, the largest country in the Scandinavian Peninsula, enjoy one of the highest standards of living in the world. The country is a representative monarchy. On the N, it has boundaries with Norway and Finland. On the E and S it has a long and greatly broken coastline on the Gulf of Bothnia and the Baltic Sea. On the SW, the Kattegat and Skagerrak separate it from Denmark. And on the W it is flanked by Norway. The northern part of the country, called *Norrland*, lies partly within the Arctic Circle; the frozen northern lands belong to *the land of the midnight sun*. The mountains that rise along almost the entire western side of Sweden slope gradually down to the coastal plains of the E and S. The N of the country has many long, narrow lakes in the mountain valleys. The S has even more lakes, some of which, such as Lake Vanern (2,140 sq miles; 5,540 sq km), cover large areas. Some offshore islands in the Baltic

Sweden. The monument of Gustavus II, Adolphus (1594-1632) in Hälsingborg. The soldier-king who led Sweden's armies through Europe was killed at the battle of Lützen, a battle in which his troops were victorious.

dant hydro-electric power, manufacturing industries are highly developed. Swedish products that have a high reputation for quality include aircraft, motor vehicles, heavy machinery, ships, precision instruments, paper, glass, and foodstuffs. The people of Sweden are related to the peoples of Denmark and Norway. The Swedish language is similarly related. The chief cities are Stockholm (the cap.), Göteborg and Malmö. Area 173,400 sq miles (449,110 sq km). For population, see Table on page 61.

SWITZERLAND

Switzerland, a small country in central Europe, is celebrated for its scenery, its prosperity, and the fact that for more than a hundred years it has succeeded in standing aloof from the wars and quarrels of the other countries of Europe. It is a federation of 22 little states called *cantons*. Switzerland has no coastline. It is bordered by West Germany on the N, Austria and Italy on the E, Italy on the S, and France on the W. Three-quarters of the country consists of mountain ranges. The Jura Mountains extend in an arc in the NW. The Alps stretch across the whole of the southern part. Between is the plateau of the Mittelland. Two of Europe's largest rivers rise in the Swiss Alps. The Rhine originates as two headstreams that join and flow into Lake Constance. The Rhône has its source in a glacier, and flows westwards through Lake Geneva to France. There are many waterfalls, some of them having falls of nearly 1,000 ft (300 m). The climate varies. The higher regions have much snow. Places on S-facing slopes have in general more equable climates than those facing N. In spring, the Mittelland has a prevailing warm SW wind called the *föhn*. Dairy farming is the chief form of agriculture; Swiss cheeses and other dairy products are known all over Europe. In the Alps, the cattle are driven up into the mountain meadows to graze in summer. Some cereals are grown, particularly in the Mittelland. On the warmer hill slopes there are vineyards and fruit trees. Swiss industry is highly developed, specializing in products that require skilled workmanship: watches, optical instruments, musical instruments, and engines. The country is the centre of international banking. Another major source of revenue is tourism. The people speak German, French, or Italian, according to the locality in which they live. Some also speak a language called *Romansh*. The chief cities are Zürich, Basel, Geneva, Bern (the capital), Lausanne, Winterthur, and St. Gallen. Area 15,940 sq miles (41,284 sq km). For population, see Table on page 61.

Switzerland. The Marktgasse in Bern, the federal capital. The clock-tower was part of the medieval city walls. Its 16th-century clock has puppet figures that parade each hour.

are part of Sweden. Despite its northerly location, much of the country has a temperate climate. This results partly from the effect of the Gulf Stream. Sweden makes good economic use of timber from the forests that cover some two-thirds of its area. Other valuable natural resources are iron, zinc, lead, and copper. The greater part of the country is unsuitable for agriculture, but there are some rich farmlands in the S, where dairy farming is important. The fishing industry, also important, is concentrated chiefly in the SW. Partly because of abun-

Turkey, which lies between the Mediterranean Sea and the Black Sea, is mainly an Asian country. But in the NW its territory extends into the Balkan Peninsula, which is part of Europe. European Turkey, called *Thrace*, contains two of the country's chief cities, Istanbul and Edirne. The two parts of Turkey are separated by the Dardanelles, the Sea of Marmara, and the Bosporus. For the main article on Turkey, see page 145.

The Union of Soviet Socialist Republics is the largest country in the world. Three-quarters of its territory is in Asia. But Moscow—the country's capital—and the majority of its other important cities are in Europe.

The U.S.S.R. is made up of 15 *union republics.* It has been said that it is really more an empire than a country, because its citizens belong to many different races and cultures. Only the United States rivals it in industrial and military power. It is the leading communist country, though its leadership is challenged by China.

The U.S.S.R. is often called the *Soviet Union* because its constitution provides for government by a system of councils known as *soviets.* It is also frequently known simply as *Russia.* The use of this term is not strictly accurate, because only some 60 per cent. of the people who live in the U.S.S.R. are Russians.

The distance across the U.S.S.R. from E to W is 5,250 miles (8,450 km). The country's area is 8,649,000 sq miles (22,400,800 sq km). The area in Europe is 2,151,000 sq miles (5,571,100 sq km) and in Asia

U.S.S.R. The church of St. Basil in Moscow, on the southern side of Red Square. It was built in the mid-1500s, during the reign of Ivan the Terrible. Each of its eight multi-coloured 'onion' domes has a different shape.

U.S.S.R. The former Winter Palace of the czars in Leningrad. The huge Baroque building now houses part of the Hermitage Museum.

U.S.S.R. A demonstration in Red Square, in Moscow. The vast square, more than half a mile (0.8 km) long, is bordered by the wall of the Kremlin. Red Square is the site of the great parades that take place on May 1 (May Day) and November 7 (the anniversary of the October Revolution).

6,498,000 sq miles (16,829,700 sq km). For population, see Table on page 61.

Land and Climate In the N the U.S.S.R. has a long coastline on the Arctic Ocean, and in the E a long coastline on the Pacific. At its most north-easterly point it is only 56 miles (90 km) from Alaska in North America, across the Bering Strait. The Kamchatka Peninsula and the Kuril Islands separate the Sea of Okhotsk from the Pacific.

On the S the U.S.S.R. has borders with China, Mongolia, Afghanistan, Iran, Turkey, and Romania. The Caspian Sea and the Black Sea lie on this southern boundary. In the W the U.S.S.R. borders Hungary, Czechoslovakia, Poland, Finland, and Norway. In the W, too, it has a coastline on the Baltic. The country has six land regions:

The East European Plain occupies most of the European part of the country. To the E it extends as far as the Ural Mountains. In the S are the Caucasus Mountains, and in the SW the Carpathians. The SW part of the plain, bordering on the Carpathians and the Black Sea, is called the *Ukraine*. Hills rise here and there, the highest being in the Kola Peninsula, between the White Sea and the Barents Sea in the N.

The Turanian Plain lies around the Aral Sea in the S. Some of it consists of plateaux and grasslands, but it also includes two barren, sandy deserts, the Kara Kum ('Black Sands') and the Kyzyl Kum ('Red Sands'). There are also steppes. In the so-

▨	Spring sown wheat, sugar beet and cattle farming
☐	Winter sown wheat, maize and other cash crops
☐	Flax, fodder and stock rearing
▨	Cereals and stock rearing
▨	Mainly forest, with some farming in river valleys
▨	Fruit and vineyards
☐	Arid and semi-arid grazing lands
☐	Winter sown cereals, sugar beet and other cash crops
▨	Intensive fruit and vegetable farming with cattle
☐	Tea and citrus fruits
☐	Upland pastures (for sheep) with mixed farming in valleys
▨	Cattle rearing with some grain farming
▨	Sheep in Central Asia with cotton, fruit and cereals
▨	Reindeer herding in the Tundra region
▨	Cattle and horse breeding
◯	Industrial regions

called Caspian Depression, on the northerly shores of the Caspian Sea, is the lowest point in the U.S.S.R., some 400 ft (120 m) below sea-level.

The Central Asian Mountains are S and E of the Turanian Plain, along the borders of Afghanistan and China. The region has many plateaux, as well as mountains of the Alai, Pamir, and Tien Shan ranges. The country's highest peak, Mt. Communism (24,590 ft; 7,495 m), is in the Pamirs.

The West Siberian Plain lies between the Ural Mountains in the W and the Yenisey River in the E. This region is the greatest stretch of continuous plain in the world, and has an area of about one million sq miles (two and a half million sq km). Vast areas of the plain are bleak and marshy; they are the Siberia of literature.

The Central Siberian Plateau extends eastwards from the Yenisey to the Lena River. In its southern half it reaches elevations of 2,000 to 3,000 ft (600 to 900 m) above sea-level, but peaks of the Sayan and Yablonovyy Ranges rise to some 10,000 ft (3,000 m).

The East Siberian Highlands consist of several mountain ranges, deep valleys, and plateaux between the Lena River and the Pacific Ocean. This cold, wild, and remote region has spectacular scenery. Some of its peaks rise to more than 10,000 ft (3,000 m). The Kamchatka Peninsula, its most westerly part, has more than 20 active volcanoes.

U.S.S.R. A new town in Siberia, the Asian part of the Soviet Union. Siberia has rich natural resources, now being exploited.

U.S.S.R.: AGRICULTURAL AND INDUSTRIAL REGIONS

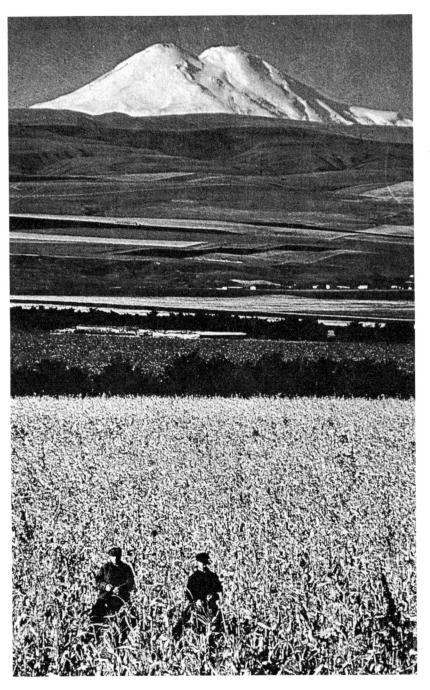

in the Altai Mountains, and flows NW and N for 2,120 miles (3,410 km) to the Gulf of Ob in the Arctic. Its banks are marshy, and flood regularly. The Yenisey, also more than 2,000 miles (3,200 km) long, rises in the Sayan Ranges and flows N to the Arctic along the edge of the Central Siberian Plateau. The Lena, the largest river of the U.S.S.R., rises near Lake Baykal and flows northwards for some 3,000 miles (4,800 km) to the Laptev Sea in the Arctic. Its delta is about 250 miles (400 km) wide.

Inland Seas and Lakes. The Caspian Sea is the world's largest inland body of water, and has an area of about 170,000 sq miles (440,000 sq km). Its waters are salt. The nearby Aral Sea, also salt, has an area of about 26,000 sq miles (67,000 sq km). Lake Ladoga, near the Baltic coast in the NW, is the largest freshwater lake in Europe. Its area is 1,700 sq miles (4,400 sq km). Lake Baykal, on the edge of the Central Siberian Plateau, has an area of about 12,000 sq miles (31,000 sq km). The largest freshwater lake in Asia, it is also the deepest body of fresh water in the world, with depths of some 5,700 ft (1,700 m).

The Climate of the U.S.S.R. varies greatly from place to place. The Turanian Plain has hot deserts, but Siberia has some of the coldest areas in the world. In general, the country has a continental climate with hot summers and cold winters.

Agriculture and Industry   Only about a quarter of the land of the U.S.S.R. is farmed, and of this less than half is cultivated. Nevertheless, the total area of farmland is greater than in any other country. Nearly all the land is organized either in collectives or in state farms. The chief crops are cereals. The U.S.S.R. is the world's greatest producer of wheat and rye. Most of the wheat is grown on the steppes. Barley and maize are also important. Other major crops are potatoes, sugar-beet, fodder, flax, and tobacco. Away from the great grain lands, mixed farming is common, and even in central and eastern Siberia there are productive pockets of agriculture. The raising of livestock plays a large part in the country's economy. In the N there are dairy farms. In the Ukraine, pig-farming is traditional. And in many areas not otherwise productive, huge flocks of sheep graze.

Fishing is a major industry. Fishing fleets operate in the domestic seas—the Baltic, Black, and Caspian—and along the coasts. Russian deep-sea trawlers are a familiar sight in many of the world's fishing grounds.

Timber from the thousands of square miles of forests is one of the most valuable

Rivers. The great rivers of the European part of the U.S.S.R. include the Volga, the Don, the Dnieper, and the Ural. All these rivers drain the steppes, the temperate grasslands extending across the U.S.S.R. just N of the Black Sea and the Caspian Sea. The Volga, the largest river in Europe, has its source in the Valdai Hills near the Baltic Sea, and flows for 2,325 miles (3,750 km) E and S to the Caspian. The wide, slow-flowing Don rises near Moscow and meanders southwards to the Sea of Azov, the northern gulf of the Black Sea. The Dnieper, which drains the Ukraine, also empties into the Black Sea. It is Europe's third longest river. The Ural flows S from the Urals to the Caspian.

The Siberian regions have several great rivers. The largest are the Ob, the Yenisey, and the Lena. The basin of the Ob is practically the whole of W Siberia. The river rises

U.S.S.R. Golden fields of crops in Georgia. In the background is Mount Elbrus (18,481 ft; 5,633 m), the highest peak in the Caucasus Mountains and in Europe.

natural resources. But the U.S.S.R. also has extremely rich mineral deposits. The chief of these are coal, iron, manganese, and petroleum. But there are also many other minerals of economic significance, including copper, aluminium, asbestos, and tungsten.

In manufacturing industry, the government has until recently concentrated on the production of heavy industrial goods—steel, rolling stock, motor vehicles, industrial plant, chemicals, and so on—as well as armaments. But an increasing proportion of industry is now devoted to consumer goods to raise the living standard of the people. Nearly three-quarters of industry is concentrated in the European part of the country.

Cities and Communications Moscow is the most heavily populated city as well as the capital. It and Leningrad are among the largest cities in Europe. Other cities with populations of more than one million are Gorki, Novosibirsk, Kuybyshev, Sverdlovsk, Kiev, Kharkov, Minsk, and Baku. In the European regions, communications are reasonably well developed. Elsewhere they are poor. Railways are the most important means of transport. One famous line, the Trans-Siberian Railway, which links Moscow with the Pacific coast, is more than 5,000 miles (8,000 km) long. Several new railways are helping to open up Siberia.

The People More than 60 per cent. of the people of the U.S.S.R. live W of the Urals. The Russians are by far the largest national group in the population. The Ukrainians are the next largest. Other large national groups are Belorussians, Uzbeks, Kazakhs, Tatars, Azerbaijani, Armenians, and Georgians. Each of these numbers several million people. There are also numerous smaller groups. The official language of the U.S.S.R. is Russian, which is taught in all schools. But about 60 different languages are spoken in various parts of the country.

UNITED KINGDOM

The United Kingdom of Great Britain and Northern Ireland comprises the whole of the British Isles except for the part of Ireland that forms the Republic of Ireland. Included in the U.K. are the Isle of Man, in the Irish Sea, and the Channel Islands, off the coast of Brittany in France. The U.K. is a constitutional monarchy. Its principal parts are the kingdoms of England and Scotland, the principality of Wales, and the province of Northern Ireland. It is the centre of the Commonwealth of Nations, the world-wide association of countries that has succeeded the former British Commonwealth and Empire. The U.K. still has a few oversea possessions.

The capital of the U.K. is London, in SE England. Three other cities have capital status, too: Edinburgh for Scotland; Cardiff for Wales; and Belfast for Northern Ireland. The area of the U.K. is 94,216 sq miles (244,018 sq km). For population, see Table on page 61.

Land and Climate The British Isles lie in the Atlantic Ocean on the continental shelf of the European mainland. Great Britain, the largest island in Europe, is sep-

United Kingdom. Above: The chalk uplands of Salisbury Plain, in Wiltshire, cover 300 sq miles (780 sq km). The plain is renowned for its beef cattle, and has long been used as a military training area. Right: The ruins of Whitby Abbey in Yorkshire, a Benedictine foundation built in 1078 on the site of an abbey dating from 657.

United Kingdom. Edinburgh Castle, a complex of buildings of several periods, stands on a hill 445 ft (136 m) above the centre of Scotland's capital city.

United Kingdom. Tower Bridge crosses the River Thames at the entrance to the Pool of London. The roadway hinges up in two sections to give large ships access to the Pool.

arated from the mainland by the North Sea and the English Channel. Between Great Britain and the island of Ireland are the North Channel, the Irish Sea, and St. George's Channel. The coastlines of both islands are deeply indented, and there are many small offshore islands, including (apart from the Isle of Man and the Channel Islands) the Shetland Islands, the Orkney Islands, the Hebrides, Anglesey, the Isles of Scilly, and the Isle of Wight.

The Atlantic Ocean has a dominating effect on the climate. The prevailing SW winds bring much rain to the higher districts of the W. But the warm currents of the North Atlantic Drift help to moderate temperatures. The lowest temperatures are experienced in Scotland and in the Pennines. The SE generally has the warmest summers.

England, the largest of the four countries of the U.K., has an area of 50,334 sq miles (130,364 sq km). In general, it is high in the W and lower in the E. Its chief mountain range is the Pennine Chain, extending from the Cheviot Hills on the Scottish border as far S as Derbyshire. The Pennines are often called 'the backbone of England'. Their highest point is Cross Fell (2,930 ft; 893 m) in the N. At their southern end is the Peak District, noted for its limestone caverns.

In the NW, between the Pennines and the coast, is the Lake District. There, in the Cumbrian Mountains, is Scafell Pike (3,210 ft; 978 m), the highest peak in England. The lakes or *meres* of the Lake District include

Windermere, the country's largest lake, which is about 10 miles (16 km) long.

Farther to the S are the Midlands, a region of smooth hills and valleys. To their E are the Fens, inland from the wide, shallow bay called the *Wash*.

The low plateau of East Anglia occupies the 'bulge' S of the Wash. To its SW is the valley of the River Thames. Between the Thames and the English Channel lies the chalk country of the Weald and the Downs. This continues W to Salisbury Plain. Beyond lies the beautiful rocky peninsula of Devon and Cornwall.

Southern England has two parallel ranges of low hills, the Cotswolds and the Chilterns. Both are in areas renowned for their beautiful countryside.

Most of England's rivers have their sources in the central uplands. Those flowing to the E coast include the Tweed, the Tyne, the Humber, and the Thames. Those flowing W include the Mersey, the Severn, and the Avon. The longest rivers are the Severn and the Thames.

Scotland has an area of 30,414 sq miles (78,772 sq km). It is renowned for its wild and haunting mountains and lochs. The land falls naturally into three regions: the Highlands, the Central Lowlands, and the Southern Uplands.

The Highlands consist of two mountain ranges, the North-West Highlands and the Grampian Mountains. Each runs roughly NE to SW. Between them, extending from coast to coast, is the great valley, Glen More. A series of lochs lying in the glen includes Loch Ness. The lochs are linked to form the Caledonian Canal. Just S of the glen, in the Grampians, is Ben Nevis (4,406 ft; 1,343 m), the highest peak in the British Isles. Some of the many lochs in the Highlands are lakes; others are coastal fiords. A great number of small rivers run through the glens and straths (flat valleys).

The Central Lowlands consist chiefly of the valleys of the Tay, Forth, and Clyde. The Tay (118 miles; 190 km)—which, like the Forth, has its firth on the North Sea—is the longest river in Scotland. The Clyde, however, is of far greater industrial importance. The Clyde flows to the North Channel. The Lowlands have Scotland's best agricultural land, and also its largest cities.

The Southern Uplands is a region of hilly moorland. Much of it consists of grazing land, but there are also pockets of intensive cultivation.

Wales has an area of 8,016 sq miles (20,761 sq km). About two-thirds of it is occupied by the Cambrian Mountains, a series of grassy plateaux interspersed with mountain ranges and peaks. The highest

United Kingdom. The Guild Chapel in Stratford-upon-Avon, built by the Guild of the Holy Cross in 1269 and rebuilt in the 1400s. William Shakespeare, who was born in Stratford-upon-Avon in 1564, attended the grammar school beside the chapel.

peaks are in the N, and include Snowdon (Eryri) which rises to 3,560 ft (1,085 m). There are many lakes, rivers, and waterfalls. The industrial areas are in the valleys of the S and in the NE; their location was dictated by the presence of coal. There are rich agricultural areas in the coastal plains and in the river valleys, such as the valley of the Wye.

Northern Ireland, in the NE of Ireland, has an area of 5,452 sq miles (14,121 sq km). It consists of six of the nine counties of Ulster. Lough Neagh (153 sq miles; 396 sq km) in the E-centre of the province is the largest lake in the British Isles. In the SW is another large lake, Lough Erne. The lowlands around Lough Neagh are bordered by hills and low plateaux. There are three mountain groups: the Sperrin Mountains in the NW, which rise to 2,240 ft (683 m) in the Sawel Mountain; the Antrim Plateau in the NE, with its beautiful valleys, the *Glens of Antrim*; and the granitic range of the

United Kingdom. Above: Punting on the river, a pastime known to generations of undergraduates at Cambridge University. Right: Blenheim Palace in Oxfordshire, built by the nation in the early 1700s for the first Duke of Marlborough. Below: Stonehenge, a group of huge stone blocks on Salisbury Plain, is believed to date from about 2000 B.C. It may have been connected with sun worship.

Mourne Mountains in the SE, where Slieve Donard rises to 2,796 ft (852 m).

Agriculture and Industry The U.K. has one of the world's most productive agricultural industries, largely because of the use of scientific farming methods. About half of the total land area is put to some agricultural use. The largest areas of arable land are in eastern England. There, farmers grow cereals—wheat, barley, and oats—sugarbeet, and vegetables. But there are rich farms in many other parts of the U.K., too. For example, some of the best farming land is in the Central Lowlands of Scotland. Fruit farming is of major importance in Kent and in Hereford and Worcester. The Fens are famous for their potato crops. Dairy farming is common in all regions, but is of greatest importance where the rainfall is high, particularly in the W. Scotland has a high reputation for its beef cattle. Many famous breeds of cattle—both beef and dairy—originated in the British Isles. Pig-breeding is widespread. In some places, sheep farming is predominant.

Britain's industrial importance was once based on its resources of coal and iron. Their declining value may be replaced in the 1980s by oil from the North Sea, where several fields are already in operation.

British industries are extremely varied. To a large extent, they depend on imported raw materials. Ranking high among them is the manufacture of steel. Much of this steel goes to other industries in the U.K. Many of these are engineering industries with long traditions of inventiveness and skilled workmanship: shipbuilding; and the manufacture of motor vehicles, rolling stock, engines of all kinds, farm machinery, industrial and construction machinery, machine tools, and electrical equipment. Other important branches of industry are concerned with chemicals, plastics, glass, paper, cutlery, and ceramics. The making of textiles, one of the oldest industries, is still one of the most important. Today, much of it is based on man-made fibres.

Cities and Communications The largest cities are London (the capital), Birmingham, Glasgow, Leeds, Liverpool, Sheffield, and Manchester. Communications are highly developed in all parts of the U.K. except in very thinly-populated areas. Although a number of unprofitable rail routes have been closed since the 1960s, much of the remaining track and rolling stock has been modernized, and a high-speed 'bullet' train has been developed. The road system has been greatly improved by the building of more than 1,000 miles (1,600 km) of motorways on the major through routes. There are several international airports.

The People The U.K. is one of the world's most densely-populated countries. The greater part of the population lives in urban areas, the chief concentrations being in the London area, the Midlands, the Central Lowlands of Scotland, and the Belfast area. Since the end of World War II more than a million people have emigrated to the U.K. from the West Indies, Africa, India, and Pakistan. Although almost everybody speaks English, some other languages are spoken, too, usually as second languages —Welsh in Wales, Gaelic in Scotland and Northern Ireland, and Manx in the Isle of Man.

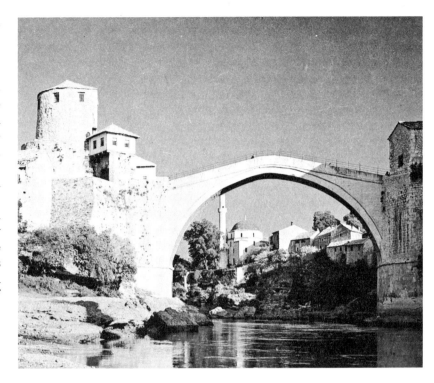

Yugoslavia. Mostar, in the east of the country, is the chief city of Hercegovina. Its stone bridge across the River Neretha was built in 1566 by the Turks.

VATICAN CITY

The smallest independent state in the world, Vatican City lies on the right bank of the River Tiber, within the city limits of Rome. It is all that remains of the large territories of the former Papal States. It is ruled by a lay governor and council on behalf of the pope. The chief buildings of Vatican City are St. Peter's Basilica, the largest church in the world, and the vast Vatican Palace, with its libraries, museums, and state apartments. The state also has jurisdiction over Castel Gandolfo, the pope's summer residence on Lake Albano, and over 13 other churches and palaces in Rome. Vatican City has its own police and postal services. It also has a radio station and a newspaper. It issues its own passports, money, and stamps, and maintains a diplomatic service. Area 108.7 acres (44 hectares). For population, see Table on page 61.

YUGOSLAVIA

Yugoslavia is a mountainous country in the Balkan Peninsula. Its name means 'the Country of the Southern Slavs', and it was formed in 1918 of a number of small Slav states, the largest of which was Serbia. Since the end of World War II it has had a communist government. Yugoslavia lies in the W of the Balkans. It is bordered by Austria and Hungary in the N, Romania and Bulgaria in the E, and Greece and Albania in the S. In the W it has a long coastline on the Adriatic Sea. Off the coast there are many islands. The coastal plain is narrow for most of its length. Behind it rises the *Karst*, a high limestone region that includes some mountain ranges, principally the Velebit Mountains and the Dinaric Alps.

Farther inland is a vast, broken plateau, rising here and there to some 8,000 ft (2,400 m) above sea-level. In the NE is the wide Pannonian lowland, a fertile plain in the basin of the River Danube. The Danube has four large tributaries in this region, the Sava, Drava, Tisza, and Morava. The Vardar, a river whose outlet is on the Aegean Sea, flows in a broad valley in the SE of the country. Yugoslavia has many lakes. The largest are Scutari and Ohrid on the Albanian border, and Prespa on the border with Albania and Greece. The climate in the Danube basin is hot in summer and cold in winter. In the mountains the climate also tends to be extreme. But coastal regions are mild and warm. Yugoslavia has rich mineral deposits, principally lead, antimony, aluminium, coal, and iron. Copper, chromium, mercury, and other metals are also found. The best farmlands are in the Danube basin. There, farmers grow wheat, maize, rye, hemp, sugar-beet, and potatoes. There are many vineyards. Other crops include tobacco, olives, and plums. The raising of livestock is extremely important. A number of new manufacturing industries have been developed in recent years, but the country also has some industries that are well established. Chief among these are shipbuilding and the making of machinery. Yugoslavia has three principal languages, Serbo-Croatian, Slovene, and Macedonian. They have a common origin and many points of similarity. The chief cities include Belgrade (the cap.), Zagreb, Skoplje, Sarajevo, and Ljubljana. Area 98,750 sq miles (255,760 sq km). For population, see Table on page 61.

ASIA

Asia, the largest of the continents, covers almost a third of the Earth's total land area. It is also the most varied of the continents. It has rich green forests; hot, arid deserts; great, slow rivers; luxuriant vegetation that in places consists of masses of perfumed and brightly-coloured blossom; vast dry windswept plains; and regions of permanent snow and ice. In it are the world's highest peak, Mount Everest, and the world's lowest land region, the deep valley around the Dead Sea. It also contains some of the richest countries, and some of the poorest.

Some countries of Asia, such as Kuwait, are small and newly independent. China, on the other hand, has more people than any other country, and a history that goes back more than 3,500 years. The world's oldest civilizations began in Asia, and many of the earliest traces of Man are there.

COUNTRIES OF ASIA
POPULATIONS AND MONETARY UNITS

Country	Population	Monetary Unit
Afghanistan	16,700,000	Afghani
Bahrain	216,000	Dinar
Bangladesh	75,000,000	Taka
Bhutan	1,012,000	Rupee
Brunei	142,000	Dollar
Burma	29,512,000	Kyat
China	800,500,000	Yuan
Cyprus	633,000	Pound
Hong Kong	4,200,000	Dollar
India	592,700,000	Rupee
Indonesia	128,000,000	Rupiah
Iran	31,000,000	Rial
Iraq	10,200,000	Dinar
Israel	3,200,000	Pound
Japan	106,800,000	Yen
Jordan	2,600,000	Dinar
Khmer Republic	7,200,000	Riel
Korea, North	15,000,000	Won
Korea, South	34,000,000	Won
Kuwait	900,000	Dinar
Laos	3,100,000	Kip
Lebanon	2,700,000	Pound
Macau	320,000	Pataca
Malaysia	11,800,000	Dollar
Maldives	114,000	Rupee
Mongolia	1,400,000	Tugrik
Nepal	11,600,000	Rupee
Oman	690,000	Rial
Pakistan	64,800,000	Rupee
Philippines	42,600,000	Peso
Qatar	150,000	Riyal
Saudi Arabia	8,300,000	Riyal
Singapore	2,200,000	Dollar
Sri Lanka	12,748,000	Rupee
Syria	6,600,000	Pound
Taiwan	15,200,000	Dollar
Thailand	36,000,000	Baht
Timor, East	620,000	Escudo
Turkey	37,600,000	Lira
United Arab Emirates	300,000	Dirham
Vietnam	42,300,000	Dong
Yemen (Aden)	1,600,000	Dinar
Yemen (San'a)	6,000,000	Riyal

Asia is bounded on the N by the Arctic Ocean. Asian lands extend for hundreds of miles beyond the Arctic Circle. To the S lies the Indian Ocean, with the Arabian Sea, the Bay of Bengal, and the South China Sea. On the E is the Pacific Ocean. The south-eastern part of Asia is made up of many islands, the southernmost of which lie close to the continent of Australia. On the W, Asia is joined to Europe: the Ural Mountains and the Caspian Sea form the boundary between the two continents. Some geographers consider that Europe and Asia are not really separate continents and that together they form one large continent, which is given the name *Eurasia*. But, historically, Europe and Asia are always treated as different continents. In the SW, Asia is bounded by the Mediterranean Sea and the Red Sea. It is linked to Africa by the narrow Isthmus of Suez, artificially cut by the Suez Canal.

The total area of Asia is 17,336,000 sq miles (44,900,000 sq km). The greatest distance N to S is about 5,400 miles (8,700 km), and W to E about 6,000 miles (9,700 km). The total coastline measures just over 80,200 miles (129,000 km). The population in the mid-1970s was estimated at about 2,221 million and increasing rapidly.

Physical Features and Climate People in Europe and North America often describe some parts of Asia as the Near East, Middle East, and Far East. These are not true geographical divisions. The *Near East* comprises Turkey and those Asian countries that border the Mediterranean Sea, plus Egypt and Libya in Africa. The *Middle East* comprises Iraq, Afghanistan, the Indian sub-continent, and Burma. But many people use this term to include the Near East countries but exclude the Indian sub-continent and Burma. The *Far East* comprises China, Japan, and the south-eastern part of the continent.

Geographically, Asia has six well-defined land divisions marked off by mountain ranges. Each of these regions differs from the others in character and climate.

Northern Asia stretches from the Ural Mountains in the W to the Pacific. It is the vast region known as Siberia, part of the U.S.S.R. The northern part of Siberia lies inside the Arctic Circle. It is a land of treeless plains, with long cold winters and short, warm summers. In summer the temperature may rise to 50°F (10°C), but even then the ground, except for a few inches near the surface, remains frozen. This permanently-frozen subsoil is called *permafrost,* and the great Arctic plains are often known as *tundra.*

PEOPLES AND POPULATIONS

Asia has peoples from all three of the main racial groups of Man—Caucasoid, Mongoloid, and Negroid. Asians have widely different cultural backgrounds, and speak innumerable languages and dialects. In one Chinese province alone, more than 100 dialects are used. There are more people in China than in Africa and North America together. The continent's population increases by some 100,000 persons every day.

Religion and Art. The famous statue called the *Daibutsu* at Kamakura, in central Honshu, Japan. The Japanese word *Daibutsu* means 'great Buddha'. The bronze statue, cast in 1252, is 42 feet (13 m) high. It depicts the Buddha in an attitude of serene contemplation.

The more southerly part of the region includes wide expanses of *steppe* (grasslands), and some forests. The southern edge of the region is marked by the Tien Shan, Sayan, and Yablonovyy ranges, and by Lake Baykal, the world's deepest lake, which has depths of some 5,700 ft (1,700 m). The three biggest rivers of northern Asia, the Ob, Yenisey, and Lena, flow to the Arctic Ocean, while the Amur flows eastwards to the Pacific. All four rivers are frozen for part of the year.

Central Asia is one of the least-explored parts of the world. The Tien Shan and Sayan ranges form its northern boundary, and it extends to the great Himalaya range in the S. It includes Mongolia and the Western Chinese provinces of Sinkiang (Chinese Turkestan) and Tibet. Mongolia and Sinkiang are areas of high, barren plateaux, with some grassland but also the Takla Makan and Gobi deserts. The Tibetan plateau is extremely high. It is an area of cold deserts, stark mountains, and deep valleys. Near the Tien Shan mountains in Sinkiang is the Turfan Depression, about 500 ft (150 m) below sea-level. It is partly occupied by salt lakes.

Eastern Asia lies between the uplands of Central Asia and the Pacific. It is a region of highlands, deep valleys, and fertile plains. Its main rivers are the Yangtze-Kiang and

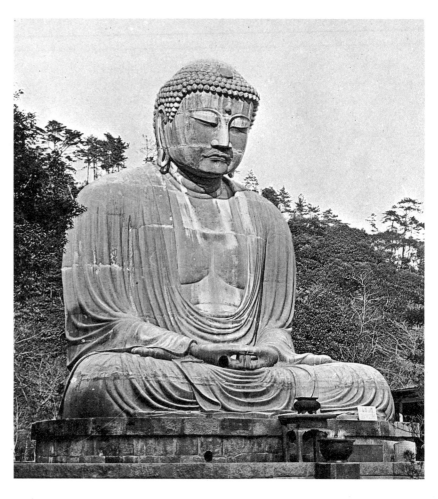

the Hwang (Yellow) Ho. Along the eastern coast of Asia are the Kamchatka Peninsula, really part of Northern Asia, and a series of islands. The three most important of these islands—Hokkaido, Honshu, and Kyushu—form the country of Japan, while farther S is the smaller island of Taiwan (Formosa). The Korean Peninsula lies opposite Japan.

South-Western Asia includes most of the land commonly called the Near East or Middle East, and consists mainly of two large peninsulas: Anatolia and the Arabian Peninsula. Anatolia runs westwards between the Black Sea and the Mediterranean, and forms the larger part of the country of Turkey. To its S lies the island country of Cyprus. The Arabian Peninsula extends between the Red Sea, the Persian Gulf and the Arabian Sea.

Anatolia, which is often called *Asia Minor,* is a rugged plateau whose many rivers and lakes are dry for much of the year. Most of the Arabian Peninsula is hot and dry, and it includes great areas of sandy desert. The land bordering the eastern Mediterranean, once commonly known as the *Levant,* is a mixture of hot, arid deserts and fertile valleys. The River Jordan flows southwards through this area in a deep cleft in the Earth's surface, ending in the Dead Sea. The Dead Sea, which is 1,268 ft (391 m) below sea-level at its surface, is the world's saltiest body of water. The Sea and the Jordan Valley, together with the Red Sea and the Great Rift Valley of Africa, are sections of a natural fault in the Earth's crust.

The Caspian Sea, to the NE of the region, also lies below sea-level, but only by some 90 ft (27 m). Its waters contain more salt than is usual in sea water, but are not as salty as those of the Dead Sea. The Caspian is the world's largest inland body of water. E of it, on the northern borders of Western Asia, lies the Aral Sea, another sea of high salinity, fed by two large rivers of central Asia, the Amu Darya and the Syr Darya.

The eastern part of South-Western Asia, comprising Iran and Afghanistan, is semi-arid and sparsely vegetated with cool winters, hot summers and only light rainfall.

Southern Asia consists of the countries of the Indian sub-continent—Pakistan, India, and Bangladesh in the S, and Bhutan and Nepal in the N. The island country of Sri Lanka (formerly Ceylon) also forms part of this region. The north-western areas are bordered by the Hindu Kush Mountains, which run N and E to a tangle of peaks and valleys called the *Pamir Knot.* Eastwards from the Pamir Knot stretch the Karakoram

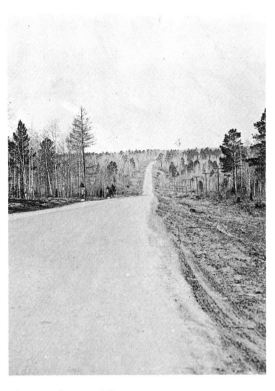

Endless Forests. The taiga, the immense forested plain that stretches across northern Asia and Europe. It makes up a quarter of all the world's forest. Until recently it was practically uninhabited, but new roads—and air communications—are now opening up some of it to settlement and industry.

Mountains and the Himalayas, the world's highest and most rugged mountain range. Mount Everest, on the frontier between Nepal and Tibet, rises to 29,028 ft (8,848 m) above sea-level. For more than a century it defied attempts of mountaineers to reach its summit. But in 1953 it was at last conquered. The Himalaya Range includes many other lofty peaks, too, of which several are

Highest Mountains. Mount Everest (29,028 ft; 8,848 m), the world's highest peak. It rises in the central Himalayas, on the border between Nepal and Tibet. It is named for Sir George Everest, who was surveyor-general of India in the 1830s.

(WINTER)

PRECIPITATION IN mm
IN ins.

| | less than 250 less than 9.85 | from 250 to 500 from 9.85 to 19.7 | from 500 to 1,000 from 19.7 to 39.4 |

from 1,000 to 2,000 from 39.4 to 78.8   more than 2,000 more than 78.8   **A** Areas of predominantly high pressure   **B** Centres of low pressure

(SUMMER)

Prevailing winds

Direction of tropical anticyclones

more than 25,000 ft (7,620 m) high and among the world's highest mountains.

The rest of Southern Asia consists of two main areas: the first is made up of a series of wide plains watered by rivers flowing from the Himalayas: the second is the Deccan, a triangular plateau that occupies most of central and southern India. The Deccan rises on its edges to two mountain ranges, known as the Eastern Ghats and the Western Ghats. In the N, the Vindhya Mountains, and other ranges, running E-W, cut the Deccan off from the rest of India. Southern Asia has two great river valleys, each of which, like the Nile in Africa, has been the focus of a civilization. One is the valley of the River Indus, which flows southwestwards through Pakistan to the Arabian Sea. The other is the valley of the River Ganges, which wanders eastwards through northern India to the Sundarbans, its huge delta in Bangladesh, on the Bay of Bengal. In Bangladesh the Ganges is joined by the Brahmaputra, which rises on the Plateau of Tibet on the northern side of the Himalayas.

Southern Asia is strongly influenced by the *monsoons,* winds that change their direction from one season to another. From about June to September the SW monsoon blows in from the Indian Ocean bringing torrential rains to large areas. The heaviest rains are in the Western Ghats and in the NE. There, they often cause severe floods, particularly in the Ganges delta in Bangladesh. From October to March, the winter monsoon, a much lighter wind, blows from the N over the Himalayas, bringing cool, dry weather.

South-Eastern Asia consists of Burma and the Indochina Peninsula (including the Malay Peninsula), lying to the E of India and the S of China, and a large number of large and small islands that were once collectively known as the *East Indies.* The mainland part of this region is, broadly, a mixture of wooded mountains in the N with wide river plains in the S. In places, the plains are well-drained and fertile; but in some places they are marshy, and have dank swamp forests. The islands tend to have similar terrain, with mountainous interiors covered by almost impenetrable woodland. The largest of these islands are Borneo, Sumatra, Java, and Celebes (Sulawesi) to the SE. N of them lies the chain of islands that forms the Philippines.

The two main rivers of mainland South-Eastern Asia are the Irrawaddy in Burma and the Mekong in Indochina.

The eastern islands, in the Pacific, lie in an area of volcanoes and frequent earthquakes. Earthquakes also occur in Anatolia

MINERAL DEPOSITS

Crude oil
Coal
Iron ore
Copper
Lead
Zinc
Manganese
Antimony
Tin
Silver
Gold
Tungsten
Uranium
Mica
Chromium

INDUSTRIES

Metal
Non-ferrous metal
Chemical
Textile
Oil refining
Hydro-electricity

IMPORTANT DATES IN THE HISTORY OF ASIA

3000s B.C. The civilization of Sumer developed in Mesopotamia.

2500s B.C. A Bronze Age culture developed in the Indus Valley.

1900s B.C. Rise of the Shiah dynasty in China; by tradition the first dynasty, but probably legendary.

1500s B.C. Rise of the Shang dynasty in China.

1500s B.C. The Aryans from central Asia invaded the Indus Valley.

1200s B.C. Moses led the Israelites out of Egypt.

1000s B.C. Chou dynasty in China. Art and learning flourished.

c. 563 B.C. The Buddha born in India.

c. 551 B.C. Confucius born in China.

336-323 B.C. Alexander the Great created an empire that extended, briefly, from south-west Asia as far as the Indus Valley.

300s B.C. Building of the Great Wall of China began.

269-232 B.C. Asoka ruled in India—a Golden Age.

c. 6 or 4 B.C. Birth of Jesus Christ in Palestine.

300s A.D. Beginnings of the Byzantine Empire in south-west Asia.

300s A.D. The Huns from Mongolia established a kingdom in northern China.

570 Muhammad born in Arabia.

618 Rule of the T'ang dynasty began in China—one of the greatest periods in China's history.

600s Muslim warriors conquered much of south-west Asia.

1206-1227 Genghis Khan and his Mongols conquered China, Persia, and part of India.

1264-1294 Kublai Khan ruled the Mongol Empire—its greatest period.

1368 Rule of the Ming dynasty began in China.

1526 Babar, the ruler of Kabul, established the Mughal Empire in India.

1500s The Ottoman Empire in south-west Asia flourished under the rule of Suleiman the Magnificent.

1500s European soldiers and traders made their first incursions into Asia.

1644 Rule of the Manchu dynasty began in China.

1757 The British East India Company ruled Bengal after Robert Clive's victory at Plassey.

1774 Warren Hastings became first British governor-general of India.

1842 China, under duress, opened five ports to British trade.

1854 An American naval force obliged the Japanese to open their ports to American trade.

1857-1858 The Indian Mutiny. After its suppression, the Mughal emperor was deposed.

1876 The British parliament declared Queen Victoria empress of India.

1904-1905 The Russo-Japanese War. Russia defeated.

1911 Revolution in China. End of the Chinese Empire. Sun Yat-sen became president.

1931 The Japanese seized Manchuria.

1937 Full-scale Japanese attack on China.

1939-1945 World War II. The Japanese occupied large areas of Asia, but were forced to relinquish their gains within a few years.

1940s-1960s End of colonial regimes in many countries and regions. Newly independent states included India, Burma, Ceylon, Malaysia, North and South Korea, the Philippines, Syria, Jordan, Indonesia, and North and South Vietnam.

1950-1953 The Korean War.

1959-1973 The Vietnamese War.

and in several other parts of South-Western Asia.

Resources and Industry Because Asia is so immense, it has almost every kind of industrial and agricultural resource in one area or another. But many of Asia's mineral resources have not yet been fully exploited, and probably there are enormous mineral riches the locations of which are still unknown.

Northern Asia—that is, the Asiatic part of the U.S.S.R.—has major coal deposits both in the E and in the W, and experts believe that there are vast untapped resources in the permafrost regions of the N. Oilfields are actively exploited to the N of the Caspian Sea and also on the long, slender island of Sakhalin, off the Pacific coast to the N of Japan. Other important minerals which the Russians are extracting include asbestos, chrome, cobalt, copper, gold, iron, mercury, manganese, nickel, silver, tin, uranium, and vanadium. In addition, the major rivers have been harnessed as a source of energy, and several industrial regions are dependent on the hydro-electric power that they provide.

Central Asia is very largely unexplored, but gold, iron, and other metals are mined there, together with coal and petroleum.

Eastern Asia is rich in coal, and is known to have small deposits of a great many metal ores, particularly those of antimony and tungsten. One of the most important resources in this region is iron ore, but surveys have so far indicated only small amounts of petroleum. Japan has good deposits of coal, copper and sulphates, but very little other mineral wealth.

CENTRAL TREATY ORGANIZATION (CENTO)

CENTO is an alliance for mutual defence of Iran, Pakistan, Turkey, and the United Kingdom. The United States co-operates with CENTO. CENTO is based on the Baghdad Pact of 1955. Its headquarters are in Ankara, Turkey.

Below left:
Patient Toil. Work on a communal farm in China Despite increasing industrialization, agriculture is still of primary importance all over Asia. The introduction of machinery and improved methods of farming have helped to increase productivity in some regions, but in most of the continent work is still carried out slowly and methodically by hand.

Below right:
Crowded Life. Hong Kong is one of Asia's most important ports and a major trading centre. It is also one of the world's most overcrowded cities. More than 100,000 of its inhabitants have no homes except the small boats called sampans.

South-Western Asia is rich in petroleum and coal. The area around the Persian Gulf is one of the world's chief sources of oil, and there is coal in Iran and Turkey. The countries along the Mediterranean Sea, apart from Turkey, are almost completely without minerals, and the known sources of other minerals elsewhere in the region are relatively few.

Southern Asia is rich in iron ore and also has valuable supplies of coal, the chief concentrations being found in India. The region also has some petroleum, and Pakistan and Bangladesh both possess natural gas deposits. But other minerals are scarce, though much of Southern Asia has yet to be surveyed. The island of Sri Lanka has few minerals.

South-Eastern Asia has good supplies of petroleum, mostly located in the islands, and there are also coalfields and beds of iron ore. Other minerals relatively plentiful in the region include bauxite (aluminium ore), gold, lead, nickel, tin, and zinc.

Japan, despite having to import nearly all of its raw materials, has become one of the world's leading industrial nations. Its factories produce a wide range of goods for export, many of them products requiring high technical knowledge and skill. The U.S.S.R., too, has many highly-developed manufacturing centres. Most other Asian countries have, by comparison, few and small manufacturing industries, and have to import many of the goods they need. China and India are both rapidly increasing their manufacturing capacity, though both have a long way to go before their industry is equal to their needs. In western Asia, the most industrialized country is Israel.

Communications In general, communications are poorly developed in Asia, and vast tracts of the continent are almost inaccessible to travellers. However, Japan, India, and Pakistan have first-rate rail systems, and the Trans-Siberian Railway spans northern Asia. Israel, Japan, and the countries of southern Asia have good roads and highways; other countries are improving their roads gradually. The fastest progress in developing communications is being made by the wealthy oil-producing countries of the Persian Gulf area. Rivers and canals are still major links in many parts of Asia, and there is an increasing use of air transport. Aircraft have been particularly useful in opening-up northern Asia. Most places are within reach of radio broadcasting, but there is little television, and newspapers reach only a comparatively few people. Few areas are now without telephone communications.

Agriculture, Fisheries, and Land Use
Two Asians out of every three live by farming. Many have their own smallholdings where they grow just enough food for themselves and their families; in bad years they often go hungry. Others work for richer neighbours, or are employed on state farms or communal farms.

The best agricultural lands are in southern Asia, south-eastern Asia, China, and the south-western part of Siberia. The two chief subsistence crops are rice and wheat. Rice needs warmth and plenty of rain; consequently, it does best in regions that are affected by the monsoon winds. The largest crops of rice are produced in China, India, and Pakistan. Wheat is grown in Siberia, China, and the dryer parts of India. Although improved agricultural methods and the use of better seeds have led to larger crop yields in recent years, many parts of Asia do not manage to grow enough food to support their inhabitants if the weather is unfavourable and have to import wheat from North America.

A number of valuable crops are grown for export. India, Sri Lanka, and China produce most of the world's tea. Malaysia and Indonesia are major producers of rubber. Rubber trees were introduced into these countries from Brazil in the 1870s. Other important cash crops include cotton, citrus fruits, jute, and tobacco. The islands of the East Indies produce most of the world's spices. In the past they were usually known as the 'Spice Islands'. It was because of the need of people in Europe to find spices to preserve food and to camouflage the taste when food was not fresh that explorers set

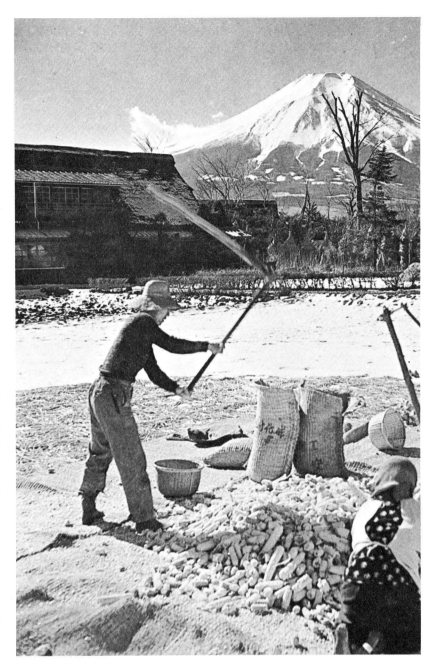

Work and Beauty. Japanese farm workers on the island of Honshu thresh the newly-gathered grain. Behind them rises the symmetrical cone of Fujiyama, the sacred mountain. Japan has relatively little farmland, and what there is is intensively cultivated. The main crop is rice.

out to find routes to the Indies in the late 1400s.

Most of central and south-western Asia is unsuitable for crop-growing. But on the scanty pasture-lands of these regions, nomadic (wandering) herdsmen tend goats, sheep, and cattle. They also raise camels, which are an important means of transport. In southern Asia cattle are little used for food; in some countries they are protected for religious reasons, and in other places they are protected because they represent wealth. But they are widely employed for ploughing and for drawing carts. In Tibet the yak, a hardy animal which can stand up to cold weather, is used both as a working animal and for food.

Many wild animals still roam the forests of Asia, but some, such as the tiger, are becoming rare and may soon be extinct. In some countries wild animals are now pro-

Legend:
- Oases and other special areas
- Intensive subsistence farming (mainly rice)
- Intensive subsistence farming (except rice)
- Plantations
- Other intensive food cultivation
- Food crops and pasturage
- Mediterranean crops
- Chiefly pasturage, some crops
- Primitive cultivation
- Nomadic grazing
- Forests and woods
- Tropical forests; scattered crops
- Tundra with seasonal pasturage
- Wild and infertile
- Limit of olive cultivation
- Limit of date palms
- Limit of sago palms
- Limit of coconut palms

tected, and there are severe penalties for killing them without licence. Elephants live wild in southern Asia and Burma, but many are domesticated and are used as work animals. Reindeer roam the tundra lands of the N; they are herded, are used for work, and are a valuable source of food in regions where there are few crops.

Forests still cover large parts of Asia, including the remote areas of Siberia. The Russians extract a great deal of softwood timber from those parts of the forests that lie near rivers or railways. In south-eastern Asia there are thick teak forests that are exploited for timber, but most of the forests of the islands, including the Indonesian islands, are untouched.

Fish are important as food in the coastal lands and islands of eastern Asia. Both China and Japan possess large fishing fleets which roam the waters of the northern Pacific in search of catches. Southern Asia, Bangladesh, Sri Lanka, and Burma all rely heavily on fish as an item of food. River and lake fishing are important in many parts of south-eastern Asia, particularly in the deltas of the great rivers, such as the

SOUTH-EAST ASIA TREATY ORGANIZA- TION (SEATO)

SEATO is based on a treaty signed in 1954 by Australia, France, New Zealand, Pakistan, the Philippines, Thailand, the United Kingdom, and the United States. Pakistan left SEATO in 1973. The members pledge mutual defence against aggression in SE Asia or the SW Pacific. SEATO headquarters are in Bangkok, Thailand.

Opposite page:
Riches of the Past. Demon figures guard the royal library in the Grand Palace in Bangkok. The Grand Palace is really the walled royal city. It is about one square mile (2·5 sq km) in extent.

Mekong. China, too, has lakes and rivers in some areas that are well stocked with fresh-water fish.

Japan possesses a major whaling fleet, which hunts not only in home waters but also in the Antarctic Ocean. Edible sea-weeds are collected by the Japanese, and some of this valuable foodstuff is exported. In Japan and the Persian Gulf, pearl fishing is an important industry. The Japanese pioneered the cultivated pearl by breeding oysters and introducing beads into their shells. This process induces the oysters to make pearls by secreting pearl substance around the bead.

Land use is very uneven in Asia. Half the world's people live in six Asian countries —China, India, Indonesia, Japan, Pakistan, and Bangladesh. But huge areas of central Asia and the Arabian Peninsula are almost uninhabited, because they are desert or semi-desert. Over most of the continent there are fewer than 25 people to each sq mile of land (less than 10 to each sq km).

People and Ways of Life Asia is not only the largest continent but also much the most

ANCIENT CIVILIZATIONS OF ASIA

Indus Valley

Excavations at Mohenjo-daro and Harappa in the north-west of the Indian peninsula (today in Pakistan) have shown that an advanced civilization existed in the Indus Valley in about 3000 B.C., and reached its peak between about 2500 and 1500 B.C. It is thought that the people of the city of Mohenjo-daro traded with the people of Sumer. They wrote in a hieroglyphic script, and had a distinctive art. Their craftsmen produced fine work in bronze. The religion of Mohenjo-daro had something in common with the later religion of Hinduism. Little is known of the reasons for the decline—apparently, a rapid decline—of Mohenjo-daro, but it may have been overrun by the Aryans.

The Aryans, a fair-skinned people from central Asia, invaded the Indus Valley in the 1500s B.C. The Sanskrit word *Aryan* means 'lord of the land'. The Aryans gradually extended their rule over most of India, and developed the Vedic civilization that underlies Hindu life and culture.

Central China

The earliest developed civilization in China was probably that of the Shang dynasty between 1500 and 1000 B.C. The people of this age were accomplished builders and had a simple, pictorial form of writing. The Shang dynasty was succeeded by the Chou dynasty, from the valley of the Yangtze Kiang. The Chou emperors built up a system of feudalism, and controlled much of China until 249 B.C. During this period art and literature were encouraged, numerous important discoveries were made in science, and there were many inventions. In the teachings of philosophers, a basis was found for attitudes of mind that the world looks upon as peculiarly Chinese. Among the great thinkers of the period were Lao Tse (c. 605 B.C.) and Confucius (551?-479 B.C.).

Mesopotamia

The Greeks gave the name Mesopotamia to the land between the Tigris and Euphrates rivers in south-western Asia. In Greek, *Mesopotamia* means 'between the rivers'.

Some of mankind's earliest civilizations developed in and around Mesopotamia. In the 3000s B.C., the Sumerians were building cities and organizing small states. They had a system of writing using a cuneiform (wedge-shaped) script. Clay tablets bearing cuneiform writing have been preserved, and tell us much about the life of the Sumerians. They were acquainted with many sciences, and were skilful farmers, architects, weavers, potters, and jewellers.

Near Sumer, another civilization developed —that of the Babylonians. They borrowed ideas and techniques from the Sumerians, but also contributed much that was original. Their language was a dialect of a Semitic language known as *Akkadian*.

Babylonia was closely linked with Assyria. The Assyrians also spoke an Akkadian dialect, and attained their greatest power and prosperity in the 800s to 600s B.C. For part of this period, they controlled Babylonia.

PEOPLES OF ASIA

113

Symbolism in Architecture. A huge torii in the Inland Sea at the island of Miyajima, Japan. A *torii* is a gateway built at the approach to a Shinto temple. Miyajima has a temple celebrated for its elusive beauty.

Left:
Wisdom of the Ages. A giant statue, one of 24 depicting men and animals that line the road to the mausoleums of the Ming emperors in Peking, in China. The Ming dynasty ruled China from 1368 to 1644.

Below:
Travelling Folk. A family caravan moves along an ancient road in Afghanistan. With it travel possessions that represent considerable wealth.

populous. It has more people than the whole of the rest of the world, and its population is increasing faster than that of any other part of the Earth. Asia's peoples include members of all the main racial groups of Man, and their ways of life are as varied as the places in which they live.

Mongoloid people live in eastern Asia and in most of central Asia. Some have emigrated to the countries of south-eastern Asia, too. Typically, they have straight, dark hair and yellowish skins. They are related to the American Indians, whose ancestors are believed to have crossed the Bering Strait into North America about 20,000 years ago.

Caucasoid people, related to the people of Europe, live in southern Asia, south-western Asia, and Siberia. Most of the Caucasoids of Europe originally came from Asia.

Negroid people, similar to the peoples of Africa S of the Sahara, live in many of the islands of south-eastern Asia. Pygmies, called *Negritos,* live in Malaya and several of the larger islands.

In most countries there are representatives of more than one racial stock. Though people of mixed race are found all over Asia, they are most numerous in the SE.

The world's principal religions began in Asia; some of them now have more members in other continents. Hinduism, the main religion of India, began nearly 4,500 years ago. From it developed Buddhism about 500 B.C. This religion spread eastwards to China and Japan. Confucianism, China's other great religion, and Shintoism, the traditional Japanese religion, began around the same time. In the W, Judaism evolved among the ancient Israelites more than 3,000 years ago. From it have come two other great religions, Christianity, founded nearly 2,000 years ago, and Islam, which dates from the A.D. 600s. Today, Islam is the main religion of south-western Asia and parts of the S and SE; India is largely Hindu; the Chinese are Buddhists, Confucianists, or a mixture of religions; and many of the people of Siberia are nominally Christians.

Religion plays a very large part in the daily lives of most Hindus and Muslims (followers of Islam). In the U.S.S.R., and still more in China, the place of religion has been largely taken over by Communism, which serves as a philosophy of life as well as a political belief.

The way people live in the various regions of Asia depends not only on tradition and religion but also on the physical characteristics of their homelands. In the belt of dry, semi-arid land that runs through south-western Asia to central Asia, many of the people are nomadic herdsmen, tending herds of such animals as goats, sheep, cattle, yaks, and camels. Much of the land is unsuitable for crop growing except in the few areas where artificial irrigation has been possible. However, in the oil-rich regions around the Persian Gulf, the people are rapidly adopting a Westernized way of life.

The most Westernized and industrial countries in the continent are Japan, which depends for its wealth largely on industry, and Israel, which is peopled largely by Jewish settlers from Europe and North America. The former European colonial territories of southern and south-eastern Asia have ways of life that reconcile tradition with Western ideas. One of the chief effects of Western influence has been ever-accelerating urbanization.

The rapid increase in Asia's population in this century has been largely confined to the Indian peninsula, China, and Indonesia. The principal cause of the increase has been a reduction in the death rate. Until comparatively recently, famine, floods, and disease took a fearful toll of human life. Some Asian countries have agonizing difficulty in finding enough food for their peoples. Millions of Asians live at subsistence level—that is, they are able to grow barely enough food to feed themselves. In Bangladesh, which is one of the world's poorest countries, countless people never have enough to eat. And every few years, the fertile delta of the Ganges—which is where most of the people live because it has soil suitable for crop-growing—is visited by devastating floods in which many die.

Over large parts of Asia, people's ways of life are centred on the village, whether it is in a forest clearing, a river valley, or an upland plain. In most villages people work together to gain a living. In China this type of co-operative effort is controlled and led by the government, which also provides technical advice and help. One of the big problems for most Asian countries is education. In a world increasingly dominated by technology, real progress is not possible until a majority of the people have mastered the arts of reading, writing, and simple mathematics.

The Uncertainty of Life. Refugees look for a new life—anywhere. With them they carry the few possessions they were able to salvage. In this century, wars and natural disasters have uprooted countless people from their homes.

115

THE COUNTRIES OF ASIA

Mountainous country of southern Asia, bordered on the N by the U.S.S.R., on the NE by China, on the S by Pakistan, and on the W by Iran. The Hindu Kush crosses the country from NE to SW. Some of its peaks rise to about 25,000 ft (7,600 m). In the SW is an arid plain, but the N has wide, fertile valleys. The largest river, the Helmand, flows from the Hindu Kush to the lakes called Hamun-i-Helmand and Hamun-i-Pusak in the SW. Afghanistan has a dry climate, with hot summers and cold winters. Farmers grow wheat, rice, fruit, and cotton, and keep goats and sheep. There are many nomads. The country is famous for its lambskins, carpets, and lapis lazuli. The chief towns are Kabul (the cap.) and Kandahar. Area 250,000 sq miles (647,000 sq km). For population, see Table on page 105.

Afghanistan. A high valley in the bleak mountains of the north-east.

Tiny but oil-rich state consisting of several islands in the W central Persian Gulf, about 15 miles (24 km) E of the Saudi Arabian coast. The largest island is Bahrain; others include Muharraq, Sitra, Umm Nasan, and the Hawar group. The islands are predominantly desert with daytime temperatures exceeding 100° F (38° C). There are large natural gas deposits as well as oilfields. The main industries are oil production and oil refining. A refinery on Sitra processes local oil as well as oil from Saudi Arabia. Bahrain also has an aluminium smelter using imported ore. Crops include dates, rice, and vegetables. Pearl fishing was once important but is now declining. The main language is Arabic. The chief towns are Manama (the cap.) and Muharraq. Area 420 sq miles (1,080 sq km). For population, see Table on page 105.

The newest and among the poorest of Asian nations, Bangladesh was formerly East Pakistan. Its name means 'the Nation of Bengal'. Its territory consists of the old province of East Bengal and part of Assam. The country has a 440-mile (710-km) coastline in the S on the Bay of Bengal. In the SE it has a short boundary with Burma. On all other sides it is bordered by India. The land is mainly alluvial plain, with many river valleys and river mouths, tropical jungles, and swamps, and a tropical monsoon climate. The chief rivers are the Ganges and the Jamuna, the main stream of the Brahmaputra before its junction with the Ganges. The mouths of the Ganges, the *Sundarbans*, form the largest delta in the world. There are reserves of natural gas and some coal deposits. The country's principal crops are jute, rice, tea, and tobacco. Jute milling is the main industry. With more than 1,300 people per sq mile, Bangladesh has one of the world's highest population densities. Most of the people are Muslim Bengalis, with a few tribal people in the SE hills. The chief cities are Dacca (the cap.) and Chittagong. Area 55,126 sq miles (142,776 sq km). For population, see Table on page 105. In 1970, the East Pakistan nationalist party won an election majority, but the Pakistani government resisted demands for autonomy in the E. Later, the Pakistani army was used to suppress a revolt. Civil war followed, and intervention by India led to the creation of the independent state of Bangladesh.

Kingdom in the Himalayas, bordered on the N by China, and on other sides by India. High mountain peaks along the N border give way to densely populated, fertile valleys in central Bhutan and forested plains in the S. There are several rivers. Most Bhutanese are farmers working their own land. The main crops are rice, wheat, and barley. The official language is Dzongka. The cap. and chief town is Thimbu. Area 18,000 sq miles (46,000 sq km). For population, see Table on page 105.

The sultanate of Brunei, on the N coast of Borneo, consists of two small enclaves within the territory of Sarawak. Brunei has a coastline on the South China Sea. Much of it is forested. Its chief sources of revenue are oil, rubber, and timber. The capital is Bandar Seri Begawan. Area 2,226 sq miles (5,765 sq km). For population, see Table on page 105.

Bangladesh. Above: The port of Chittagong, near the mouth of the Karnaphuli River on the Bay of Bengal. It is an important commercial and industrial town.

The Union of Burma, in SE Asia, has coastlines on the Bay of Bengal and the Andaman Sea. It is bordered on the NE by China, on the E by Laos and Thailand, and on the W by Bangladesh and India. The Arakan Mountains mark the boundary with India and extend southwards into the country. They include the Naga, Chin, and Lushai Hills. Most of eastern Burma is a vast plateau, the Shan Massif. Between the mountains and the plateau is a wide central lowland, drained by the Irrawaddy and Sittang Rivers. A narrow range of low mountains, the Pegu Yoma, rises between the valleys of the two rivers. The Irrawaddy is more than 1,300 miles (2,100 km) long, and has formed a wide delta where it flows into the Bay of Bengal. Its largest tributary is the Chindwin. To the E of the Shan Massif is another great river, the Salween. A long, narrow strip of Burmese territory stretches southwards to the Malay Peninsula. It is called the *Tenasserim Coast.* The Burmese climate is monsoonal. Coastal districts are hot and wet, but hilly areas are often cool. The country has rich mineral resources, of which petroleum is the most important. Lead, tungsten, silver, zinc, wolfram, and tin are found too. Manufacturing industries produce building materials, chemicals, textiles, and other goods. Most people live by farming. The chief crop is rice. Other crops are wheat, maize, millet, ground-nuts, tea, tobacco, rubber, and sugar-cane. About three-quarters of the people are Burmans. The chief cities are Rangoon (the cap.), Mandalay, Moulmein, Bassein, and Pagan. Area 262,000 sq miles (679,000 sq km). For population, see Table on page 105.

Burma. The Shwe Dagon pagoda in Rangoon, built over a shrine believed to contain eight of the Buddha's hairs. The pagoda stands on a hill, and is 368 ft (112m) high. It is built of brick, covered in gold. By tradition, it was founded some 2,500 years ago, but in its present form it dates from the 1500s.

CHINA

More than one-fifth of all the people in the world live in China, though in area China is only the third largest of the world's countries, after the U.S.S.R. and Canada. The official name of the country is *Chung-Hua Jen-Min Kung-Ho Kuo*; in English it is called *The People's Republic of China*. The name 'China' comes from the name of a dynasty, the Ch'in, that ruled China for a brief period in the 200s B.C.

China has one of the world's oldest and most highly-developed civilizations, with writings, philosophies, and traditions going back for more than 3,500 years. It has contributed greatly to the culture of neighbouring countries, including Japan, and its influence in thought, arts, and invention has been felt in every continent. In the 1900s, China has undergone violent change, brought about by revolution and war. Today it is a leading communist nation.

Most of the people of China live in small villages and work on the land. Large parts of the country, where the land is too poor for successful farming, are very thinly popu-

lated. But China also has many of the world's largest cities. Most of them are in the eastern third of the country.

China is bordered on the N by the U.S.S.R. and Mongolia, and on the NE by North Korea. On the S it borders on Vietnam, Laos, Burma, India, Bhutan, and Nepal. In the W it has a short border with Afghanistan. On the SE coast there are two tiny enclaves ruled by other countries: Hong Kong (United Kingdom), and Macao (Portugal).

In the E, China has a 4,250-mile (6,800-km) coastline on three seas that are part of the Pacific Ocean. From N to S, these are the Yellow Sea, the East China Sea, and the South China Sea. Between the East and South China Seas is the island of Taiwan (Formosa), which is a separate country sometimes called 'Nationalist China'.

The capital of China is Peking, which is in the NE. The area of the country is estimated at about 3,700,000 sq miles (about 9,600,000 sq km). For population, see Table on page 105.

Land and Climate China, occupying more than one-fifth of Asia, is cut off from the rest of the continent by natural barriers —high mountain ranges that fringe the country on the N, S, and W. The Great Plain of eastern China is the largest lowland and the part of the country in which most people live. It has many waterways. Tibet, in the SW, is a high plateau or tableland lying between the world's greatest mountain range, the Himalayas, in the S, and the Kunlun Mountains in the N. Northwards from the Kunluns stretches the sandy Takla Makan Desert, bordered on the N by the lofty Tien Shan range. This part of the country, Sinkiang, is sometimes known as Chinese Turkestan.

The flat, low plains of Manchuria occupy NE China. They are bounded by the Khingan Mountains, which—on the W—separate Manchuria from Mongolia. On the country's N-central border in Inner Mongolia (the Chinese part of Mongolia) is the vast Gobi Desert. Southern China includes the high Yunnan Plateau, crossed by the Red and Mekong Rivers.

Eastern China is 'China proper'—that part of the country that is enclosed by the Great Wall of China. 'China proper' excludes Tibet, Chinese Turkestan, Inner Mongolia, and Manchuria. It is divided into northern and southern parts by the Chin Ling or Central Mountains, which stretch from W to E and separate the great river plains of the Hwang Ho, or Yellow River, to the N, and the Yangtze-Kiang to the S. Both of these rivers rise in the mountains of the W and flow in a generally eastern direction to the sea. Each of them is liable to frequent flooding, caused by melting mountain snows. Over thousands of years, these floods have levelled the lower Yangtze and Hwang Ho valleys into broad plains. The Hwang Ho gets its name from the large amount of yellow mud it carries. This often causes the river to silt up, burst over its banks, and change course. The Hwang Ho mud is also responsible for the slightly yellow tinge of the Yellow Sea.

The Si or West River is SE China's largest river. It empties into the South China Sea near Macao and Hong Kong.

China has several large lakes, mostly in the vicinity of the Yangtze-Kiang. Two of the main lakes are Tungting and Poyang, both in the SE. The Tungting covers some 4,000 sq miles (10,000 sq km) in summer, but contracts in winter.

Because of China's great size, climatic conditions vary considerably. In the Tibetan Plateau and Inner Mongolia, the climate is very dry and marked by great extremes of

Opposite page: China. The Great Wall winds across China in an east-west direction from the Gulf of Khihli, on the Yellow Sea, to a point some 1,500 miles (2,400 km) inland. It was built in the 300s and 200s B.C., and has been reconstructed many times since. Its purpose was to protect China against attacks by Mongols from the north. The total length of the wall is probably about 3,000 miles (4,800 km), because it is not constructed in a straight line but follows the natural contours of the land.

China. A cheerful family meal in a busy city street. The food is being eaten with kwaitsze—chopsticks. The word *kwaitsze* means 'quick ones'.

heat and coldness. This is due to the influence of the great, waterless land mass of central Asia.

China's two great river basins, being nearer to the ocean, have a wetter, more equable climate. In this part of China the dry season lasts from November to February. The rest of the year, particularly May, is extremely wet. In the southern coastal regions, the climate is sub-tropical, with intense summer heat and intermittent typhoons, which cause much destruction.

Agriculture and Industry The fertile soil of the lowland regions is one of China's principal natural resources. The richest farmland is found in the Yangtze-Kiang plain. Farther N, in the Hwang Ho plain and in Manchuria, the soil is also fertile. But cold winters and irregular rainfall limit agriculture particularly in the most northerly areas. Rich, red soil is found in S-central China. This soil is particularly suitable for grain crops. North-western China is covered with windblown *loess*, the fine yellow particles of which give the Hwang Ho its colour. The thick deposits of loess are good for crops, but lack of rain in this region makes farming difficult.

Farming is the basis of the Chinese economy, and most of the people live on the land. Because there is so little level land, and because so many people have to live on it, every usable patch of soil is planted with crops. Especially in the S, hillsides are terraced far up their slopes to prevent erosion of the soil. Fertilizers, both natural and

Soya beans, sorghum, spring sown wheat
Woods, grasslands, marshes
Winter wheat, soya beans, sorghum, cotton
Winter wheat, millet
Spring sown wheat, millet
Steppe: stock rearing, pockets of cultivation
Desert, semi-desert, crops in oases
Woods, pasturage
Winter wheat, rice
Rice, winter wheat, sweet potatoes
Rice, tea, sweet potatoes

Sub-tropical rice-growing region (two crops)
Tropical rice-growing region (three crops)
Rice and maize
Summer rice growing
Grazing, cultivation in watered spots
Main crop-growing area of Tibet (barley)
Cold desert or steppe
Japan: mixed farming
Japan: rice growing
(one crop)
(one crop) mixed farming in some areas
Boundary between areas of spring & winter wheat
Limit of intensive rice cultivation
Northern limit, sub-tropical fruits
Northern limit, tea
Northern limit, bananas and cinnamon

artificial, are widely used to increase fertility.

The most important food crops are rice and wheat. Rice needs a hot climate and wet soil, the latter requirement meaning that there must be a plentiful supply of water for irrigation. These conditions are found in southern China, where most of the country's rice is grown. Nearly all the level land and some of the terraced fields are planted with rice. Because the seedlings must be planted in water, the fields are banked up into small, pond-like paddies. The farmers wade into the paddies to plant the seedlings. The paddies must be kept flooded for several weeks after planting.

Other crops cultivated in the warm S include tea (formerly the most important export crop) and sugar, cotton, sweet potatoes, fruit, and vegetables. In order to give as much as possible of the level land to rice growing, these other crops are planted mainly on the terraced hillsides. The mulberry tree, the leaves of which are used to feed silkworms, is an important crop in southern China.

Wheat, which needs far less water than rice, is the main crop in northern China. Other crops grown in this region include millet, maize, sorghum, soya beans, fruit, and vegetables.

Few Chinese farmers raise cattle, because the land that would be needed to grow cattle feed can be better used for crops destined for human consumption. Water buffaloes are raised as draught animals. Nearly every farmer keeps chickens, ducks, and pigs. Most of China's sheep, horses, and cattle are raised on the grasslands of Manchuria and Inner Mongolia.

China has considerable mineral wealth, particularly in the S, where there are deposits of copper, tin, antimony, lead, and zinc. In the N there are quite large coal deposits, and iron is found in many places. Petroleum has been found in northern, central, and western China.

Each of China's major port cities is a centre of light industries, including the production of cotton and silk textiles, and food processing. The main heavy industries include the manufacture of iron and steel, heavy machinery, motor vehicles, and farm machinery. Other industries include the production of chemicals and fertilizers.

Cities and Communications China's largest city and chief port, Shanghai, is situated near the mouth of the Yangtze-Kiang. Some cities on the Yangtze, which is navigable by large ships, are also major ports. They include Nanking and Wuhan (really three cities—Hankow, Wuchang, and Hanyang—in one).

N of Shanghai, at the northern end of the Hwang Ho lowland, is the port city of Tientsin, located at the mouth of the Pei River. It is the main port for the Hwang Ho plain. Manchuria's chief port is Lü-Ta (Dairen). S China's leading port is Canton, on the bank of the Pearl River (Chu Kiang) and in the delta of the Si-Kiang.

Other large cities important as political or industrial and commercial centres include Harbin and Mukden in Manchuria, and Peking, the national capital, located in the northern part of the Hwang Ho plain.

Transport chiefly depends on rivers or roads, though the railway system is being extended, especially in the N and NW. There, centres of heavy industry are being built up on the basis of oil and iron deposits. The port of Lü-Ta is the eastern terminus of the famous Trans-Siberian Railway. Coastal shipping is important along the China and Yellow Seas. Many port cities depend almost entirely on this kind of transport. In the villages, goods are carried in carts or wheelbarrows, or on pack animals or carrying poles.

China has a total of about 300,000 miles (480,000 km) of surfaced roads, and about 24,000 miles (38,000 km) of railways.

The People More than 90 per cent. of the people are Han Chinese. The remainder

China. The Forbidden City, in Peking, within the Inner City. It was once the residence of the emperor, and only members of the imperial court could enter it. Today, the Forbidden City, with its palace, temples, gardens, and priceless works of art, is the site of several museums and is open to the public.

include many small groups, of which the most important are Uigur, Hui, Yi, Tibetan, Miao, Manchu, Mongol, Chuang, and Korean. The bulk of the people are crowded into about a third of the country's area, mainly in the farms and cities of the great river plains. The minority nationalities live mainly in China's outlying regions.

The Chinese language is spoken in many different regional dialects, but is written in a uniform script. The government is promoting a national language based on the Mandarin dialect of the N. The main southern dialects include Wu, Cantonese, and Hakku. Many other languages are spoken by minority groups. Tibetans have a distinctive language and script. The Uigurs of Sinkiang speak a Turkic language.

CHINA AND MONGOLIA: MINERALS AND INDUSTRY

CYPRUS

The republic of Cyprus occupies the Mediterranean's third largest island, and lies S of Turkey and W of Syria. It has been independent of British rule since 1960, but has been torn by strife between the Greek majority of its people and the Turkish minority. In 1974, troops from Turkey invaded Cyprus and forcefully partitioned the island, expelling many Greeks from the NE sector. The dominant features of the Cyprus landscape are the Troodos Mountains in the western centre and the Kyrenia Range along the N coast. The highest point in Cyprus is Mount Olympus (6,408 ft; 1,953 m) in the Troodos Mountains. The Mesaoria, the island's central plain, lies between the two ranges. The largest of the many rivers is the Pedieos. The climate is a warm Mediterranean one. Minerals account for a third of Cyprus's exports; mineral deposits include copper, asbestos, iron pyrites, and chromite. The chief agricultural products are wine, raisins, cattle, cotton, and barley. Nearly four-fifths of the people are of Greek descent, and one-fifth of Turkish. Greek and Turkish are the official languages, and English is also widely spoken. The chief towns are Nicosia (the cap.), Limassol, Larnaca, Paphos, and Famagusta. Area 3,572 sq miles (9,251 sq km). For population, see Table on page 105.

HONG KONG

Top:
Cyprus. The splendid ritual of a service in an Orthodox church.

Below:
Hong Kong. The harbour, one of the world's finest. The name *Hong Kong* means 'Fragrant Harbour'.

Small British colony on the southern coast of China, about 100 miles (160 km) SE of Canton. It includes Hong Kong Island and some 230 other islands, and Kowloon and the New Territories on the mainland. Part of the territory is leased from China. Most of the land is swampy or rocky and hilly. The climate is sub-tropical. Hong Kong is a major shipping, banking, commercial, and industrial centre, and an important entrepot for China. It has textile and other light industries. Farming and fishing are intensive, but much food is imported from China. China also supplies water. Almost all the people are Chinese. Chinese is generally spoken, but the official language is English. The capital is Victoria, on Hong Kong Island. Area 398 sq miles (1,031 sq km). For population, see Table on page 105.

The republic of India occupies the greater part of the Indian sub-continent, the large triangular peninsula in Asia that projects southwards into the Indian Ocean. The rest of the peninsula is occupied by Pakistan and Bangladesh. On the N, India is bordered by China (Tibet), Nepal, and Bhutan, and on the E by Burma. At its southern tip, it is separated from the island of Sri Lanka (Ceylon) by the Palk Strait. The part of the Indian Ocean that washes India's western coast is called the Arabian Sea; that on the E is the Bay of Bengal.

INDIA

India has one of the world's longest histories of civilization. The ruins of the cities of Mohenjo-daro and Harappa (now in Pakistan) testify to an Indian civilization that dates back to 2500 B.C. In the India of today, ancient traditions and habits exist side-by-side with modern technology.

In landscape, India is extremely varied. It is often spoken of as a land of vast plains, high mountains, and great rivers. But it has regions that are almost park-like in their scenery, and it also has deserts, swamps, and jungles.

Its people are as varied as its countryside. They belong, racially, to several different stocks, and speak scores of different languages. Some have great wealth, some have not enough to eat. Because of the great pressure of population—India has more people than any other country except China, and its population is increasing rapidly —one of the government's most difficult problems is providing enough food. The use of modern farming methods is encouraged in order to increase the productivity of the soil. But when India's two worst enemies, drought and floods, ruin the harvest, many people still die of starvation.

India's capital is New Delhi, in Delhi Territory in the N. The country has an area of 1,262,000 sq miles (3,268,400 sq km). Its greatest length from N to S is about 1,900 miles (3,060 km), and its greatest width E to W about 1,700 miles (2,740 km). For population, see Table on page 105.

Land and Climate The land is divided naturally into three main regions: the triangular Deccan plateau, which comprises the greater part of southern and central India; the fertile northern plains; and the Himalaya Mountains.

The Deccan Plateau is the largest single region. It is bounded on the N by a number of mountain ranges, of which the most important is the Vindhya Range, which extends ENE across the country as far as the valley of the Ganges River. The highest peaks in this range reach about 5,000 ft

India. 13th-century Kesava temple at Somnathpur, Mysore. It is typical of the Vesara style of Hindu architecture: star-shaped and having a raised platform, pyramidal towers, and lavish ornament. The style influenced architecture in other countries apart from India, notably Cambodia (the Khmer Republic).

(1,500 m). The plateau is highest in the S and W, and is tilted downwards towards the E. But on both its western and its eastern edges it rises to ranges of mountains called the Ghats, which meet near Cape Comorin at the southern tip of the peninsula. The Western Ghats rise to about 5,000 ft (1,500 m), and extend for about 800 miles (1,300 km) as far N as the Tapti River. In the S they meet the higher Nilgiris—the 'Blue Hills'. Between the Western Ghats and the sea is a narrow coastal plain on which is situated the city of Bombay. The Eastern Ghats are much lower, their greatest height being only about 2,000 ft (600 m). They are about 500 miles (800 km) in length, and slope down to a wide coastal plain.

The principal rivers of the Deccan follow the slope of the plateau. They rise in the Western Ghats, and flow eastwards to empty into the Bay of Bengal. The largest of them are the Godavari, the Krishna (or Kistna), and the Cauvery. Most of them are flooded during the rainy season.

The Northern Plains stretch right across the sub-continent, from sea to sea. They are crossed by some of the world's greatest rivers, and they comprise the largest area of alluvial lowland in the world. Their width is some 1,500 miles (2,400 km) and they have an average depth of about 200 miles (320 km). The soil of this region is extremely fertile, and it produces a major part of the country's crops. It is the most densely-populated part of India.

The chief rivers of the northern plains are the Indus in the W, and the Ganges and

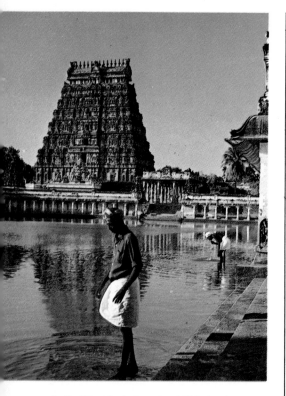

India. The Hindu temple of Chindambaram with its tank, the reservoir of water that is so important a feature of village life. In areas that may have to go for several months without rainfall, water is a precious commodity that must be carefully gathered and protected.

Map legend:

Desert
Rice, jute and sugar cane
Wheat and cotton
Jowar, bajra and cotton
Rice and coconut palms
Plantations of india rubber trees
Tea plantations
Rice and wheat
Wheat
Unproductive mountain areas
Mountain areas with some farming

Coal
Iron ore
Crude oil
Bauxite
Manganese
Tungsten
Magnesium
Industrial centres
Iron and steel industry centres
Railways

Brahmaputra in the E. The Indus, some 1,800 miles (2,900 km) long, rises in Tibet on the northern slopes of the Himalayas, and flows through NW India before entering Pakistan and swinging SW to its mouths on the Arabian Sea. It was on its banks that the civilization of Mohenjo-daro and Harappa developed. The Ganges, the holy river of the Hindus, rises on the southern slopes of the Himalayas, and flows eastwards into Bangladesh, and thence to the Bay of Bengal. The Brahmaputra also rises in the Himalayas, on the northern slopes. It flows eastwards through Tibet before turning SW to flow through NE India and into Bangladesh, where it joins the Ganges at its delta.

The Himalaya Mountains, the world's greatest mountain range, tower over some 1,500 miles (2,400 km) of India's northern border. However, for the greater part of the way only the foothills extend into India. The highest peak within Indian territory is Nanda Devi, which is 25,645 ft (7,817 m) high. Several other peaks rise to more than 20,000 ft (6,100 m). The Himalayas are cut by the Indus River in the W and the Brahmaputra in the E. Spurs of the Himalayas that project into the northern plains are the Sulaiman and Siwalik Ranges, and, in the E, the Naga Hills.

Because of India's enormous size and the diversity of its land features, its climate varies considerably from place to place. The Himalayas have a major effect on climate, shielding the country from the influence of the great land mass of central Asia.

The hot season lasts from March to June. The highest temperatures are in the N; though places in the mountains remain cool, and the so-called 'hill stations' or 'hill towns' owe their popularity to this fact. The rainy season begins in about the middle of June, and goes on until the end of September. This is the period of the SW *monsoon,* when rain-bearing winds blowing in from the Indian Ocean bring torrential rainstorms to much of the country. The rains are heaviest in the Western Ghats and, particularly, in the NE, where they often cause disastrous flooding —though the farmers in many areas depend upon them for the growth of their crops. In October, the rains ease off, temperatures rise, and the air becomes extremely humid. The third climatic season, the cool season, lasts from October to February. In most areas the weather is mild, generally somewhat warmer in the S than in the N. But the mountains of the N experience extreme cold at this time, temperatures dropping lower and lower as altitude increases.

Except in desert regions and on parts of the plateau, the vegetation of India tends to be thick and lush. On the wet western coast and on the middle slopes of the northern mountains there are dense evergreen tropical rain forests. Other mountain forests include evergreen oaks and cedars. Covering many other areas of the peninsula are deciduous or monsoon forests made up of such trees as sal and teak.

Agriculture and Industry Most of the people of India live by farming, many of them at subsistence or near-subsistence level. Many farmers rely on age-old methods of tilling the land, with the result that the yield is low. But government plans to increase productivity have had considerable success. Farmers have been taught to use modern tools, have been supplied with fertilizers and high-yield seeds, and have been given grants to buy machinery. A wide range of crops is grown. Rice is the most important cereal, and is the staple food of southern and eastern India. In the drier parts of northern India, wheat is the principal grain crop. Millet, maize, and barley are also grown. Sugar-cane is grown in the NW and in the Ganges valley. Oilseed crops, including ground-nuts, sesame, rapeseed, and linseed, are an important source of income in many parts. Cotton thrives on the soils of the W-central Deccan, and is the basis of a major industry. It grows also in NW India, where the plantations are artificially irrigated. Tea is the chief cash crop in the NE, in the foothills of the Himalayas, and in the hills of Assam. There are also extensive tea plantations in the Nilgiri Hills of southern India. Jute is the main crop of the Ganges delta region. The forests produce teak and other valuable woods.

India has millions of cattle, including buffalo, but most of them are kept as draught animals, not for their meat, because to Hindus the cow is a sacred animal. The ox-drawn cart or plough is one of the country's commonest sights. Sheep and goats are kept for food, both on the plains and in the hilly regions, and there are vast numbers of poultry. The ox is not the only animal raised as a beast of burden. Elephants, camels, mules, and horses are used too.

In a country subject to such extremes and uncertainties of rainfall, irrigation and drainage are very important. Canals, tanks (ponds), and wells provide irrigation.

India has enormous mineral resources, though they are not yet fully exploited. It has some of the world's richest deposits of iron ore, much of it of high grade. Some iron ore is exported, and some, together with home-produced coal (of which there are huge deposits), forms the basis of the country's

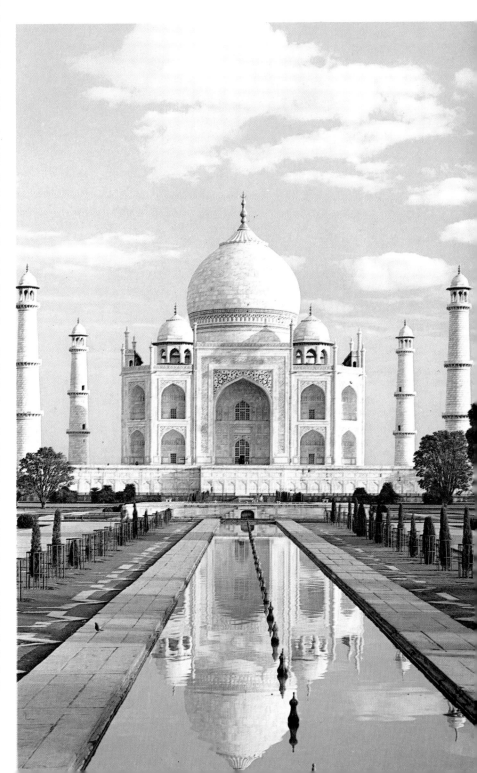

India. The Taj Mahal at Agra, often described as the world's most beautiful building. It was built in the 1660s by the Mughal Emperor Shah Jahan as a mausoleum for his wife Mumtaz-i-Mahal. The Taj Mahal is constructed of white marble inlaid with semi-precious stones, and rests on an octagonal base of red sandstone. Its dome is 120 ft (36.5 m) high. The bodies of the Shah and his wife lie in a vault beneath the centre of the building.

steel industry, which is centred at Jamshedpur, near Calcutta. Other important minerals include manganese, bauxite, copper, chromite, barytes, antimony, and gold. The country produces about nine-tenths of the world's supply of mica.

The chief industry is the manufacture of textiles, principally cotton, silk, and wool materials. The leading centres are Bombay, Nagpur, Jabalpur, Amritsar, and Varanasi (Banaras). Other major industries are the manufacture of cement, paper, burlap products (from jute), chemicals, plastics, and leather goods. The engineering industry is being progressively developed, and the manufacture of machine tools and rolling stock is important. In towns and villages many people work in cottage industries, producing goods in their own homes.

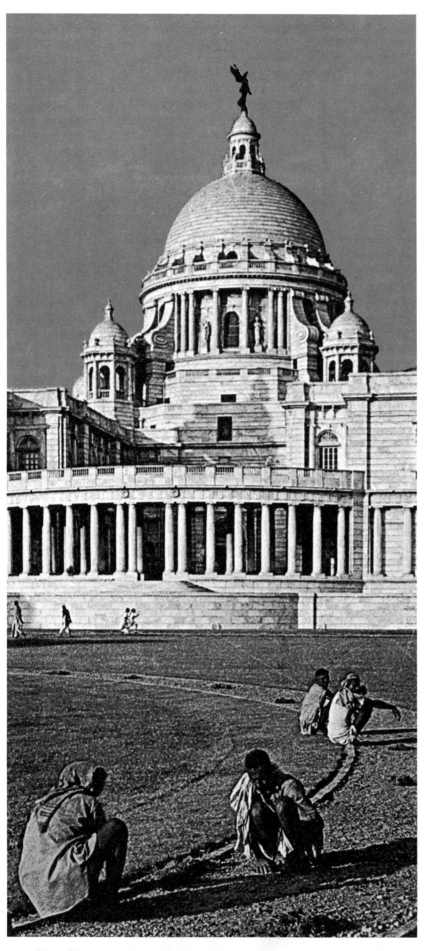

India. The Victoria Memorial in Calcutta, built to commemorate the reign of Queen Victoria, the first British monarch to be described as ruler of an Indian empire. It was opened in 1921. Built of white marble from Jodhpur, the Memorial houses a museum of art and history. The revolving figure of 'Victory' which surmounts the dome is 16 ft (5 m) high.

Cities and Communications Less than 20 per cent. of India's enormous population lives in cities. The largest cities are the ports of Bombay on the W coast, Calcutta on the NE coast, and Madras on the E. Delhi, the capital, is also a large city, and other major centres of population include Ahmadabad, Bangalore, Kanpur, and Poona.

India has about 200,000 miles (320,000 km) of surfaced roads, and some 38,000 miles (60,000 km) of railway. Apart from Bombay, Calcutta, and Madras, there are also many other busy ports, including Vizagapatam, Cochin, and Kandla.

The People India has people of many different races. The two major groups are the Indo-Aryans, who form the majority and who live chiefly in the northern part of the country, and the Dravidians, who are most numerous in the S. Generally, the Indo-Aryans are taller and lighter skinned than the Dravidians. There is also a large aboriginal or tribal population of forest and hill peoples, such as the Maria Gond, the Bhils, and the Nagas. In the extreme NE, many of the people are of Mongoloid stock, related to the peoples of China.

The official language of the country is Hindi, which is spoken by about 35 per cent. of the population. English is also widely spoken. In all, there are about 60 languages and dialects, including 14 national languages recognized by the Indian Constitution. Some of the more important languages are Telugu, Bengali, Marathi, Tamil, Urdu, Gujarati, Kanarese, Punjabi, Oriya, and Malayalam.

More than four-fifths of the people are Hindus, and the *caste* (class) system has a strong influence on their lives. A person remains for life in the caste into which he was born. The four main castes are the *Brahmins* (priests and scholars), *Kshatriyas* (warriors), *Vaishyas* (merchants), and *Sudras* (those whose function is to perform menial tasks). Some people, the 'untouchables' (called *harijans,* 'children of God', by Mahatma Gandhi), are traditionally outside the caste system altogether. Since 1950, any form of discrimination based on caste has been illegal in India.

India also has a large Muslim community, living mainly in the N. Smaller religious minorities include Christians, Sikhs, Buddhists, Jains, and Parsees (Zoroastrians). The Jains believe that all life is sacred, and some take elaborate precautions to avoid the destruction of any living creature. The Parsees, who are believed to have emigrated to India from Persia in the A.D. 700s, live mainly in or near Bombay.

INDONESIA

Indonesia, the largest country of South-East Asia, consists of about 3,000 islands lying between the Pacific and Indian Oceans. The more northerly islands are crossed by the equator. Malaysia and the Philippines are N of Indonesia, and Australia is to the SE. The capital is Djakarta, on the island of Java. Area 735,000 sq miles (1,904,000 sq km). For population, see Table on page 105.

Land and Climate The three largest wholly Indonesian islands are Sumatra (Sumatera), Celebes (Sulawesi), and Java (Djawa). In addition, two large islands are shared between Indonesia and other countries: Borneo with Malaysia; and New Guinea with Papua New Guinea. The Indonesian part of Borneo is called *Kalimantan*; and the Indonesian part of New Guinea, *West Irian*. Between Celebes and West Irian are the Moluccas (Maluku), once famous as the *Spice Islands.*

Java, the most densely populated island, has a fertile alluvial plain in the N-centre, heavily cultivated and supporting most of the island's people. A belt of volcanic mountains runs the length of Java. Sumatra is similar, with a volcanic mountain range in the W, and a swampy alluvial plain inland from the north-eastern coast. Kalimantan is thinly peopled and mostly covered with primeval forest and swamps. The island is mountainous in the centre. Celebes has a deeply indented coast and a volcanic, mountainous, and thickly forested interior. West Irian is mostly forested, with vast mountain ranges. The large islands have many rivers.

The climate is generally hot and wet. The southernmost islands, such as Java, are less hot than equatorial Indonesia, and have alternate wet and dry seasons. All the islands are subjected to the monsoons.

Agriculture and Industry About 70 per cent. of the people live by farming. Much of the farming is at a subsistence level, but there are also many plantations. Crops include rice, maize, tapioca, sago, bananas, beans, cassava, and spices. The chief plantation crops are rubber, coffee, tea, sugar, and sisal. Forest products include quinine, copra, palm oil, hardwoods, and bamboo.

India. Above: The 15th-century Varadarajaswami Vaishnavite temple at Kanchipuram, one of the seven Hindu sacred cities. It is built in the Dravidian style. Below: The great temple of Tiruvannamalai, some 50 miles (80 km) from Vellore, dedicated to the god Siva. In modern times, a famed holy man, Sri Ramana Maharshi, lived nearby.

Indonesia. Rice growing in terraced *paddies* or *cuts*, small fields surrounded by dikes. Because the young plants need a great amount of water, the paddies are kept flooded. Towards harvest time, the water is allowed to drain away.

Indonesia. A temple dancer of Bali, one of the Lesser Sunda Islands. Balinese culture is highly developed, and the music, dance, and drama of the island are famous. Most islanders are Hindus.

Indonesia has large deposits of oil, tin, bauxite, and other minerals. Oil and tin, together with rubber, are the main exports. The main industries are oil refining and food processing. Java is well-known for *batik* —fabric printed by means of a wax process.

Cities and Communications More than 60 per cent. of the people live on Java. The chief cities are Djakarta, Surabaja, Semarang, Bandung, and Surakarta in Java; Palembang, Bukit Tinggi, and Djambi in Sumatra; Bandjarmasin, Balikpapan, and Pontianak in Kalimantan; and Makassar in Celebes. Indonesia has about 50,000 miles (80,500 km) of surfaced roads and 3,500 miles (5,600 km) of railways.

The People The main population groups are Malays and Papuans. Chinese form the largest minority. In the more remote jungle and mountain regions there are groups of tribal peoples. Bahasa Indonesia, a form of Malay, is the official language. English is taught in secondary schools as a second language. About four-fifths of the people are Muslims and most of the rest are Christians. Bali is mainly Hindu.

The Imperial Kingdom of Iran in SW Asia has a history that goes back to the Persian Empire of 2,500 years ago. Today, Iran is still often called *Persia*. The country has a coastline in the S on the Persian Gulf and the Arabian Sea, and in the N on the Caspian Sea. Its land borders are with the U.S.S.R. in the N; Afghanistan and Pakistan in the E; and Iraq and Turkey in the W. Most of the country consists of a plateau averaging some 4,000 ft (1,220 m) above sea-level. In the N the plateau is rimmed by the Elburz Mountains, the highest peak of which is Mount Damavand (18,933 ft; 5,771 m). Another mountain system rims the plateau on the SW. These mountains, the Zagros, rise in several parallel ranges, their highest point, Zardeh Kuh, reaching 14,921 ft (4,548 m). To their E is another series of ranges, the Central Mountains. The central plateau consists of two great deserts: the Dasht-i-Kavir (Great Salt Desert) in the N; and the Dasht-i-Lut (Great Sand Desert) in the centre and E. In the cool season small rivers flow into the Dasht-i-Kavir from the mountains, forming shallow salt-water lakes and marshes. When they dry out during the hot season, much of the surface of the desert is covered with salt deposits. Lake Urmia, in the NW of this desert, covers some 2,400 sq miles (6,200 sq km) in the wet season, but loses a third of its area in the dry season. A low, fertile strip lies between the Elburz Mountains and the Caspian Sea. This is one of the country's few agricultural regions. Another is the Khuzestan plain at the head of the Persian Gulf, which is really part of the Mesopotamian lowlands—mainly in Iraq. Khuzestan is the fount of much of the country's wealth because of its oil-fields. In the interior, the climate is one of very hot summers and very cold winters. Towards the S, the climate becomes more temperate, but the autumn period is still extremely hot. Iran is one of the world's leading petroleum producers. Other mineral resources of importance include iron, copper, coal, chrome, lead, and zinc. Despite the small proportion of cultivable land, more than half the people live by farming. Crops include wheat, barley, rice, fruits, and cotton. The most widely-used language is Farsi. The chief cities are Tehran (the cap.), Tabriz, Esfahan, Mashhad, and Shiraz. Area 628,000 sq miles (1,626,500 sq km). For population, see Table on page 105.

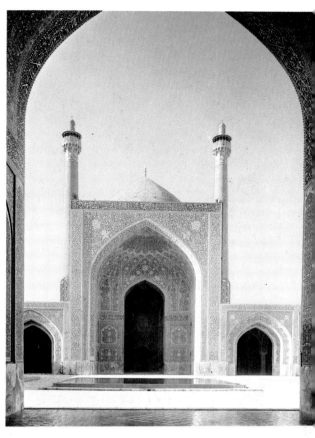

Iran. The exquisite imperial mosque at Esfahan, one of the world's greatest architectural treasures. Its construction was begun in 1612, during the reign of Shah Abbas I. The dome and minarets are faced with turquoise-coloured majolica, and much of the building is decorated with intricate and subtle mosaics.

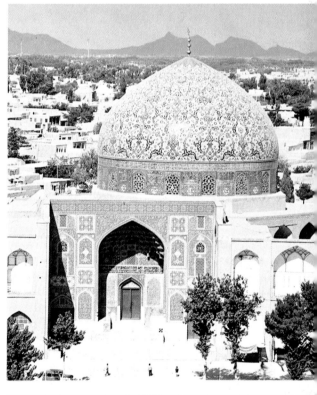

Iran. The mosque of Shaykh Lutf Allah at Esfahan, begun in 1601, was built by Shah Abbas as a private oratory. Like the imperial mosque it stands on the city's great maidan, scene of splendid parades.

Iran. The tomb of Cyrus the Great at Pasargadae. Cyrus, who founded the ancient Persian Empire, reigned from 559 to 529 B.C.. He invaded and conquered many lands, including Babylonia, and called himself *King of the World*.

Arab republic of SW Asia at the head of the Persian Gulf. It includes the region of Mesopotamia where Sumeria, one of the world's first civilizations, was founded. Iraq is bordered on the N by Turkey, on the E by Iran, on the S by Kuwait and Saudi Arabia, and on the W by Jordan and Syria. On the S it has a very short coastline on the Persian

IRAQ

Gulf. Sparsely-wooded highlands in the NE and the arid Syrian Desert in the W are separated by the broad lowlands and plateaux crossed by the Tigris and Euphrates rivers. The region bounded by the rivers is the ancient *Mesopotamia*—'the land between the rivers'. At a point some 120 miles (190 km) from the Persian Gulf, the rivers join to form the River Shatt al Arab. The NE highland region is part of *Kurdistan*—the land of the Kurds. The climate is dry and hot, especially in the desert region. The country is an important producer of petroleum—the chief source of its wealth. Other minerals include sulphur, phosphates, salt, and gypsum. More than half the people live by farming. Their crops include dates, wheat, barley, rice, tobacco, and cotton. Livestock provide wool and hides. Industries include food processing, and the manufacture of cement, soap, and textiles. Efforts are being made to develop industry and agriculture by hydro-electric and irrigation works. Most of the people are Arabs and the rest are mostly Kurds. Arabic is the official language. The people are nearly all Muslims, divided two to one between the Shi'a and Sunni sects. There is a small Christian minority. The chief cities are Baghdad (the cap.), Basra, and Mosul. Area 172,000 sq miles (445,500 sq km). For population, see Table on page 105.

Iraq. Refuelling the desert bus at Baghdad. The powerful, air-conditioned, and specially equipped bus makes a 500-mile (800-km) journey to Damascus in Syria. For most of the way, its route crosses the hot, barren limestone hills and sandy wastes of the Syrian Desert.

ISRAEL

Located in SW Asia at the E end of the Mediterranean, Israel is bordered on the N by Lebanon, on the E by Syria and Jordan, and on the SW by Egypt. It has a long Mediterranean coastline. About half of Israel consists of the Negev desert in the S. The rest includes a hilly region in the N and a narrow coastal strip in the centre. The climate is hot and dry. The chief rivers are the Jordan and Kishon. The country's mineral resources include potash, phosphates, copper, iron, and gypsum. Through irrigation and careful cultivation, much formerly arid land has been made productive, and Israel is a major exporter of citrus fruits and vegetables. Industries include food processing, and the manufacture of machinery, textiles, and chemicals. Almost all the people are Jews, with the rest mostly Arabs. Hebrew is the official language but English is widely used in business. The chief cities are Tel Aviv-Yafo, Jerusalem, Haifa (the main port), and Ramat Gan. The Israeli parliament declared Jerusalem the cap. in 1950, but this is not recognized by the United Nations. Area 7,992 sq miles (20,700 sq km). For population, see Table on page 105.

Israel. Tel Aviv, the country's largest city. It lies in the Plain of Sharon, on the Mediterranean coast. The city was founded in 1909, and is laid out in a regular pattern, with modern buildings arranged in rectangular blocks. It has a university, and is an important centre of industry and communications.

The island nation of Japan is a long archipelago off Asia's eastern coast. The Sea of Japan separates it from the Asian mainland, and to its E is the Pacific Ocean. Japan's capital is Tokyo, on the island of Honshu. The country's area is 142,810 sq miles (369,880 sq km). For population, see Table on page 105.

JAPAN

Japan. The sacred mountain of Fujiyama or Fujisan, on Honshu. Its snow-capped peak rises to 12,388 ft (3,776 m). Fujiyama, about 70 miles (112 km) from Tokyo, is in Fuji-Hakone National Park, and its beauty has provided inspiration for many Japanese artists and poets.

Land and Climate Japan consists of four main islands and more than 3,000 smaller ones. The four largest, in order of size, are Honshu (area 87,800 sq miles; 227,400 sq km); Hokkaido, the northernmost major island (30,140 sq miles; 78,060 sq km); Kyushu, the southernmost major island (14,110 sq miles; 36,447 sq km); and Shikoku (7,050 sq miles; 18,260 sq km). The chief groups of smaller islands are the Ryukyus and Bonins.

All the larger islands are extremely mountainous. The highest mountains are the Japanese Alps in central Honshu. Japan has nearly 200 volcanoes, some of them active. Mount Fuji, or Fujiyama, on Honshu, is Japan's highest peak, rising 12,388 ft (3,776 m). It is an inactive volcano.

Japan's many rapid mountain rivers are an important source of hydro-electric power. The main rivers are the Ishikari (Hokkaido); and the Shinano, Tone, and Kitakami (Honshu). The islands have many beautiful lakes, including Biwa and Inawashiro (Honshu) and Shokotsu (Hokkaido).

Japan's climate ranges from cool temperate in the N to warm temperate in the S. The variation is caused by the influence of the warm Japan Current in the S and the cold Oyeshio Current in the N. All Japan has plentiful rain. Many of its mountain peaks are snow-covered in the winter.

Agriculture and Industry Only about 15 per cent. of Japan's land can be farmed. As a result, the available farmland has to be intensively cultivated. The main crops are rice, wheat, barley, fruit, and potatoes and other vegetables. There being so little lowland, many hills and mountainsides are cultivated by means of terracing.

Most of Japan's mountains are covered with forests, providing valuable timber and charcoal, which is widely used as a domestic fuel. Japan is among the world's leading fishing nations, catching about $8\frac{1}{2}$ million tons of fish every year.

The country is poor in natural resources. Only coal is mined in large amounts, and most of it is low grade. There are small amounts of copper, zinc, bauxite, lead, and nickel. Despite the shortage of resources, Japan is one of the world's great industrial and Kobe, on Honshu; Fukuoka on Kyushu; and Sapporo on Hokkaido.

Japan has about 80,000 miles (130,000 km) of surfaced roads, and a rail network of 15,000 miles (24,000 km) covering the four main islands. The major ports are Tokyo, Yokohama, Nagoya, Kobe, and Osaka.

The People The Japanese are a Mongoloid people, closely related to other groups of eastern Asia, but with some mixture of Malaysian and Caucasoid strains. The only major minority is about 600,000 Koreans. In Hokkaido there is a small minority of about 15,000 Ainu, a Caucasoid people who were probably Japan's original inhabitants.

Japanese is the official and universal language. Its structure is similar to Korean, and both languages resemble those of the Ural-Altaic family, which also includes Turkish and Finnish. Written Japanese is based on Chinese characters, some of them in simplified form. Buddhism and Shintoism are the main religions. Almost all Japanese follow one or the other, and many people practise both.

Japan. The ancient city of Kyoto, until 1868 the country's capital. Above: Cherry-blossom time at the temple of Dai-goji. Left: A garden without trees or flowers. This peaceful spot is a place of contemplation for Zen Buddhists. Below: The temple of Heian-Jingu. The city has hundreds of temples and shrines.

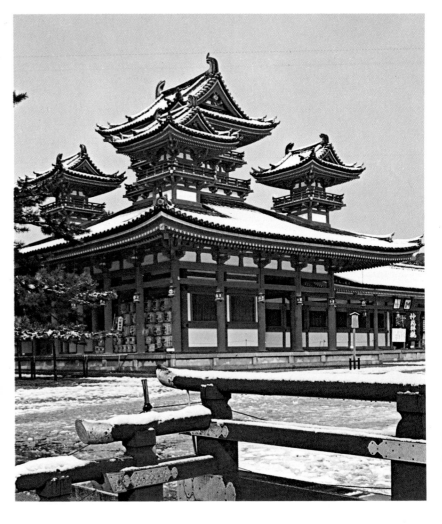

countries, importing almost all the raw materials it needs. It is one of the leading steel producers, and leads the world in shipbuilding. Other heavy industries include the construction of locomotives and industrial machinery. The country has a high reputation for the manufacture of textiles, motor vehicles, optical instruments, electronic equipment, chemicals, and precision tools.

Cities and Communications More than two-thirds of the people live in cities or other urban communities. The chief cities include Tokyo, Osaka, Yokohama, Kyoto, Nagoya,

JORDAN

Arab kingdom of SW Asia. It lies near the eastern coast of the Mediterranean, but its only outlet to the sea is at the port of Aqaba on the Gulf of Aqaba leading to the Red Sea. It is bordered on the N by Syria, on the NE by Iraq, on the E and S by Saudi Arabia, and on the W by Israel. In the W the River Jordan flows N-S through the country. At its S end is the Dead Sea, a salt lake, which is the earth's lowest surface point—1,286 ft (391 m) below sea-level. W of the Jordan are uplands, including the Judean Hills and Samaria. E of the river, the land is also high. These arid, stony uplands form the greater part of the country, and slope away to the Syrian Desert in NE Jordan. The climate is hot and dry. The chief river, apart from the Jordan, is the Yarmuk. Exploited minerals include phosphates, potash, and marble; there are also deposits of manganese, iron, sulphur, and copper. About a third of the people live by farming. The chief crops are wheat, barley, citrus fruits, olives, and vegetables. Industries include oil refining, tanning, and the manufacture of electrical goods, cement, and soap. Almost all the people are Arabs, divided between Bedouin to the E of the Jordan and Palestinians to the W. The largest cities are Amman (the cap.), Irbid, and Nablus. The Jordanian sector of Jerusalem was occupied by Israel in 1967. Area 37,700 sq miles (97,600 sq km). For population, see Table on page 105.

Khmer Republic. Right: Angkor Wat, a ruined Hindu temple in the north of the country, believed to be the largest religious structure in the world. Below right: Cultivating rice in a paddy field, a farming technique that is probably 5,000 years old.

KHMER REPUBLIC

Country of SE Asia, in the Indochina Peninsula. It was formerly the kingdom of Cambodia. The Khmer Republic is bordered on the N by Thailand and Laos, and on the E and SE by Vietnam. In the SW it has a coastline on the Gulf of Siam, an inlet of the South China Sea. The E of the country lies in the alluvial basin of the lower Mekong River, with extensive jungles and the large Tonle Sap ('Great Lake') in the W part. The Phanom Dong Rak mountain range runs along the N border, and the peaks of the Cardamom Range rise just inland from the coast. The climate is tropical and affected by the monsoon. The country's mineral resources are little exploited; they include phosphates, iron, and limestone. Agriculture and fishing are the two main occupations. Many farmers live at a subsistence level, growing, chiefly, rice. Other crops include rubber, tobacco, maize, and sugar. Nearly 90 per cent. of the people are Khmers. Minorities include Vietnamese, Chinese, and Chams. Khmer (or Cambodian) is the official language. The chief cities are Phnom Penh (the cap.), Battambang, Kompong Cham, and Kampot. Area 70,000 sq miles (181,000 sq km). For population, see Table on page 105.

KOREA, SOUTH

South Korea. The harbour at Pusan, the country's largest port. The city lies on the Korea Strait, and is an important industrial centre.

KOREA, NORTH

The Democratic People's Republic of Korea occupies the N part of the Korean Peninsula on the NE coast of China. It is bordered on the N by China, on the extreme NE by the U.S.S.R., and on the S by South Korea. It has coastlines on the Yellow Sea in the W and the Sea of Japan in the E. North Korea is a mountainous land with thousands of peaks. The highest mountains are in the N, where some rise to more than 8,000 ft (2,440 m). Another impressive mountain range extends along the E coast. The only flat and low-lying land is in the W. The chief rivers are the Tumen, Yalu, Ch'ongch'on, and Taedong. The climate is subject to the monsoon, but is more temperate and somewhat cooler and drier than in South Korea. The country is rich in minerals. There are large deposits of coal, iron, lead, zinc, copper, gold, graphite, silver, and tungsten. More than half the people work in mining and industry. The main industrial products are steel, machinery, and textiles. Only a fifth of the country's area is suitable for farming, and less than one-tenth of the people live on the land. The chief crops are rice, maize, barley, wheat, and cotton. The people of North Korea are related to the Mongolians with some Chinese mixture. Korean is the official language. The chief cities are Pyongyang, Chongjin, Hungnam, Hamhung, and Kaesong. Area 46,800 sq miles (121,200 sq km). For population, see Table on page 105. See also *South Korea*.

South Korea. The Eastern Gate of the city of Seoul, one of the three surviving gates in the ancient city walls. As well as being the country's capital, Seoul is the cultural centre and has large industries.

The Republic of Korea occupies the southern part of the Korean Peninsula on the NE coast of China. It is bordered on the N by North Korea. In the E it has a coastline on the Sea of Japan, in the S on the Korea Strait, and in the W on the Yellow Sea. There are thousands of off-shore islands to the S and W. Most of the land is mountainous, the highest peak being Chiri-San in the S which rises to 6,281 ft (1,914 m). There is limited arable land in the lowlands and river valleys of the S and SW. The chief rivers are the Naktong and the Han. The climate is temperate and somewhat warmer and wetter than in North Korea. South Korea has less than one-third of the peninsula's estimated deposits of gold, silver, and tungsten, and about one-tenth of its coal and iron ore. The main crops include rice, barley, wheat, millet, soya beans, tobacco, and sweet potatoes. Livestock raising is also important. Industries include food processing, and the manufacture of textiles, wood products, ceramics, glass, and electronic goods. As in N Korea, the people are almost all of Korean stock, thought to be related to Mongolians with some Chinese mixture. Korean is the official language. Traditional religions include Buddhism and Confucianism. There is also a Christian minority of more than a million and a half. The chief cities are Seoul (the cap.), Pusan, Taegu, and Inchon. Area 38,400 sq miles (99,460 sq km). For population, see Table on page 105. Annexed by Japan in 1910, Korea was divided into Russian and American occupation zones in 1945 after Japan's defeat in World War II. This division became the basis of two states, communist North Korea and non-communist South Korea.

KUWAIT

Located at the NE corner of the Arabian Peninsula, at the head of the Persian Gulf, the tiny, oil-rich state of Kuwait is bordered on the N and W by Iraq and on the S by Saudi Arabia. The country is almost all flat desert, with a few fertile oases. There are no rivers, and the climate is hot and dry. Only a very small part of the land is cultivated, producing fruit and vegetables. The economy is almost totally based on the country's oil deposits, amounting to nearly 20 per cent. of the world's known resources. Natural gas is also produced. Kuwait's oil revenues give it almost the world's highest per capita income. The population is mainly Arab, but less than half are Kuwaitis. Arabic is the official language and English is widely spoken. The capital is Kuwait City. Area 7,500 sq miles (19,000 sq km). For population, see Table on page 105.

LAOS

A landlocked country of SE Asia, Laos is bordered on the N by China, on the E by Vietnam, on the S by the Khmer Republic, on the W by Thailand, and on the NW by Burma. The northern part of the country consists of jungle-covered mountains and plateaux, cut by narrow river valleys. In this region is Phu Bia, the highest peak in Laos, rising to 9,240 ft (2,816 m). The S part has sparsely-forested, arid limestone terraces, descending to the valley of the River Mekong. The Mekong is the largest and most important river. Like surrounding lands, Laos has mainly a tropical monsoon climate. The country's chief mineral resource is tin. It also has deposits of iron, gold, copper, and manganese, but these are largely unexploited. Laos is one of the least developed countries in SE Asia. Most of the people live by subsistence farming. But cash crops are grown, too. The main crops are

Above:
Laos. Houses built on piles line a swampy creek in Vientiane, the country's capital. The city extends along 3 miles (5 km) of the left bank of the Mekong River.

Top:
Kuwait. Part of El Sabaah hospital, one of the most modern and lavishly-equipped medical centres in the world. Much of the state's large income from oil has been spent on hospitals, schools, roads, and houses.

rice, maize, coffee, cotton, tea, and tobacco. The country's physical features hamper development. And the lack of an adequate road network makes communication difficult between the Mekong basin in the W, where most of the people live, and the rest of the country. About half of the people are Lao, descendants of Chinese settlers of the 1200s. Various scattered hill tribes live in the rest of the country. They include peoples of Chinese, Tibeto-Burman, Thai, and Indonesian descent. There are also other small minorities from many parts of Asia. Lao is the official and main language. French is widely used in business. Theravada Buddhism is the official religion. The chief cities are Vientiane (the administrative cap.), Pakse, Thakhek, and Luang-Prabang (the ancient royal cap.). Area 91,400 sq miles (236,725 sq km). For population, see Table on page 105.

The smallest country of SW Asia, Lebanon is located on the E coast of the Mediterranean, and is bordered on the N and E by Syria and on the S by Israel. Its main physical features are a narrow coastal plain, behind which rise the Lebanon Mountains. Inland is the fertile Bekaa Valley, behind which rise the Anti-Lebanon Mountains, along the Syrian border. Mount Hermon in the Anti-Lebanons reaches 9,232 ft (2,814 m). The largest river is the Litani. Lebanon has a hot, Mediterranean climate, but has fairly heavy rainfall and snow in the mountain regions. The country's mineral resources include deposits of bitumen, iron, limestone, and salt. Its industries include food processing, oil refining, and the manu-

LEBANON

Malaysia. Below: The national mosque in Kuala Lumpur. The city has many fine and imaginative modern buildings. Bottom: A village in Sarawak, on the island of Borneo.

facture of textiles and cement. About a quarter of the land is cultivated. The main crops are citrus fruits, vegetables, and tobacco. There are also large numbers of livestock. Much of Lebanon's income is derived from commerce, finance, and banking. Most Lebanese are Arabs, but, unusually for SW Asia, they are divided nearly equally between Muslims and Christians. Arabic is the official language, but French and English are widely spoken in business and commerce. Beirut, the cap. and largest city, is the financial cap. of the Middle East. Other cities include Tarabulus (Tripoli), Saida (Sidon), and Sur (Tyre). Area 4,300 sq miles (11,000 sq km). For population, see Table on page 105.

MALAYSIA

The Federation of Malaysia in South-East Asia consists of the southern part of the Malay Peninsula—known as West Malaysia—and most of the northern part of the island of Borneo—known as East Malaysia. West Malaysia is bordered by Thailand on the N. On the E it has a coastline on the South China Sea. On the S, the Johore Strait separates it from Singapore. And on the W it has a coastline on the Andaman Sea and the Strait of Malacca. East Malaysia consists of two territories, Sarawak and Sabah. They are bordered by Indonesian Borneo (Kalimantan) on the S, and have coastlines on the South China Sea, Sulu Sea, and Celebes Sea. Malaysia's capital is Kuala Lumpur, in West Malaysia. The country's area is 130,000 sq miles (337,000 sq km). For population, see Table on page 105.

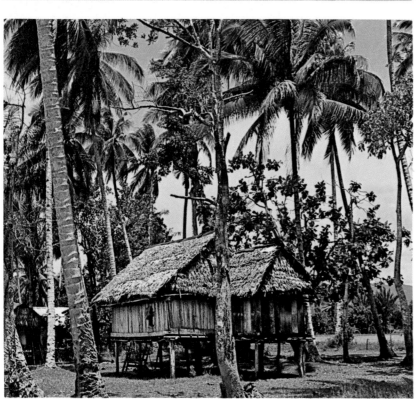

Land and Climate Malaysia lies just N of the equator. More than two-thirds of its land area is covered by dense tropical rain forest. Forested mountains form the backbone of the Malay Peninsula. They are flanked by coastal plains, more extensive in the W than in the E. Much of the population is concentrated in the W. In Sabah and Sarawak, coastal plains rise to forested, mountainous, and sparsely populated inland areas.

The chief rivers are the Perak and Pahang in West Malaysia, the Rajang in Sarawak, and the Kinabatangan in Sabah. The cli-

Legend:
- Rice fields
- Rubber plantations
- Teak forests
- Plantations (tropical produce)
- Tribal and village farming
- Coal
- Petroleum
- Iron
- Nickel
- Manganese
- Lead
- Zinc
- Tungsten
- Chrome
- Tin
- Copper
- Railways
- Oil pipelines

mate is hot and humid, with heavy rainfall, and is affected by the monsoons.

Agriculture and Industry Malaysia's main natural resources are rubber and tin, which with timber and palm oil account for about three-quarters of the country's exports. With Indonesia, Malaysia is the world's leading natural rubber producer, its climate being ideally suited to rubber production. The rich tin deposits in the Malay Peninsula make Malaysia the world's main producer of this metal. Other mineral deposits include oil, coal, and bauxite. The dense forests produce valuable hardwoods.

Most Malaysians live by farming. The main food crops are rice, palm oil, coconut oil, fruits, sago, sugar, tea, and pepper.

Many industries are based on the processing of food, metals, and rubber. There are also manufactures of consumer goods, chemicals, and machinery.

Cities and Communications Nearly 40 per cent. of Malaysia's people live in cities. Apart from Kuala Lumpur, the capital, the largest cities are Penang and Ipoh, both in the western part of the Malay Peninsula. Kuching is the chief town of Sarawak, and Kota Kinabalu of Sabah. Malaysia has extensive road and rail networks and its

MACAU

The tiny Portuguese possession of Macau lies at the mouth of the Pearl River, on the SE coast of China. It is about 40 miles (64 km) W of Hong Kong. It lives by trade. Area 5 sq miles (8 sq km).

MALDIVES

main population centres are linked by internal air services. The chief ports are Penang, Port Swettenham, Dungun, Talok Anson, Malacca, and Port Dickson.

The People Malays (45 per cent.) and Chinese (35 per cent.) are the two largest population groups. About one-tenth of the people belong to non-Malay tribal groups. They include the Dyaks in Sarawak and the closely related Dasuns in Sabah. There are also other sizeable minorities, chiefly Indian and Pakistani.

Malay (Bahasa Malaysia) is the official language. Several Chinese and Indian dialects, as well as English, are also spoken. Almost all the Malays, as well as the Pakistani minority, are Muslims.

The Maldive Islands are an archipelago of some 2,000 coral islands in the Indian Ocean, about 400 miles (640 km) SW of Sri Lanka. The islands are grouped into a dozen atolls. About 220 islands are inhabited. The climate is hot and humid. Fishing and copra and coconut oil production are sources of revenue. The people are Muslim, and speak Maldivian, a Sinhalese dialect. The capital is Malé. Area 115 sq miles (298 sq km). For population, see Table on page 105.

MONGOLIA

Mongolia, in central Asia, is the world's most sparsely populated country. It lies between the U.S.S.R. to the N, and China to the S. Mongolia is mainly a vast, high plateau fringed by mountains in the NW, N, and E, and with much of the Gobi Desert in the SE. Its climate ranges from extremely hot in the desert to very cold in higher regions. Coal and oil are the main mineral deposits. Most of the people are Mongols, many of them herdsmen. Ulan Bator is the cap. and largest city. Area 600,000 sq miles (1,550,000 sq km). For population, see Table on page 105.

NEPAL

Top:
Mongolia. A herdsman on the great plains that form the eastern part of the country. Mongol horsemen have been famed for their skill and fearlessness since the 1100s A.D. Then, under Genghis Khan, they swept over Asia and eastern Europe and established the largest empire in history.

Centre:
Nepal. Onlookers mount decorated 'beasts' for a better view of a religious procession. The majority religion in Nepal is Hinduism.

Bottom:
Nepal. A fertile valley on the southern slopes of the Himalayas, near the city of Katmandu.

The Hindu kingdom of Nepal is situated in the Himalayas. It is bordered on the N by Tibet (China), and on the E, S, and W by India. Nearly three-quarters of Nepal is mountainous, and it has dense swampy jungles. The southern Terai region consists of plain and swamps. The central part of the country is traversed by high, fertile valleys and sharp mountain ridges. The High Himalaya Range lies along the northern border, and includes Mount Everest (29,028 ft; 8,848 m), the world's highest peak. There are several small rivers. The climate, mild along the Indian border, is bitterly cold in the High Himalayas. Nepal's mineral resources include coal and iron, but are largely unexploited. Nearly all the people live by farming. The chief crops include rice, maize, wheat, millet, and jute. Forestry is also important. The main industries are the weaving of textiles, and village crafts. The people are descended from Indian, Tibetan, and central Asian settlers. There are numerous population groups, of which the Gurkhas (the largest) and the Sherpas are the best known. Nepali is the official language. Hinduism, the state religion, is the principal religion in the S; Buddhism is followed in the less populous N. The chief cities include Katmandu (the cap.), Morang, and Lalitpur. Area 54,362 sq miles (140,797 sq km). For population, see Table on page 105.

OMAN

Oman, in the SE of the Arabian Peninsula, is bordered on the N by the United Arab Emirates and Saudi Arabia, and on the W by Yemen (Aden). It has coastlines on the Arabian Sea and the Gulf of Oman. High, barren hills in the NW are separated from the Dhofar Plateau in the SW by hundreds of miles of desert. The principal source of revenue is petroleum. The capital is Muscat. Area 82,000 sq miles (212,000 sq km). For population, see Table on page 105.

The Islamic Republic of Pakistan, the north-western part of the Indian sub-continent, lies on the northern shore of the Arabian Sea. It is bordered by Iran on the W, Afghanistan on the N, and India on the E. Its NE frontier is in Jammu and Kashmir, a territory whose ownership it disputes with India. Until 1972, Pakistan also included the country that is now Bangladesh. Pakistan's capital is Islamabad. The country's area is about 310,400 sq miles (804,000 sq km). For population, see Table on page 105.

PAKISTAN

Pakistan. Right: A busy street in Rawalpindi, in the Punjab. The city lies in the shadow of the Himalayas, and is a major industrial centre.

Land and Climate Pakistan includes all or part of several historic lands: Baluchistan, North-West Frontier Province, Sind, and the Punjab. It has a mountain wall in the N and NW which includes the towering peaks of the Himalaya, Hindu Kush, and Karakoram Ranges. Nanga Parbat in the W Himalayas rises to 26,660 ft (8,126 m) above sea-level. On the NW frontier with Afghanistan is the famous Khyber Pass. Other mountain ranges include the Sulaiman Range in central Pakistan and the Makran and Kirthar Ranges in the S.

Pakistan's main river, the Indus, flows generally SW through the country. It waters the fertile plains of the Punjab region in the NE and the alluvial plain of its estuary. Its main tributaries include the Sutlej, Jhelum, and Chenab. The Thar Desert of NW India extends into SE Pakistan.

Pakistan's climate ranges from hot and dry in the desert regions to cold and wet in mountainous areas. Rainfall varies greatly. The climate of the whole country is strongly affected by the monsoons.

Agriculture and Industry Pakistan has deposits of a number of minerals, including coal, petroleum, iron, chromite, gypsum, sulphur, and antimony. Dams on the Indus and its tributaries provide hydro-electric power, and water for irrigation.

About two-thirds of the people of Pakistan live by farming. They grow wheat, maize, millet, barley, and fruit. Other important crops are cotton, sugar, tobacco, and tea. Most farm work is still done by hand with simple implements. Yet Pakistan is normally self-sufficient in food production.

Pakistan's industries include food processing, and the manufacture of cotton and other textiles, chemicals, cement, paper, and metal goods.

Cities and Communications The chief cities are Karachi, Lahore, Lyallpur, Hyderabad, Multan, Rawalpindi, and Islamabad (the cap.). Pakistan possesses well-developed road and rail systems,

Pakistan. The valley of the River Swat, on the historic North-West Frontier. Much of the region is stark and barren. The people hold tenaciously to their own way of life.

though communications are difficult in mountainous and desert regions. Air transport provides access to some remote areas.

The People The chief population groups are Pathans, Punjabis, Sindhis, and Baluchis. There are also various tribal peoples living in remote mountain areas. Urdu is the official language, but English is widely used in government and business. Most of the people are Muslims, and Islam is the state religion.

139

The republic of the Philippines is an archipelago of more than 7,000 islands in South-East Asia. It stretches for more than 1,000 miles (1,600 km) between the Pacific Ocean and the South China Sea. The northernmost islands are about 330 miles (530 km) from the coast of China and the southern tip of the chain is about 30 miles (50 km)

PHILIPPINES

from the island of Borneo. The capital of the Philippines is Quezon City on Luzon Island. The area of the Philippines is 115,800 sq miles (299,900 sq km). For population, see Table on page 105.

Land and Climate The three largest islands are Luzon (41,765 sq miles; 108,170 sq km) in the N, Mindanao (38,344 sq miles; 99,310 sq km) in the S, and Palawan (5,751 sq miles; 14,895 sq km) in the W. Other major islands and groups include Negros, Bohol, Cebu, Leyte, Masbate, Mindoro, Panay, Samar, and the Sulu Archipelago. In all, more than 700 islands are inhabited.

Most of the larger islands are mountainous. All have coastal lowlands, and some have fertile river valleys. The country's largest lowland area is in central Luzon. Volcanic peaks rise in many of the islands, and there are numerous hot springs.

The Philippines lie just N of the equator. The climate is warm and humid, with little seasonal variation in temperature. But rainfall varies considerably. The northern islands of the Philippines are frequently hit by typhoons, and earthquakes occur from time to time especially in Mindanao.

Agriculture and Industry More than half of the people live by farming. They grow rice, maize, sweet potatoes, manioc, and tropical fruits. Sugar, tobacco, Manila hemp (abaca), and coconuts are important cash crops. Many goats, sheep, pigs, and water buffalo are raised. Forest products and fishing are also important.

The Philippines are rich in minerals. The main metal is gold. Other minerals include silver, iron, copper, chromite, manganese, limestone, coal, and petroleum. Industries include oil refining, food processing, and the manufacture of cement, textiles, chemicals, and rope and fibre products.

Cities and Communications About a third of the people live in cities. The chief cities are Manila, Quezon City, Cebu City, Davao City and Isabela. The country has good roads, and a railway system.

The People The majority of the people are of Malay stock, with Indonesian and Mongoloid strains. There are about half a million Chinese and smaller minorities of Americans and Europeans. There are also about 60,000 pygmy Negritos.

The national language, Filipino, is based on Tagalog, one of the several Malayo-Polynesian languages spoken in the Philippines. English and Spanish are also official languages and are widely used.

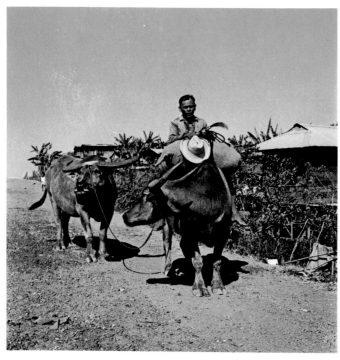

Philippines. Above: The slow, strong *carabao*, the water-buffalo, is the most common beast of burden in the Philippines. It can work in flooded rice fields where other animals would be unable to walk. Below: The port of Manila, the country's largest city.

Qatar, on the E coast of the Arabian Peninsula, occupies a 120-mile (190-km) long peninsula extending into the Persian Gulf. It is bordered by Saudi Arabia and the United Arab Emirates. The country is a low desert plain. Its climate is hot and dry. The basis of the country's prosperity is its petroleum. Revenues from the production of petroleum have been used to develop roads, water supplies, and schools and welfare services. Natural gas is also produced, and fishing is important. Vegetables are grown, and there are herds of goats and camels. Most of the population lives in and around the capital, Doha. Area 4,000 sq miles (10,300 sq km). For population, see Table on page 105.

The kingdom of Saudi Arabia occupies most of the Arabian Peninsula in south-western Asia. It is bordered on the N by Jordan, Iraq, and Kuwait; on the E by Qatar and the United Arab Emirates; on the SE by Oman; and on the S and SW by Yemen (Aden) and Yemen (San'a). In the W it has a coastline on the Red Sea, in the E on the Persian Gulf, and in the S on the Arabian Sea and Gulf of Aden. The capital city is Riyadh, though the administrative centre is Jidda. In the W of the country is Mecca, the holy city of Islam. Saudi Arabia's area is about 920,000 sq miles (2,383,000 sq km). For population, see Table on page 105.

Land and Climate Saudi Arabia is a country of rocky plains, deserts, and mountains. Stretching along the Red Sea coastline is an arid, narrow plain, the Tihama. Mountains extend along the W coast: the Hejaz Range in the N and the Asir Range in the S. Inland is the great central plateau, the Nejd. This barren region of rocky and sandy deserts covers most of Saudi Arabia. The largest deserts are the Rub'al Khali, or Empty Quarter, in the S, and the An Nafud in the N. At the eastern edge of the plateau is a long, narrow belt of sandy desert, the Dahana. The eastern coast lands along the Persian Gulf form a low plain, fringed by mud flats, lagoons, and sand banks. There are no permanent rivers or lakes.

Saudi Arabia's climate is among the world's hottest, with summer temperatures reaching 125°F (52°C) in the deserts and coastal regions. On the other hand, winter temperatures in mountainous areas can drop to freezing. The coastal regions are humid, but the interior has almost no rain.

Agriculture and Industry Over half the people are farmers, growing crops in desert

Saudi Arabia. Oil refinery at Mina Saud, on the Persian Gulf. Oil is the source of the country's wealth.

oases and other scattered localities. They depend heavily on artificial irrigation. The main crops are dates, alfalfa, barley, citrus fruits, millet, and wheat. Nomadic Bedouin tend herds of camels, sheep, and goats.

Saudi Arabia's main source of wealth is its immense oil deposits, which make it the world's fifth largest oil producer. Much of the oil is transported by pipeline to Bahrain and the Lebanon to be refined. There is a major oil refinery at Ras Tannura on the Persian Gulf coast.

Cities and Communications About one-tenth of the population live in cities. The main cities, apart from Mecca, Jidda, and Riyadh, are Al Hufuf, Medina, Al Khobar, and Dammam.

Camel and truck caravans are the main means of transport. There are about 10,000 miles (16,000 km) of roads, mostly unsurfaced. The only railway is a 350-mile (560-km) link between Riyadh and Dammam.

The People Almost all the people are Arabs, divided between nomadic Bedouin and the city and town dwellers and farmers. The official language is Arabic, and Islam is the official religion.

SINGAPORE

Singapore. Barges line the banks of the Singapore River. Singapore's splendid harbour is crowded with craft of all sizes, from great liners to sampans.

Island republic in SE Asia, off the southern tip of the Malay Peninsula, from which it is separated by the Johore Strait. The island is linked to Malaysia by a causeway. Most of Singapore is low lying. Its highest point is Bukit Timah (581 ft; 177 m). There are several small rivers. Singapore's climate is hot and humid. Singapore City has one of SE Asia's few natural harbours and is the world's fourth largest port. The economy is based on the international trade passing through the port and on the processing of raw materials from other countries. More than three-quarters of the people are Chinese. The cap. is Singapore City. Area 224 sq miles (580 sq km). For population, see Table on page 105.

SRI LANKA

Formerly called *Ceylon*, the island republic of Sri Lanka lies in the Indian Ocean off India's southern tip, from which it is separated by the Palk Strait. The S central part of the island is mountainous and richly forested. The rest of the land is a fertile, low-lying plain. The chief river is the Mahaweli Ganga. Sri Lanka has a hot, humid, tropical monsoon climate. More than half the people work on farms and plantations. The chief products are tea, rice, rubber, and coconuts. Mineral deposits include graphite and semi-precious and precious stones. Industries include food processing and the manufacture of chemicals, cement, paper, and textiles. Three-quarters of the people are Sinhalese and are Buddhists. But more than 20 per cent. are Tamils, and are Hindus. The official language is Sinhala; Tamil is used, too, and many people speak English. The chief cities are Colombo (the cap.), Jaffna, Kandy, Galle, Negombo, and Trincomalee. Area 25,332 sq miles (65,610 sq km). For population, see Table on page 105.

SYRIA

Arab republic of SW Asia on the eastern coast of the Mediterranean. It is bordered on the N by Turkey, on the E and SE by Iraq, on the S by Jordan, and on the SW by Israel and Lebanon. The Syrian Desert occupies the southern part of the country, fringed on the SW by the Anti-Lebanon Mountains. Near their highest peak, Mount Hermon (9,232 ft; 2,814 m), is a source of the River Jordan. N of the desert is the fertile valley of the River Euphrates, and beyond that a region of rough plains. Parallel with the coast in the W is the valley of the River Orontes. The climate is mild along the coast, with plenty of rain. Inland it ranges from very hot to very cold, with little rain. Syria is self-sufficient in food production and more than half the people live by farming. The chief crops are wheat, barley, cotton, tobacco, and fruit. Cattle and sheep raising is also important. In desert areas there are many nomadic herdsmen. Industries include food processing, and the manufacture of textiles and cement. A dam on the Euphrates provides hydro-electric power and aids irrigation. Oil pipelines from Iraq cross the country to Mediterranean ports. Most Syrians are Arabs. The largest cities are Damascus (the cap.), Aleppo, and Homs. Area 71,800 sq miles (185,960 sq km). For population, see Table on page 105.

Syria. The construction of a pipeline across the Syrian Desert. Pipelines carry oil from Iraq for hundreds of miles across Syria to ports on the Persian Gulf.

THAILAND

TAIWAN

The island of Taiwan or *Formosa* lies about 100 miles (160 km) off the SE coast of China, from which it is separated by the Formosa Strait. Officially called the *Republic of China*, Taiwan is the seat of the Nationalist government of China which was driven from the mainland in 1949. The territory of the republic includes the Pescadores Islands and some small islands off the Chinese mainland. About two-thirds of Taiwan is mountainous. The W part is a flat coastal plain that supports most of the population. Taiwan has few minerals, but industry, including textiles, chemicals, food-processing, and electrical goods, is well developed. The chief towns are Taipei (the cap.), Kaohsiung, and Tainan. Area 13,885 sq miles (35,962 sq km). For population, see Table on page 105.

Formerly known as *Siam,* Thailand, in SE Asia, is bordered on the NE by Laos, on the SE by the Khmer Republic, and on the S by Malaysia. It has a long coastline on the Gulf of Siam in the S, and a short one on the Andaman Sea in the SW. The northern part of the country is on the SE Asian mainland, and the southern part stretches down the Malay Peninsula. The country can be divided into four main natural regions. The mountainous northern region lies between the Salween River in the NW and the Mekong River in the NE. The eastern region is a high and arid plateau, stretching from the mountains to the border with Laos. Central Thailand consists of a low, flat alluvial plain formed by the Chao Phraya and other rivers. Southern Thailand is long, narrow, and mountainous. As well as the Chao Phraya, Mekong, and Salween, Thailand's many rivers include the Chi, Nan, Pattani, Ping, Wang, and Yom. Most of them are navigable, and they form an important means of transportation. In southern Thailand, just N of the Malaysia border, is a large lagoon called Thale Luang. About half of the country is covered by forests that provide valuable teak, iron-wood, rattan, and bamboo. The forests are also known for their wildlife, including tigers, panthers, monkeys, and wild boars. There are more than 1,000 species of birds in Thailand. Most of the people of the country live by farming. Rice, the main crop, is chiefly grown in the central region. It is the staple food and main export. Rubber is the second most important crop. Other crops include maize, sugar, pepper, coconuts, jute, tapioca, and fruits. Fishing, for local con-sumption and for export, is important. Tin, tungsten, and sapphires are mined. There are also considerable deposits of other min-erals, including coal, gold, and lead, but as yet these have been little exploited. The few

Thailand. The Grand Palace—the walled royal city—in Bangkok. In the centre background is the magnificent *Wat Phra Keo*, the Temple of the Emerald Buddha. The chief of its treasures is a jasper figure of the Buddha.

TIMOR

The island of Timor in the Malay Archipelago is the most easterly of the Lesser Sunda Islands. The western part of the island was a Dutch colony until 1949, when it became part of Indonesia. The eastern part was col-onized by the Por-tuguese as early as the 1600s. The chief town in the eastern section is Dili.

---

industries include food and rubber proces-sing, and there are traditional handcraft industries such as the making of cotton and silk cloth and lacquerware. Most of the people are Thais, the descendants of Mongol invaders and settlers from China. A large Chinese minority, numbering about three million, dominates Thailand's retail trade. The official language is Thai, which belongs to the Sino-Tibetan group. It is written in a script derived from Sanskrit. Most of the people are Buddhists, but there are some Christians, Confucians, and Muslims. The country has nearly 20,000 Buddhist tem-ples. Traditionally, Thai men who are Buddhists spend at least a short period of their lives as monks. Most of the people live in farming villages. Each village has a central group of buildings called a *wat*, which is used as a religious and social centre. The chief cities are Bangkok (the cap.), Chiang Mai, and Lampang. Area 198,460 sq miles (513,810 sq km). For popu-lation, see Table on page 105.

The republic of Turkey is mainly in SW Asia ('Asia Minor'). But a small part of it is in Europe. European Turkey (called *Thrace*), about 3 per cent. of the total area, is separated from Asiatic Turkey (the greater part of which is called *Anatolia*) by the Dardanelles, the Sea of Marmara, and the Bosporus. Turkey has long coastlines on the Black Sea in the N, and on the Mediterranean in the S. Part of the Mediterranean coastline is on the Aegean Sea. In the NE the country borders on Greece and Bulgaria, in the NW on the U.S.S.R., and in the SE on Iraq and Syria. The European part has rolling farmland, and is hilly in the NE. Asiatic Turkey has fertile coastal strips. The semi-arid Anatolian Plateau in the centre is fringed by hills and mountains, including the Pontic Mountains and the Taurus Range. The chief rivers include the Kizil, Seyhan, Menderes (Maeander), Tigris, and Euphrates. The climate ranges from mild and moist along the coasts to harsh and dry on the inland plateau. Turkey has good mineral deposits, but they are mostly unexploited. More than two-thirds of the people live by farming. The chief crops are cereals, tobacco, fruit, and cotton. Animal husbandry is also important. Industries include the production of iron and steel, sugar refining, and the manufacture of textiles. Most people are Turks. Minorities include about two million Kurds, and small groups of Greeks, Arabs, Armenians, and Jews. The chief cities are Ankara (the cap.), Istanbul (Constantinople), Izmir (Smyrna), Adana, and Bursa. Area 301,300 sq miles (780,360 sq km). For population, see Table on page 105.

The U.S.S.R., the world's largest country, lies mainly in Asia. About three-quarters of its territory is E of the Ural Mountains, the traditional dividing line between Europe

Above: Thailand. A floating market in Bangkok. The city lies on a river, the Chao Phraya or Menam, and is intersected by numerous canals. Below: Turkey. The mosque of Sultan Ahmed in Istanbul, popularly called the *Blue Mosque*. It has 6 minarets instead of the usual 4.

United Arab Emirates. The port of Dubai on the southern coast of the Persian Gulf. It was for long a port of call for coastal trade in the Gulf, and has today a large import trade for the seven states.

Opposite page: Vietnam. A girl measures cereals in a city street. The light and practical 'lampshade' hat is the typical headgear of the country.

and Asia. But the majority of its important cities—including the capital city, Moscow—are in the European section. So are more than 60 per cent. of its people. For the main article on the U.S.S.R., see page 95.

UNITED ARAB EMIRATES

A federation of seven emirates or sheikhdoms, the United Arab Emirates is located on the Persian Gulf, in the SE 'horn' of the Arabian Peninsula. The area was previously known as the *Trucial States* and as *Trucial Oman*. The sheikhdoms are Abu Dhabi, Dubai, Sharjah, Ajman, Umm al Qaiwain, Ras al Khaimah, and Fujairah. The U.A.E. is bordered by Oman, Saudi Arabia, and Qatar. It consists mostly of low, flat desertland, with some coastal hills. There are vast oil deposits in Abu Dhabi, and oil has also been found in Dubai. Many people live by herding or fishing, and dates are grown. The people are mainly Arabs. The chief cities are Abu Dhabi (the cap.) and Dubai. Area 32,000 sq miles (82,900 sq km). For population, see Table on page 105.

VIETNAM

Located on the W coast of the South China Sea, Vietnam is bordered on the N by China, and on the W by the Khmer Republic and Laos. On the W it also has a short coastline on the Gulf of Siam. Once part of French Indochina, from 1954 Vietnam was divided between a communist government in the N and a non-communist government in the S. In 1975, communist forces defeated the S Vietnam government and took control of the whole country. Vietnam is about 1,000 miles (1,600 km) long from N to S, and only about 40 miles (55 km) wide at its narrowest point. The country can be divided into three main natural regions: the fertile delta of the Red River in the N; the long, narrow coastal region of central Vietnam; and the delta of the Mekong River in the S. There are high mountains along the northern borders, sloping down to the lowlands of the Red River delta. Central Vietnam rises from a coastal plain to a plateau region and then to wooded mountains. Most of the S is marshy and flat, with fertile alluvial soil. Nearly half the area of Vietnam is forested. The country has a humid tropical monsoon climate, with a hot rainy season in the summer and a cooler, drier season in the winter. Vietnam's mineral resources include coal, iron, manganese, phosphates, tin, and zinc in the N; and coal, lead, limestone, and molybdenum in the S. Important forest products include cinnamon, quinine, and bamboo. Most Vietnamese are farmers. Rice, grown mainly in the delta regions, is by far the most important crop. Rubber is the main export crop. Other crops include coffee, tea, maize, manioc, ground-nuts, coconuts, sugar, tobacco, and sweet potatoes. Vietnam's industries include coal mining, sugar refining, rubber processing, and the manufacture of textiles and cement. Vietnamese is the official language in both parts of the country, and French is widely spoken. Only about 10 per cent. of the population live in cities. The main cities are Saigon (the southern cap., now renamed Ho Chi Minh City), Hanoi (the northern cap.), Da Nang, Gia Dinh, Hué, and Haiphong. Area 128,400 sq miles (332,560 sq km). For population, see Table on page 105.

YEMEN (ADEN)

The People's Democratic Republic of Yemen, or South Yemen, lies on the SW coast of the Arabian Peninsula, on the Gulf of Aden. It is bordered on the N by Saudi Arabia, on the E by Oman, and on the NW by Yemen (San'a). The country consists of a narrow, sandy coastal plain, giving way to a mountainous interior cut by deep valleys. The mountains in turn give way to the desert that occupies much of the peninsula. The climate is hot and dry. There are few natural resources and almost no industries apart from an oil refinery. Most people live by subsistence farming or fishing. The main crops are sorghum, millet, and wheat. Some cotton is grown for export. The people are mainly Arabs. Arabic is the official language, and English is widely spoken. The capital is Aden. Area 180,000 sq miles (466,000 sq km). For population, see Table on page 105.

YEMEN (SAN'A)

The Yemen Arab Republic, in the SW of the Arabian Peninsula, is bordered on the N and E by Saudi Arabia, and on the S by Yemen (Aden). In the W it has a coastline on the Red Sea. The country consists of two main regions: a hot, arid coastal strip along the Red Sea, and, in contrast, a mountainous interior which receives abundant rain. Little is known of Yemen's mineral resources: salt is the only mineral produced in any quantity. Most of the people live by farming or as herdsmen. The main crops are millet, maize, sorghum, fruit and dates. Coffee and qat, a narcotic, are the chief cash crops. Other exports include hides, other skins, and salt. The people are chiefly Arabs, with some African mixture along the coast. The chief cities are San'a (the cap.), Ta'iz, Ibb, and the coffee-exporting port of Moka. Area 75,000 sq miles (194,000 sq km). For population, see Table on page 105.

AFRICA

Africa, the second largest of the continents, is a land of great contrasts. It was the home of one of man's oldest civilizations, but today some of its peoples have still scarcely progressed beyond the most primitive kind of existence.

Africa has burning deserts and luxurious tropical rain forests. In places it also has some of the world's most equable climates. Its forests and plains are rich in animal life.

Politically the continent is divided into the Arab states of the N, and 'Black Africa' S of the Sahara. Two of its countries are ruled by white minority governments.

Africa lies due S of Europe. The Mediterranean Sea forms its northern boundary. At the Strait of Gibraltar Africa is only 9 miles (14 km) from Europe. In the NE, the Isthmus of Suez joins Africa to SW Asia. The Suez Canal has been cut through the 72-

COUNTRIES OF AFRICA
POPULATIONS AND MONETARY UNITS

Country	Population	Monetary Unit
Afars and Issas	101,000	Franc
Algeria	15,200,000	Dinar
Angola	5,700,000	Escudo
Botswana	650,000	Rand
Burundi	3,700,000	Franc
Cameroon	6,100,000	Franc
Central African Empire	1,800,000	Franc
Chad	4,000,000	Franc
Congo	1,900,000	Franc
Dahomey (Benin)	2,940,000	Franc
Egypt	35,500,000	Pound
Equatorial Guinea	300,000	Ekpwele
Ethiopia	26,000,000	Dollar
Gabon	500,000	Franc
Gambia	493,000	Dalasi
Ghana	9,900,000	Cedi
Guinea	4,100,000	Franc
Guinea-Bissau	580,000	Escudo
Ivory Coast	4,600,000	Franc
Kenya	12,900,000	Shilling
Lesotho	1,170,000	Rand
Liberia	1,300,000	Dollar
Libya	2,250,000	Dinar
Malagasy Republic	7,200,000	Franc
Malawi	4,800,000	Kwacha
Mali	5,100,000	Franc
Mauritania	1,250,000	Ouguiya
Mauritius	861,000	Rupee
Morocco	16,500,000	Dirham
Mozambique	8,233,000	Escudo
Niger	4,300,000	Franc
Nigeria	79,700,000	Naira
Rhodesia	5,800,000	Dollar
Rwanda	3,800,000	Franc
Senegal	4,000,000	Franc
Seychelles	52,700	Rupee
Sierra Leone	2,600,000	Franc
Somalia	2,900,000	Shilling
South Africa	21,500,000	Rand
Sudan	17,000,000	Pound
Swaziland	400,000	Rand
Tanzania	14,000,000	Shilling
Togo	2,100,000	Franc
Tunisia	5,600,000	Dinar
Uganda	10,600,000	Shilling
Upper Volta	5,550,000	Franc
Zaire	18,600,000	Zaire
Zambia	4,700,000	Kwacha

Ice on the Equator. An elephant feeds on the tropical grasslands near Kilimanjaro, Africa's highest mountain. Although the mountain is only about 200 miles (320 km) from the equator, snow lies on its upper slopes. Kilimanjaro is an extinct volcano and has two peaks, Kibo and Mawenzi. Kibo, the higher, has a vast snow-filled crater, the source of a great glacier.

mile (116-km) wide Isthmus to provide a passage for ships from the Indian Ocean and the Red Sea to the Mediterranean and the Atlantic Ocean.

The coastline of Africa is mostly smooth, with comparatively few bays and inlets. There are a few islands off the coast, the largest being Madagascar (the Malagasy Democratic Republic) in the Indian Ocean. The Madeira, Canary, and Cape Verde islands are small groups off the NW coast.

The total area of Africa is about 11,700,000 sq miles (30,300,000 sq km). The greatest distance N to S is about 5,000 miles (8,000 km), and W to E 4,700 miles (7,600 km). The population in the mid-1970s was more than 380 million.

Since the end of World War II greater political changes have taken place in Africa than in any other continent. In 1945, all but four of Africa's countries were ruled as colonies of European countries. Now there are hardly any colonial territories left in Africa.

Physical Features and Climate The equator crosses the middle of Africa, and most of the continent lies between the tropics of Cancer and Capricorn. The greater part of Africa is a vast plateau, lying more than 500 ft (150 m) above sea-level. The coastal plain, all round the continent, is narrow in most places, but broadens in the NE, along parts of the Indian Ocean coast, and in the NW.

The great continental plateau lies at several levels, with steep, step-like escarpments between them. On each 'step' there are wide tracts of land with few hills to break the surface.

Almost two-fifths of the continent is desert. The Sahara, the world's largest desert, stretches across the northern part of the continent, where Africa is at its widest, from the Red Sea almost to the Atlantic. It forms a barrier between the coastal lands along the Mediterranean Sea and the rest of Africa. As a result, the peoples N and S of the Sahara differ in their cultures and ways of life.

In the far S, the Kalahari Desert occupies a large area of Botswana and South Africa. Another large desert, the Namib, lies along the SW coast.

In the eastern part of the continent is the Great Rift Valley, a natural fault in the Earth's crust which extends from Mozambique in the S up through the Red Sea and into Asia, where it continues as the Dead Sea and the valley of the River Jordan.

Africa's highest mountains also lie in the E. Mount Kilimanjaro, the highest peak, is almost on the equator. Its summit, 19,340 ft (5,895 m) above sea-level, is always snow covered. Some of the mountains in the Rift Valley region are volcanic, and a few of the volcanoes are still active.

Another mountain range is the Atlas Mountains, a long chain which cuts off the coastlands of Morocco, Algeria, and Tunisia from the Sahara. The Drakensberg range of South Africa has several peaks of more than 10,000 ft (3,050 m) above sea-level. There are two smaller ranges of mountains in the Sahara: the Hoggar Mountains in southern Algeria, of which the highest peak is Tahat (9,557 ft; 2,913 m); and the Tibesti Massif in northern Chad, whose highest peak is Emi Koussi, an extinct volcano.

The region of the Rift Valley contains Africa's chief lakes. The largest, Lake Victoria, has an area of 26,828 sq miles (69,484 sq km); the only larger freshwater lake is Lake Superior in North America. Victoria lies between two arms of the Rift Valley. One of the other lakes in the region, Lake Tanganyika, is 420 miles (680 km) in length and nearly 5,000 ft (1,500 m) deep.

Africa has four of the world's mightiest rivers. The Nile, the world's longest river, flows 4,160 miles (6,695 km) northwards from the lakes of the Rift Valley to the Mediterranean Sea. Water from the Nile irrigates all the fertile land in Egypt and Sudan. The Aswan High Dam, near the border between the two countries, helps to control the river's flow, and has created a huge artificial lake, Lake Nasser.

The Niger is the main river of western Africa. It rises in the hills close to the eastern boundary of Sierra Leone and flows E and N through Mali before swinging through a right-angle and coming out in a large delta to the Atlantic Ocean in the Gulf of Guinea. Its length is 2,600 miles (4,180 km). It is gradually changing course: Timbuktu, which was built on its banks 900 years ago, now lies 7 miles (11 km) from it.

The Zaire, often still called the Congo, rises in SE Zaire, and flows N and W in a wide curve to the Atlantic. At its source it is called the Lualaba. With its tributaries it drains an area almost one-eighth of the entire continent. The Zambezi rises in NW Zambia, and flows over an S-shaped course for 1,600 miles (2,575 km) to the Indian Ocean. Its many waterfalls include the Victoria Falls, which are among the world's largest.

Many parts of Africa, particularly central Africa, are thickly forested. Some forest regions have not yet been thoroughly explored.

Africa is, in general, a hot continent; the central regions are hot all the year round.

A Lofty View. Giraffes, the world's tallest animals, in a national park in Kenya. Africa has a richer variety of animal life than any other continent, but there are only one-twentieth as many animals today as there were fifty years ago. In many countries, those that remain are protected by law, and conservation areas have been established where they can live in safety.

The Boiling Pot. At the Victoria Falls, the waters of the Zambezi River drop spectacularly into a narrow gorge, and then surge through a rock-bound cleft called 'the boiling pot'. The Falls, about 1 mile (1.6 km) wide, were reached by the explorer David Livingstone in 1855, and named in honour of Queen Victoria.

Legend:

Older rocks, often metamorphosed
Younger rocks, mainly sedimentary

◆ Bauxite   ▽ Nickel
☆ Silver   ☆ Gold
● Coal   ⊟ Crude oil
▼ Chrome   ◻ Lead
◇ Diamonds   ★ Platinum
◆ Felspar   ★ Copper
● Iron ore   ⋀ Salt
▣ Phosphates   ▼ Tin
△ Graphite   ○ Tungsten
◼ Magnesium   ✳ Uranium
◆ Manganese   ◉ Vanadium
✛ Mercury   ◼ Zinc
▲ Mica   ◇ Sulphur
✖ Other minerals

The Zaire River basin has a tropical climate with heavy rainfall, but the rest of central Africa has moderate rainfall. The Sahara has little rain, and is hot nearly all the time, though some areas occasionally have frost at night. Southern Africa, outside the desert areas, has moderate or light rainfall. Most of southern Africa is covered by the *veld*, grasslands which in places are very thick and lush.

Because temperatures decrease by about 1°C for every 100 m (330 ft) of height above sea-level, many areas of Africa, even areas close to the equator, have moderate temperatures. For example, Mombasa on the coast of Kenya is hot and humid all the year round, but Nairobi, the capital, which is about 5,500 ft (1,675 m) above sea-level, is seldom more than pleasantly warm.

Resources and Industry Africa is rich in mineral wealth, but so far only a small fraction of it has been found and exploited. The countries along the N African coast have large deposits of petroleum and some natural gas. Most of the world's diamonds come from the countries of southern Africa, and industrial diamonds—those too small for use as jewellery—are found in several

countries in the western part of Africa.

Gold is one of the assets of South Africa, where there are also deposits of other minerals including coal, chromium, manganese, and uranium. Rhodesia has important copper mines, and so has Zaire. Uranium and other valuable metals are found in South Africa and Zaire. There are important deposits of iron in Liberia and South Africa. But even with all the known mineral wealth, there are vast areas of Africa, including most of the Sahara and the forests of central Africa, that have never yet been surveyed by prospectors.

Many African countries are striving to increase their industrial output. Some have planned and built hydro-electric installations to provide power. Among these schemes are the Aswan High Dam in Egypt; the Kariba Dam on the Zambezi, between Rhodesia and Zambia; and the Volta Dam on the River Volta in Ghana.

Communications Communication is a major problem in Africa. Broad rivers, dry, wind-swept deserts, and thick forests make travelling difficult over large areas of the continent. Some of the rivers can be used for transport, but not over their whole length because of rapids and waterfalls. These natural barriers have slowed up development. A large part of the continent has been explored only in the past hundred years, and then with difficulty. The first E-W crossing was made in 1877 by Sir Henry Morton Stanley. It took him nearly three years.

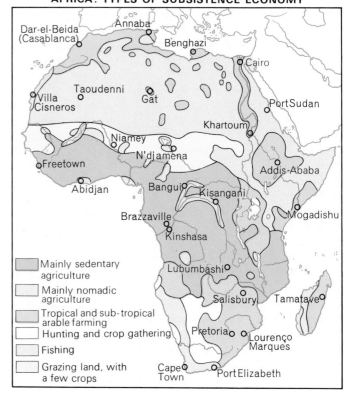

AFRICA: TYPES OF SUBSISTENCE ECONOMY

Legend:

Mainly sedentary agriculture
Mainly nomadic agriculture
Tropical and sub-tropical arable farming
Hunting and crop gathering
Fishing
Grazing land, with a few crops

AFRICA: AREAS OF AGRICULTURAL PRODUCTION AND CHIEF TRANSPORT ROUTES

Areas producing cash crops

Mining area of Zaïre (Katanga)

—— Railways

—— Inland waterways

○ Ports

Railways have gradually been built across Africa, but only South Africa, Morocco, Algeria, and Tunisia have anything like good rail networks. Northern Egypt and the Nile Valley also have an efficient rail system. But many countries have only one or two rail lines, and some have none at all. Away from the cities there are few good roads. Most roads are dirt tracks, dusty in summer and deep in mud in winter, while floods frequently wash away large stretches. Fortunately, the development of air transport is beginning to open up the more remote areas.

Agriculture, Fisheries, and Land Use The main work of Africa's people is farming of one kind or another. In the more remote areas, the people are subsistence farmers, growing just enough food for their own needs, with perhaps a little over to barter for other goods.

Subsistence Farming is the rule mainly in tropical Africa. Shifting cultivation is the most usual form. Tribesmen clear a part of the forest by felling trees and burning brushwood. They cultivate the land for several years until the soil is exhausted, then they move off to a fresh site and start all over again. The land has to be left for about 10 years to recover its fertility. In places where there are too many people to allow

AFRICA
TERRITORIES NOT COMPLETELY INDEPENDENT

Name	Location, etc.	Area sq. miles	sq. km	Population	Capital
Azores (Port.)	Atlantic Ocean. Group of 9 islands.	893	2,313	290,000	Ponta Delgada
British Indian Ocean Territory (U.K.)	Indian Ocean. Group of scattered coral islands and atolls.				
Canary Islands (Spain)	Atlantic Ocean. Group of 7 islands and 6 uninhabited islands.	2,807	7,270	1,165,000	Las Palmas
Cape Verde Islands (Port.)	Atlantic Ocean. Group of 10 islands and 5 islets.	1,550	4,014	270,000	Praia
Comoro Archipelago (Fr.)	Indian Ocean, at northern end of the Mozambique Channel. Group of 4 islands and several islets. In 1975 the islands, except for Mayott, declared themselves independent.	838	2,170	280,000	Moroni
Madeira (Port.)	Atlantic Ocean. Group of 2 islands and several island groups.	308	798	255,000	Funchal
La Réunion (Fr.)	Indian Ocean. Island formerly called Île de Bourbon. Administers several other islands.	970	2,512	450,000	St. Denis
Saint Helena (U.K.)	Atlantic Ocean. Mountainous island. Administers other islands, including Ascension Island and Tristan da Cunha.	47	122	5,200	Jamestown
São Tomé e Principe (Port.)	Gulf of Guinea. Group of 2 islands and 2 islets.	372	963	70,000	São Tomé
South-West Africa (S. Africa)	Formerly called German South-West Africa. Placed under South African administration by League of Nations mandate. Sometimes called Namibia.	318,260	824,290	670,000	Windhoek
Western Sahara	North-western coast. Formerly called Spanish Sahara.	120,000	311,000	63,000	El Aiún

migration, farmers practise crop rotation.

Cultivation is carried out with hoes. In many regions farmers cannot keep horses or cattle because of diseases carried by the tsetse fly, so ploughing is out of the question. In any case, ploughing is not suitable in some places because the broken soil washes away in the heavy rains. Crops grown by subsistence farmers include yams, maize, rice, cassava, beans, and ground-nuts.

Tree Crops are important in the tropical forests of the Zaire River basin, on either side of the equator. The trees cultivated include the oil palm, which yields vegetable oil; the cacao tree, which produces cocoa beans; the banana palm; the rubber tree; and the coffee tree.

Savanna Crops are those grown on the savannas or grasslands—including the veld of South Africa—that cover two-fifths of the continent. Farmers in savanna regions grow all the same crops as those who practise shifting cultivation, plus cotton, millet, and

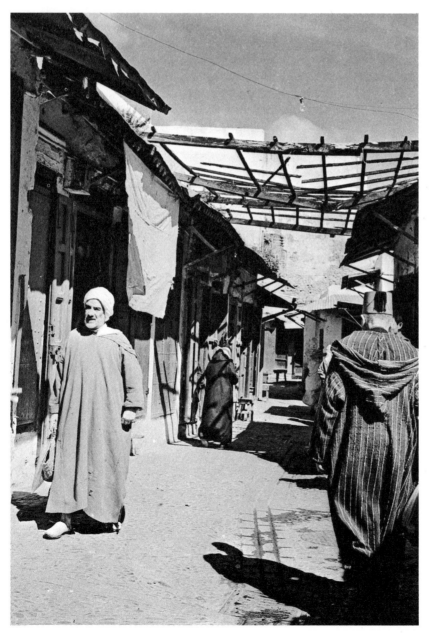

Narrow Sun-Baked Streets. A *casbah* or bazaar in a Moroccan city, typical of bazaars in the Arab cities of northern Africa. Jewellers and other craftsmen sell the objects they have fashioned, and traders offer fruit, vegetables, cooked dishes, cloth, leatherwork, and baskets for sale.

tobacco. In many areas the only cereal that will grow is maize, which forms the main food of the people there. Ground-nuts are produced as a *cash crop*—that is, one purely for sale.

Oasis Crops are grown in the oases of the Sahara. The main crop is dates from the date palm. The sap from this tree is used to make palm wine, while its leaves provide material for matting and roofing. Cotton is another important oasis crop. In addition, the people grow fruit and vegetables for their own use.

Mediterranean and Cape Crops depend on the mild winters and warm summers of these two coastal regions. They include grapes from which wine is made—particularly in South Africa and Algeria—wheat, and barley. Oranges, lemons, and grapefruit are also grown. Along the Mediterranean coast, olives are an important crop, mainly for the oil they yield.

River Valley Crops depend on the annual flooding of various rivers, particularly the Nile, which results in the deposit of rich soil. Such crops include fruit, vegetables, and cereals, and in the Nile valley cotton. Many flooded areas are ideal for growing rice.

Livestock cannot be reared in the areas of tropical Africa where the tsetse fly is prevalent. But livestock is an important part of farming in other regions. Sheep are reared in the Mediterranean lands and in South Africa, and cattle are grazed in the lands immediately S of the Sahara, and over large areas of southern Africa. In the Sahara, the *nomadic* (wandering) Bedouin tribes keep camels, for milk as well as for use as beasts of burden.

Fishing is carried on off the coasts of southern and south-western Africa, in the Atlantic to the NW, and in the Mediterranean. Several countries are seeking to develop fisheries on the big lakes of eastern Africa, and on the rivers.

People and Ways of Life The oldest known skeletons of prehistoric man have been found in Kenya, on the shores of Lake Rudolf. They have been dated at 2,800,000 years ago. But scientists do not know whether Africa was the home of the first men.

Of the present-day Africans, the oldest inhabitants are the Bushmen of the Kalahari Desert, and the Pygmy Negrillos of the Zaire River basin forests. All these peoples are shorter than average.

Generally speaking, the Sahara divides the peoples of Africa into two distinct types. The Negro peoples live S of the Sahara, and this part of the continent is often referred to

154

as 'Black Africa'. N of the Sahara are the Arabs and other related peoples.

The Negroes fall into two main groups. The Sudanese Negroes, also called the W African Negroes, live in the region of western and central Africa that lies just S of the Sahara. Some scientists regard them as the purest Negro type. They include the Yoruba and Ibo peoples of Nigeria, the Mandinka of Guinea, and the Wolof of Senegal. Some of their tribes are very small.

In the northern part of this region are a number of tribes that are a mixture of Negro peoples with peoples from N of the Sahara. This intermixture has resulted from centuries of trading across the desert.

The other main group of Negroes are the Bantus of the southern half of the continent. The Bantu Negroes all speak languages that are of the same general type. They, too, are thought to be the result of intermarriage between pure-blooded Negroes and peoples from N of the Sahara. The Bantu peoples moved into the areas they now occupy during the past 2,000 years. This colonization process was still taking place 200 years ago.

The Caucasian Peoples are those living mainly N of the Sahara. They are basically of the same stock as the peoples of India and Europe. The two main groups are the Hamites and the Semites. They are both swarthy in complexion, with black hair.

The Hamites include the fellaheen of Egypt, and the nomadic Tuareg and Fulani tribes of the Sahara. The Fulani of today are found in Niger and northern Nigeria. In the NW are the Berbers, a much lighter-skinned group similar to the people of Spain and Italy in appearance. Morocco, Algeria, and Tunisia are often called the *Barbary States* after the Berbers. Most of the Hamites are what are loosely called 'Arabs' today.

The Semites, also called 'Arabs', live in the lands E of the Barbary States, and in the lands in the eastern 'horn' of Africa. Like the Hamites they are Muslims.

Polynesian Peoples live in the island of Madagascar (the Malagasy Republic). They apparently settled in the island from southeastern Asia, and they speak a language similar to that spoken by the Malays and by the Polynesians of the Pacific islands.

Later Settlers include the white people from Europe who colonized Africa in the 1800s. They are mostly found in South Africa and Rhodesia. Many of those in other countries, such as Kenya and Uganda, have left or been ejected since these countries became independent. There are also a number of settlers from India and China. But many have left or have been forced to leave in the recent past.

A Face from the Past. Right: The Great Sphinx, near Gizeh in Egypt, has looked out across the Nile valley for nearly 5,000 years, seeing several civilizations come and go. It has the head of a pharaoh—one of the kings of ancient Egypt—and a lion's body. Near it is the Great Pyramid.

Roman Remains. Above: The ruins of the basilica at Apollonia (Marsa Susa) in Libya, dating from the 400s A.D. Hundreds of such ruined buildings in northern Africa testify to the years of Greek and Roman rule.

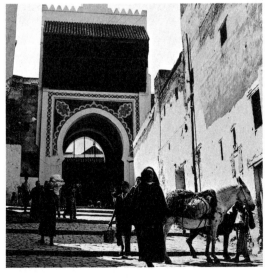

Moorish Arches. Right: A beautiful arched gateway, one of the many that pierce the ancient walls of the city of Fès in Morocco.

THE COUNTRIES OF AFRICA

AFARS AND ISSAS, TERRITORY OF

Small country of eastern Africa, lying between Ethiopia and Somalia. It has a coast-line on the Gulf of Aden, at the entrance to the Red Sea. Until 1967 it was called French Somaliland; it gained complete independence in 1976. Most of the country is stony desert. Crops grow only in the limited areas where irrigation is possible, but there are large numbers of sheep, goats, and camels, many of them tended by nomads. Of the country's two main population groups, the Afars live mainly in the N and the Issas in the S. Djibouti, the cap., is a busy port, much of its trade being with Ethiopia. Area 9,000 sq miles (23,300 sq km). For population, see Table on page 149.

ALGERIA

Algeria, in NW Africa, is the continent's second largest country; only Sudan is bigger. It has had many foreign rulers including, from 1847 to 1962, the French. In the N it has a coastline on the Mediterranean. On the E it is bordered by Tunisia and Libya; on the S by Niger, Mali, and Mauritania; and on the W by Morocco. The country has two distinct natural regions: the Mediterranean region of the N, which lies in the Atlas Mountains; and the vast wastes of the Sahara. Most of Algeria's people and towns are in the Mediterranean region. The coastal part of this region, the Tell, has hills and woods and fertile soil. It is one of the richest farming areas in Africa, with fields of grain,

Algeria. The rich, productive land of parts of northern Algeria contrasts with the barren desert wastes that make up most of the country. In the fertile areas near the Mediterranean coast, vineyards lie beside olive groves, broad fields of cereals, and plantations of fruit and vegetables. One of the farmers' most difficult problems is soil erosion. Government plans to deal with it include the planting of more tree crops and bush crops.

vegetable gardens, olive groves, and vineyards. Just to the S of the Tell are high plateau grasslands, bounded on their southern side by the peaks of the Saharan Atlas. These include the ranges of the Ksour, the Ouled Nail, and the Aurès Mountains. The highest peaks are often snow-covered in winter. Beyond, to the S, are the sands of the Sahara, occupying nine-tenths of the country. The desert has almost no surface water except in the oases. There, villagers grow maize and dates. Elsewhere in the desert, the sparse population consists of nomadic herdsmen. In the S of the country, rising above the desert sands, are the spectacular Hoggar Mountains. Northern Algeria has a Mediterranean climate; on the coast, winters are mild, but in the mountains they are cold. The desert is extremely hot, and in some places has no rain at all for many years. Algeria is rich in petroleum and natural gas. It also has iron and phosphates. There are chemical, steel, and food-processing industries. But the majority of the people live by farming. The chief towns are the capital Algiers (El Djezair), Oran (Ouahran), Constantine, Annaba, and Blida. Area 918,000 sq miles (2,378,000 sq km). For population, see Table on page 149.

ANGOLA

The large country of Angola was an overseas province of Portugal until 1975. It lies in SW Africa, and has a long coastline on the Atlantic. On the N and NE it is bordered by Zaire, on the E by Zambia, and on the S by South-West Africa (Namibia). A plain extends the whole way along the coast; it is narrow in the S and centre but up to 100 miles (160 km) wide in the N. The plain rises to the Bié Plateau, which occupies most of the country. The many large rivers include the Congo, the Cuango, the Cuanza, the Kasai, and the Zambezi. The typical vegetation is savanna, but there are areas of desert in the S. The coastal plain has a moderate climate. Inland there is much rain. An enclave of Angola, called Cabinda, lies on the coast between Congo and Zaire, N of the River Congo. Most of the people live by subsistence farming, growing chiefly millet or maize. The most important cash crop is coffee. Others are cotton and palms for oil. There are some mineral resources, and timber is exported. The chief towns are Luanda (the cap.) and Lobito. Area 481,350 sq miles (1,246,700 sq km). For population, see Table on page 149.

Algeria. The sands of the Sahara near Beni-Abbès, once famous for its connexion with the French Foreign Legion.

BOTSWANA

Landlocked republic in southern Africa, formerly the British protectorate of Bechuanaland. In the NE it borders on Rhodesia, in the SE and S on South Africa, and in the W and NW on South-West Africa (Namibia). The country is an arid plateau averaging some 3,000 ft (900 m) above sea-level. Much of it is occupied by the Kalahari Desert. In the N, the Okovanggo River from Angola forms a vast swamp called the Okovanggo Basin. The Makarikari Salt Pan occupies a large area in the NE. The typical vegetation is savanna. The northern part of the country is relatively well-watered for part of the year. Botswana has deposits of coal, gold, copper, manganese, and other minerals, but most of the people live by keeping livestock. They belong, chiefly, to Bantu groups; Bushmen live in the Kalahari. The cap. is Gaberones; other towns include Francistown and Lobatse. Area 220,000 sq miles (570,000 sq km). For population, see Table on page 149.

Botswana. The Bushmen, most of whom live in the Kalahari Desert, are among the oldest of the peoples of Africa. Once, they were spread all over the southern part of the continent, but gradually they were driven into regions where few other people could survive. These tiny people, less than 5 ft (1.5m) tall, eat roots and such animals as they can kill with their bows and arrows. Their way of life has remained unchanged for hundreds of years.

Cameroon. Village gathering house at Foumbot, in the west of the country. The building has a raised floor, walls constructed of poles, and a thatched roof.

Country of E-central Africa, formerly part of the U.N. trust territory of Ruanda-Urundi. It is small but densely populated. Burundi lies some 150 miles (240 km) S of the equator. The greater part of its western boundary runs through Lake Tanganyika. In the N it borders on Rwanda, in the E and S on Tanzania, and in the W on Zaire. In the western half of the country, a range of mountains extending N to S marks the edge of the Great African Rift Valley (in which Lake Tanganyika lies). This range separates the basins of the Rivers Congo and Nile. The rest of the country consists of broken plateaux, with many rivers in steep-sided valleys. The higher regions are warm and often extremely wet. At lower levels there are higher temperatures, but less rain. Most of the people live by subsistence farming, growing rice, cassava, barley, and wheat. The main cash crops are coffee and cotton. The chief towns are Bujumbura (the cap.) —formerly Usumbura—and Gitega. Area 10,747 sq miles (27,835 sq km). For population, see Table on page 149.

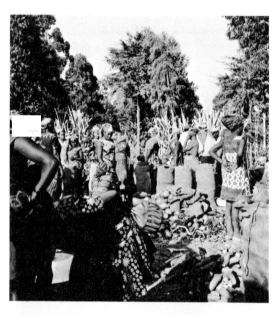

Cameroon. A busy market place, with traders buying and selling vegetables, fruit, and other local produce.

CAMEROON

Country of western Africa, on the Gulf of Guinea, formed by the union of two former trust territories, one French and one British. It is bordered on the E by Chad and the Central African Empire, on the S by Congo, Gabon, and Equatorial Guinea, and on the NW by Nigeria. The narrow coastal plain is swampy. Inland it rises to a plateau that occupies the greater part of the country. The southern region of the plateau is covered by thick rain forest. The central region, called the Adamawa Highlands, is forested, too, in many places. It averages some 1,500 to 2,000 ft (460 to 600 m) above sea-level, but isolated points reach 6,000 ft (1,800 m). In the N there is savanna. The most northerly point of the country lies in the Lake Chad basin. Mountains rise along the western border: the Mandara Mountains in the NW, the Ladamaoua Mountains in the centre, and the Bambuto Mountains in the SW. The highest peak in the country is Mount Cameroon (13,353 ft; 4,070 m), near the coast. The coast is hot and humid. The plateau is cooler and drier. Cameroon has a few mineral resources, and a little industry. But the bulk of the people live by farming. For most it is subsistence farming. The most important cash crops are cocoa and coffee. Other exports include cotton, ground-nuts, rubber, palm oil, timber, and bananas. Some aluminium is exported. The chief towns are Yaoundé (the cap.) and Douala. Area 183,570 sq miles (475,444 sq km). For population, see Table on page 149.

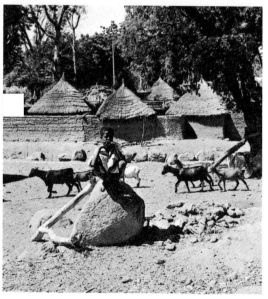

Cameroon. Village life. Most of the villagers spend the whole of the daytime in the open air. There, the daily chores are carried out, food is cooked, and families meet to eat. At night, they find shelter in huts of dried mud. Sometimes, a number of huts, the property of one family, are grouped together behind a wall.

Opposite page.
Chad. A gathering of tribal· chiefs and other notabilities at N'djamena.

Country lying just N of the equator, formerly part of French Equatorial Africa. It is completely landlocked. On the N it is bordered by Chad, on the E by the Sudan, on the S by Zaire and Congo, and on the W by Cameroon. Most of the country is a hilly plateau, rising in the NE to more than 4,500 ft (1,300 m). The northern part of the country is savanna, the southern part forested.

CENTRAL AFRICAN EMPIRE

There are several rivers: the Oubangui and its tributary the Bomu form part of the southern boundary. All of the country is hot. The S has heavy seasonal rains. Most of the people live by subsistence farming. Exports include diamonds, gold, cotton, and coffee. The chief town is Bangui (the cap.). Area 234,000 sq miles (606,000 sq km). For population, see Table on page 149.

Country in N-central Africa, formerly a French colony. It is named after the large, marshy Lake Chad on its western boundary. Chad has no coastline. On the N it is bordered by Libya, on the E by the Sudan, on the S by the Central African Empire, and on the W by Cameroon, Nigeria, and Niger. The northern part of the country is in the Sahara. The Tibesti Mountains in this region rise to 11,204 ft (3,415 m) in Emi Koussi. To the S of the mountains is a vast low-lying area called the Bodélé Depression. In the E are the high Ennedi Plateau and the Quad-

CHAD

dai Massif. The typical vegetation in the southern part of the country is savanna. All of Chad is hot. The S has moderate rainfall. The northern desert region is sparsely populated, and most of the Arab and Hamitic peoples there are nomadic herdsmen. Many of the Negro peoples of the S live by subsistence farming, but important cash crops are grown too. The chief are cotton, ground-nuts, and sugar. The chief towns are N'djamena (Fort-Lamy) and Sarh. Area 495,750 sq miles (1,284,000 sq km). For population, see Table on page 149.

Country in western Africa, on the equator. It was formerly a French colony. Congo is bordered on the N by the Central African Empire, on the E and S by Zaire, on the W by Gabon, and on the NW by Cameroon. In the SW is the small Angolan enclave of Cabinda, and also a short coastline on the Atlantic. A low coastal plain extends NE to the Mayombe Mountains. Beyond is broken upland, and the savanna of the Batéké

CONGO

Plateau. In the N is tropical forest. The rivers Congo and Oubangui form the country's eastern boundary. Many tributaries flow through the country. Congo has some mineral resources, but most people live by farming. Exports include timber, bananas, ground-nuts, and tobacco. The chief towns are Brazzaville (the cap.) and Pointe Noire. Area 129,000 sq miles (334,000 sq km). For population, see Table on page 149.

Small country of W Africa with a short sandy coast in the S on the Gulf of Guinea. Inland the country is at first savanna, rising to a high plateau, and then dropping again towards the basin of the River Niger. The other main river is the Oueme, which has a sluggish sea outlet in the southern lagoons. The climate of the S is hot and wet. Northwards, the rainfall decreases. Dahomey has some mineral deposits, but its principal source of revenue is palm oil from the palm plantations of the S, where most of the people live. Before the country became independent in 1958, it was part of French West Africa. Its main towns are Cotonou, Porto Nova (the cap.), and Abomey. Area 44,696 sq miles (115,762 sq km). For population, see Table on page 149.

Dahomey. A fishing village on the Gulf of Guinea. Several similar villages, built on stilts, lie in the lagoons behind tongues of land on Dahomey's coast.

Right:
Egypt. The golden mask that covered the face of the mummy of Tutankhamon, the boy-king who reigned in about 1350 B.C. It was among the treasures discovered when his tomb in the Valley of the Tombs of the Kings was opened in 1922. Tutankhamon succeeded to the throne when he was eight or nine years old, but died less than ten years later.

The Arab Republic of Egypt, the leading Arab country, is the home of one of the world's oldest civilizations. The greater part of Egypt is in north-eastern Africa, but a small part, the Sinai Peninsula, is in south-western Asia. The country has two coasts: that in the N is on the Mediterranean Sea, and that in the E on the Red Sea. On the E it has a land boundary with Israel, on the S with the Sudan, and on the W with Libya. The Suez Canal cuts through Egypt's territory, joining the Mediterranean with the Red Sea and the Indian Ocean, and providing a valuable trade route for shipping between Europe and Asia. Egypt is called 'the land of the Desert and the River.' The Desert covers most of the country. The River is the Nile, the longest river in the world. Its valley extends for some 960 miles (1,550 km) from the Sudan border in the S to the Mediterranean in the N. In the whole of this distance, the river receives no tributaries. The southern part of the valley has been turned into a great lake, Lake Nasser, by the building of the Aswan High Dam for irrigation and the production of hydro-electric power. Farther N, the river flows between limestone cliffs. In this section it is some 10 to 14 miles (16 to 22 km) wide. Just below Cairo, the Nile enters its delta, where it forms several branches. The main eastern branch reaches the Mediterranean at Damietta (Dumyat) and the main western branch at Rosetta (Rashid). The delta, a triangular plain, is built up of silt carried by the river. From Cairo to the sea it measures some 100 miles (160 km), and at its widest extent on the coast some 150 miles (240 km). Most of Egypt's people live in the delta and the river valley. The Nile is bounded by desert on both sides. The Western Desert is part of the Libyan Desert (in turn, part of the Sahara). Its southern end is extremely bare and windswept. In the N is the vast Qattara Depression, which sinks to 440 ft (134 m) below sea-level. The Eastern Desert or Arabian Desert, between the river and the Red Sea, rises to the Red Sea Mountains along the coast. The triangular Sinai Peninsula lies on the Asian side of the Gulf of Suez. It is the highest part of the country: the Gebel Katherina in the S rises to 8,651 ft (2,636 m). Egypt is a hot and dry country, but in Mediterranean areas the weather is often moderate and pleasant. The country has considerable mineral resources, including petroleum, phosphates, and iron. The government is working to expand industry: manufactures include textiles, chemicals, cement, and processed foods. But Egypt is predominantly agricultural. The main cash crops are cotton, rice, citrus fruits, onions, sugar-cane, and potatoes. The chief cities are Cairo (the cap.), Alexandria, Ismailia, Port Said, Asyut, El Faiyum, and Suez. Area 386,000 sq miles (999,700 sq km). For population, see Table on page 149.

EQUATORIAL GUINEA

Small country in western Africa, formerly a Spanish possession. It consists of the mainland territory of Rio Muni (lying between Cameroon and Gabon), the offshore Corisco Islands, and the mountainous island of Fernando Póo (Macias Nguema) in the Gulf of Guinea. Exports include timber, coffee, and cocoa. The chief towns are Santa Isabel (Malabo), on Fernando Póo, and Bata, in Rio Muni. Area 11,000 sq miles (28,000 sq km). For population, see Table on page 149.

ETHIOPIA

Ethiopia, sometimes called *Abyssinia*, is one of the oldest independent countries in the world. Until 1974 it was ruled by an emperor, but in that year he was deposed. The country, which is in north-eastern Africa, had no coastline until 1952, when the former Italian colony of Eritrea, on the shore of the Red Sea, became part of it. On the E, Ethiopia is bordered by the Territory of the Afars and Issas and by the Somali Republic, on the S by Kenya, and on the W by the Sudan. It is a country of mountainous

plateaux, separated into two main sections by the Great African Rift Valley. The Valley, which crosses Ethiopia from NE to SW, contains several large lakes, including Lakes Zwai, Shala, and Abaya. Another lake, Tana, lying in the N, is the largest in the country, with an area of 1,150 sq miles (2,980 sq km). It is the source of the Blue Nile. The Ethiopian Highlands, or the Central Plateau, extends W of the Rift Valley. Its surface is extremely rugged and broken and varies generally in elevation from 7,000 to

Egypt. The Great Pyramid and the Sphinx near Gizeh, in the Nile valley. The Great Pyramid, about 450 ft (140m) high, was built by the Pharaoh Khufu during the 2600s B.C. It was one of the wonders of the ancient world. The pyramids were constructed as royal tombs.

THE EAST AFRICAN RIFT VALLEYS

— Faults

Rift Valleys

▲ Main active volcanoes

Areas with young volcanic landforms

Lakes

161

Ethiopia. Priests of the Coptic Church in Addis Ababa. Christianity has strongly influenced life in Ethiopia since the 300s A.D. when the kings of Axum, in the north-east of present-day Ethiopia, were converted by Christian missionaries. Today, although Christians are in the majority, there is also a large Muslim minority and there are smaller minorities of other religions.

14,000 ft (2,135 to 4,270 m). Ras Dashan in the N reaches 15,158 ft (4,620 m). Several river valleys cut through the Highlands. The valley of the Blue Nile is some 4,750 ft (1,450 m) deep at one point. The Somali Plateau, E of the Rift Valley, is less mountainous, though it, too, rises to 14,000 ft (4,270 m) in places. The low-lying regions in the N and SE are mostly desert. The northern desert, the Danakil, is inhabited by the nomadic Danakil or Afar people. The southeastern desert, the Ogaden, is occupied by Somali herdsmen. Eritrea has a narrow coastal plain and many offshore islands. The climate in the lower grasslands is hot and dry. With increasing altitude, the climate becomes progressively more temperate, up to about 10,500 ft (3,200 m). The higher levels are too cold for normal agriculture. Ethiopia has some mineral resources, such as salt, potash, manganese, and gold. But mining contributes little to the country's economy. There are still vast resources of timber, particularly in the rain forests of the SW, but in the Ethiopian Highlands much of the original forest has been cleared to allow for agriculture. Nearly all the people live by farming, often at a precarious subsistence level and in constant danger of famine. The main cash crop is coffee. Other crops include cotton, sugar-cane, wheat, maize, barley, corn, tobacco, and fruit. In most inhabited areas, there are herds of cattle, sheep, and goats. The chief towns are Addis Ababa (the cap.), Asmera, Dire Dawa, Massawa, and Assab. Area 450,000 sq miles (1,165,000 sq km). For population, see Table on page 149.

GABON

Country in western Africa, formerly a French possession, lying on the equator. In the W it has a coastline on the Atlantic. On the N it is bordered by Equatorial Guinea and Cameroon, and on the E and S by Congo. The coastal areas are low-lying, and there is a broad plain at the estuary of the River Ogooué, which crosses the country NE to W. Inland, and occupying most of the country, is a hilly plateau. Much of Gabon is covered by tropical forests. All regions have an equatorial climate and are hot and wet. Gabon has valuable mineral resources, particularly manganese, iron, uranium, and oil. Its forests provide timber for export, the most important woods being mahogany and okoumé, which is used for making plywood. Cash crops include coffee, cocoa, palm kernels, and ground-nuts, but most people live by subsistence farming. There are few all-year roads. The chief towns are Libreville (the cap.), Port Gentil, and Lambaréné. Area 103,100 sq miles (267,000 sq km). For population, see Table on page 149.

THE GAMBIA

The Gambia consists of a narrow strip of land stretching for some 300 miles (480 km) along the course of the River Gambia in western Africa. It was formerly a British possession. On the W it has a short coastline on the Atlantic. All its land borders are with Senegal. At its estuary the river flows through mangrove swamps. Behind the swamps are the *banto faros*, grasslands that are under water for part of the year. The inner part of the country is a sandstone plateau. The climate of the Gambia is tropical. It has few mineral resources. Its only important export is its ground-nut crop. The chief food crop is millet; maize and rice are grown, too. The river provides the chief means of communication, but there is also one major road. The capital is the port of Banjul. Area 4,000 sq miles (10,500 sq km). For population, see Table on page 149.

GHANA

Ghana, on the Gulf of Guinea in western Africa, was formerly the British colony of the Gold Coast. On the N it is bordered by Volta, on the E by Togo, and on the W by Ivory Coast. There is some high ground in the E, and a high ridge, the Kwahu escarpment, crosses the country. But most of Ghana is low-lying. In the S there is a broad coastal plain, and almost the whole of the eastern part of the country lies in the basin of the Volta River system. The White Volta flows southwards through the centre of the country. The Black Volta forms part of the western boundary before turning eastwards to join the White Volta. By the construction of a dam across the river at Akosombo, some 70 miles (180 km) from its estuary, a vast lake has been created, several hundred miles in length and covering an area of 3,270 sq miles (8,470 sq km). This great sheet of water, the largest artificial lake in the world, is used for irrigation and the production of hydro-electricity. The climate is hot in northern Ghana, and warm in the centre. The N and E often suffer from long dry spells. The coastal plain is hot and wet. Ghana has some mineral resources. Gold, which gave the country its former name, is still mined in sizeable quantities. Other minerals include diamonds, manganese, bauxite, and graphite. Some of the bauxite is used to feed the important aluminium-smelting industry that has been developed using power from the Volta Dam installations. The bulk of the people live by agriculture. The chief food crops are rice, maize, cassava, and yams. Cash crops include cocoa—the country's main export—coffee, copra, and palm kernels. The chief towns are Accra (the cap.), Kumasi, Tamale, and Sekondi-Takoradi. Area 92,100 sq miles (238,538 sq km). For population, see Table on page 149.

GUINEA

Ghana. Preparing the dinner: cassava being pounded in a wooden mortar by a two-man—or, in this case, a two-woman—pestle. In most places, vegetable products are the chief items of diet.

Country in western Africa, formerly a French colony, with a coastline on the Atlantic. In the N it borders on Guinea-Bissau, Senegal, and Mali, in the E on Ivory Coast, and in the S on Liberia and Sierra Leone. Its coastal plain, about 50 miles (80 km) wide, is fringed by mangrove swamps. Inland is the hilly plateau of Fouta Djallon, which is more than 3,000 ft (900 m) above sea-level in places. The forested Guinea Highlands in the SE rise to 5,748 ft (1,752 m) in the Nimba Mountains. The savanna lands of the E are drained by the River Niger and its tributaries. The Niger has its source in Guinea, as has the River Gambia. In most parts of the country the climate is hot and wet. Guinea is rich in mineral resources, the chief being bauxite and iron. But most of the people live by farming. Their crops include rice, bananas, maize, coffee, ground-nuts, palm kernels, and citrus fruits. The chief towns are Conakry (the cap.), Kankan, and Kindia. Area 96,900 sq miles (250,970 sq km). For population, see Table on page 149.

GUINEA-BISSAU

Country in the NW bulge of Africa, lying between Senegal and Guinea. It has an Atlantic coastline and there are many off-shore islands. Formerly it was called Portuguese Guinea. The country is generally low-lying, but it rises in the SE to the hills of the Fouta Djallon. There are several rivers. Guinea-Bissau is hot and wet. The chief crop is rice; other crops include ground-nuts and cassava. The capital is Bissau. Area 14,000 sq miles (36,000 sq km). For population, see Table on page 149.

Kenya. Tribal life. Drummers wait to provide the music at a celebration.

Opposite page.
Kenya. A Kikuyu village, showing the typical red laterite soil of some regions of equatorial Africa. The Kikuyu, one of Kenya's largest tribes, live in the highlands north-east of Nairobi. They are farmers, though many of them exist at a subsistence level.

Kenya. Dancers at a festival in a prosperous village. Africa's superb heritage of dance and music is endangered by the growth of industry and by the increasing modernization of towns and villages. But many people are aware of the need to preserve cultural traditions that can never be regained once they are lost.

IVORY COAST

One of the richest countries of western Africa, Ivory Coast was formerly a French colony. In the S it has a coastline on the Atlantic. Mali and Volta border it on the N; Ghana on the E; and Liberia and Guinea on the W. A low, forested plain extends inland from the coast for some 150 miles (240 km). It rises to a plateau, averaging some 1,000 ft (300 m) above sea-level, that occupies the northern two-thirds of the country. The plateau vegetation is savanna. In the W the Guinea Highlands extend into Ivory Coast. The principal rivers, flowing N to S, are the Cavalla, Sassandra, Bandama, and Comoé. Natural resources include timber, manganese, iron, and diamonds. The most important crop is cocoa, grown mainly in the SE. Coffee, too, is produced in large quantities. Other crops include bananas, rice, maize, yams, and cassava. The chief towns are Abidjan (the cap.), Bouaké, and Grand-Bassam. Area 125,000 sq miles (323,700 sq km). For population, see Table on page 149.

KENYA

Country of eastern Africa, on the equator. It was formerly a British colony. Kenya has a coastline on the Indian Ocean in the SE. On the N it is bordered by the Sudan and Ethiopia, on the E by the Somali Republic, on the S by Tanzania, and on the W by Uganda. The land is very varied. The coastal plain in the SE is narrow at its southern end but broad near the equator. Much of it is fertile, but it also has patches of tropical rain forest and mangrove swamp. Inland, the land rises to a vast, dry, grass-covered plateau, which occupies most of the northern part of the country and some of the S. The Kenya Highlands, in the SW, are the most heavily populated region. They have Kenya's two highest mountains: Mount Kenya (17,057 ft; 5,199 m), the second highest peak in Africa; and Mount Elgon (14,177 ft; 4,321 m). Both are extinct volcanoes. W of the Highlands is the Nyanza Plateau, extending to the shores of Lake Victoria. This is one of the richest farming regions because of its volcanic soils. One branch of the Great African Rift Valley cuts through Kenya from N to S. In the N it encloses Lake Rudolf, and in the S, where it passes through the Highlands, it encloses the small Lake Naivasha. Kenya's climate is largely dependent on altitude. The coastal plain is hot. The Highlands are relatively cool. Most of the country is dry, but the W, around Lake Victoria, has moderately heavy rainfall. The country is famous for its wild life, and has a number of national parks. Kenya has very little in the way of mineral resources. Most of the people live by agriculture. Half of the country is suitable only for livestock raising. Elsewhere people grow maize, beans, sorghum, wheat, cassava, and other food crops. The most important cash crops are coffee, tea, and sisal. Timber is also a valuable source of revenue. The chief towns are Nairobi (the cap.), Mombasa, Nakuru, and Kisumu. Area 224,960 sq miles (582,644 sq km). For population, see Table on page 149.

The landlocked country of Lesotho was formerly the British colony of Basutoland. It lies in the extreme SE of the continent, and is entirely surrounded by the territory of South Africa. It has close customs and trading relations with South Africa. The country is high and mountainous. The greater part of it is more than 6,000 ft (1,800 m) above

LESOTHO

sea-level. In the E, peaks of the Drakensberg Mountains rise to more than 10,000 ft (3,050 m). Lesotho's many rivers include several tributaries of the Orange, one of Africa's greatest rivers, which has its source in the country. In general, the climate is moderate. But temperatures vary considerably according to altitude, and rainfall tends to be higher in mountainous areas. The greatest concentration of population is in the S. Most people live by farming, working in organized agricultural communities. They grow wheat, maize, barley, peas, and beans. Many of them are herdsmen, and the rearing of sheep, goats, and cattle is important to Lesotho's economy. Thousands of people find work in South Africa. The capital and largest town is Maseru. Area 11,716 sq miles (30,344 sq km). For population, see Table on page 149.

Libya. The ruins of Leptis Magna, on the Mediterranean coast to the east of Tripoli. The city was founded by the Phoenicians in about 600 B.C. Later, under the Romans, it was an important port. Its ruined temples, houses, baths, and forums make up the most impressive collection of Roman remains in Africa.

The oldest independent country in western Africa, the republic of Liberia was formed in 1847 from settlements for Negroes freed from slavery in the United States. Liberia has a 315-mile (507-km) coastline on the Atlantic in the W. On the N it borders Sierra Leone and Guinea, and on the E Ivory Coast. Much of the country is covered with tropical forest, and the coastal plain is swampy. Inland there are plateaux and low hills extending to the Guinea Highlands in the

LIBERIA

NE. On the plateaux, the vegetation is savanna. Several rivers flow NE to SW across the country from the hills. Almost all areas are hot and wet. There are valuable deposits of iron ore, and exports include rubber, coffee, and cocoa. But many people live by subsistence farming. The chief towns are Monrovia (the cap.), Buchanan, Greenville, and Harper. Area 43,000 sq miles (111,000 sq km). For population, see Table on page 149.

LIBYA

Country of northern Africa, with a 1,000-mile (1,600-km) coastline on the Mediterranean Sea. It is one of the continent's largest but most sparsely populated countries. Libya contains three historic regions: Tripolitania in the W; Cyrenaica in the E; and Fezzan in the S. It is bordered on the E by Egypt, on the SE by the Sudan, on the S by Chad and Niger, and on the W by Algeria and Tunisia. At its centre, the coast has a broad indentation, the Gulf of Sidra. On both sides of the gulf there are upland areas. That in the E, the Jebel el Akhdar, rises to some 2,000 ft (600 m). Most of the fertile land in Libya lies along the coast. The rest of the country is in the Sahara: nine-tenths of it is desert or semi-desert. Just inland from the coastal strip in the W is a broad, arid plateau. Beyond lie great wastes of sand seas and barren rocks. In places there are gaunt mountains and escarpments. Several

Libya. The port of Benghazi in the north-east of the country, on the Gulf of Sidra. It stands on the site of the ancient city of Berenice, named in honour of the queen of the Pharaoh Ptolemy III.

caravan routes and motor roads cross parts of the desert, mostly connecting oases with the coastal towns. Important oases include Ghat in the W, Jofra and Al Fuqaha in the centre, and Jalo and Kufra in the Libyan Desert in the E. Temperatures in the desert are often extremely high. The coastal regions have a moderate Mediterranean climate. Libya's great source of wealth is petroleum; the country is the chief oil producer in Africa. Agriculture is confined mainly to the coastal areas and the oases. The chief crops are cereals, ground-nuts, olives, tobacco, citrus fruits, and vegetables. In the desert, there are thousands of nomadic herdsmen. The chief towns are Tripoli (the cap.), Misurata, Benghazi, Homs-Cussabat, and Zawia. Area 679,360 sq miles (1,759,534 sq km). For population, see Table on page 149.

MALAGASY DEMOCRATIC REPUBLIC

Island country in the Indian Ocean whose people are of mixed Asian and African descent. It occupies the island of Madagascar (and some small off-shore islands) about 250 miles (400 km) from the SE coast of Africa, across the Mozambique Channel. On its western side the island has a broad coastal plain. The plain rises to a vast, high, central plateau, which is greatly broken up. Lofty mountains rise in the N, centre, and S. The highest point in the island is Mount Maromokotro (9,449 ft; 2,880 m) in the Tsaratanana Massif in the N. The escarpment of the plateau is much steeper on the eastern side, and the coastal plain is narrow. On this coast a shipping canal several hundred miles long has been constructed by connecting the many coastal lagoons. The island has a large number of rivers, the longest being the Betsiboka. Eastern regions are generally hot and wet. In the higher areas the climate is milder. Madagascar has deposits of chrome, graphite, phosphates, and mica, and has textile industries. But most people live by farming. Rice, millet, maize, and cassava are grown for food. The chief cash crops are coffee, cloves, and vanilla. Large numbers of cattle are reared. The chief towns are Tananarive (the cap.), Tamatave, Majunga, and Fianarantsoa. Area 230,000 sq miles (595,700 sq km). For population, see Table on page 149.

MALAWI

Malawi. A village on the open plateaux. The wide 'mushroom' roofs of the circular mud huts protect the walls from destructive rains during the wet season.

Malawi, in southern Africa, was formerly the British protectorate of Nyasaland. It has no seacoast, but for hundreds of miles its eastern border runs through Lake Malawi (Lake Nyasa), the third largest lake in Africa. Malawi is bordered on the N by Tanzania, on the E, S, and SW by Mozambique, and on the W by Zambia. The country is long and narrow, and the Great African Rift Valley runs through it for its whole length. The western side of the Rift Valley rises steeply from Lake Malawi in an area of wild beauty. In the SE is Mount Mlanje, the highest point in Malawi, which rises to 9,843 ft (3,000 m) above sea-level. Between are broken plateaux, generally averaging some 3,500 ft (1,070 m) above sea-level, but rising in places to more than 8,000 ft (2,450 m). In addition to Lake Malawi there are three smaller but still sizeable lakes —Malombe, Chiuta, and Chilwa. Temperatures are high in most places, and rainfall is high in the extreme N and the S. Most of the people live by farming. Important cash crops are tobacco, tea, cotton, sugar, and ground-nuts. Food crops include maize, millet, and cassava. The chief towns are Zomba (the cap.), Blantyre, Lilongwe, and Mzuzu. Area 45,500 sq miles (117,850 sq km). For population, see Table on page 149.

MALI

Mali. Launderers at work beside the River Niger, near Bamako. Clothes are scrubbed and beaten in the river, and then spread out on the bank to dry.

Large, inland country in western Africa, formerly the French colony of Soudan. It is bordered by Mauritania and Algeria on the N; Niger on the E; Volta, Ivory Coast, and Guinea on the S; and Senegal on the W. In general the country is a flat, savanna-covered plateau, the northern third of which lies in the Sahara. But the S is mountainous, and in the NE there are the highlands of the Adrar dea Iforas. In the N, in the desert, is the great oasis of Taoudenni, where salt is produced. The River Niger enters the country from Guinea, and flows north-eastwards and then south-eastwards in a great arc. In its central section it forms a number of swampy lakes. Another large river, the Senegal, waters the low-lying and relatively fertile areas of the SW. The highest temperatures are in the N, and the heaviest rainfall in the S. Mali has some mineral resources but they have not been exploited. Nearly all the people live by farming or fishing. Most of the Berbers, Fulani, and Tuareg of the N are nomadic herdsmen. The Negroes of the more densely-populated S are mainly subsistence farmers, growing rice, millet, or maize. But some cash crops, such as cotton and ground-nuts, are produced. Cattle are among Mali's exports. The chief towns include Bamako (the cap.), Gao, Kayes, and Timbuktu. Area 465,000 sq miles (1,204,000 sq km). For population, see Table on page 149.

MAURITANIA

The Islamic Republic of Mauritania was formerly part of French West Africa. It is bounded on the NW by Western Sahara, on the NE by Algeria, on the E and SE by Mali, and on the SW by Senegal. In the W it has a coastline on the Atlantic. Much of the country consists of desert plateaux. There is only one permanent river, the Senegal, which forms the south-western border. The climate is hot and dry. The flood plain of the Senegal is fertile, and there farmers grow maize, millet, ground-nuts, and other crops. In other parts of the country, the people are predominantly nomads. There are large deposits of iron and copper. The chief towns include Nouakchott (the cap.) and Atar. Area 419,000 sq miles (1,085,000 sq km). For population, see Table on page 149.

MAURITIUS

Republic consisting of a group of islands in the Indian Ocean some 500 miles (800 km) E of Madagascar. The largest islands are Mauritius and Rodrigues. Mauritius Island has a central plateau bordered by three mountain ranges. The chief industry is the production of cane sugar, but tea and tobacco are grown too. The islands are densely populated. The chief towns include Port Louis (the cap.), Beau Bassin, and Rose Hill. Area (total) 800 sq miles (2,000 sq km). For population, see Table on page 149.

MOROCCO

Moghreb-el-Aksa, 'the Farthest West'. Kingdom in NW Africa. Its most northerly part, the peninsula of Tangier, is only some 9 miles (14 km) from Europe across the Strait of Gibraltar. Morocco has a coastline on the Mediterranean Sea in the N, and on the Atlantic in the W. The Mediterranean coast is rocky. The Atlantic coast is sandy, and inland from it there is a broad coastal strip. Parallel with the coastal strip extends the

168

long, high range of the Atlas Mountains, rising to 13,661 ft (4,164 m) in Mt Toubkal. The mountains are dry and rocky, and on their E side they slope down to the hot sands of the Sahara. Several rivers flow W from the Atlas range to the Atlantic. Those that flow down in an easterly direction dry up as they reach the desert sands. The country's longest river, the Moulouya, is on the desert side of the mountains, but it flows N to the Mediterranean. On the Atlantic coastal plain the climate is moderate. In the N the summers are hot, and in the desert areas they are extremely hot as well as long. The NW parts of the Atlas range have severe winters, with much snow. Most of the people live by farming, the productive areas being the coastal plains and the Atlantic slopes of the Atlas. The chief crops include wheat, barley, maize, beans, dates, nuts, olives, oranges, lemons, and apricots. On the lower mountain slopes there are huge vineyards. Farmers and nomadic herdsmen keep sheep and goats, and there are also many cattle. Forestry and fisheries are important. Many industries are concerned with food processing. There are valuable mineral deposits, and Morocco is a leading producer of phosphates. The people of Morocco are mainly Arabs, Berbers, and Moors. Their language is Arabic. The chief cities are Casablanca, Marrakesh, Rabat (the cap.), Fès, Meknès, and Tangier. Area 171,000 sq miles (442,888 sq km). For population, see Table on page 149.

Morocco. The ancient city of Marrakesh, at the foot of the Atlas Mountains.

with heavy seasonal rains. Subsistence farming is general, but there are good soils in the river valleys. Crops include cashew nuts, sisal, cotton, maize, sugar, and tea. The chief towns are the capital Can Phumo (Lourenço Marques), Beira, and Mozambique. Area 303,000 sq miles (705,000 sq km). For population, see Table on page 149.

Morocco. Below: Nomads camp in the desert. Once, the nomads ruled the Saharan wastes. Today, they are herdsmen or merchants.

MOZAMBIQUE

The People's Republic of Mozambique in SE Africa was, until 1975, an overseas province of Portugal. It has a 1,500-mile (2,400-km) coastline on the Indian Ocean. Through its railways and ports, Mozambique provides access to the sea for the three inland countries of Rhodesia, Zambia, and Malawi to its W. On the N it is bordered by Tanzania, and on the SW by South Africa and Swaziland. Much of the country consists of a broad coastal plain, which is sandy and infertile in places. Many rivers flow eastwards across the country from the mountains and highlands of the W. They include the Zambezi and the Limpopo. The great Lake Malawi is on the NW border. The climate is tropical,

Large, landlocked country in north-western Africa, formerly part of French West Africa. It is bordered on the N by Algeria and Libya, on the E by Chad, on the S by Nigeria and Dahomey, and on the W by Volta and Mali. Most of the country is a plateau. The northern two-thirds is in the Sahara. It is high in places, principally in the Djado Plateau in the N and in the Aïr (Azbine) region in the centre, where some hills rise to about 6,000 ft (1,800 m) above sea-level. Although the Aïr is in the desert, it has valleys and plains where livestock can be pastured and millet and other crops grown. Southern Niger is covered with savanna. In the SW there is fertile, low-lying land in the valley of the River Niger. The SE is also low around Lake Chad. The whole of the country is hot and dry. In the S agriculture is aided by seasonal rains and seasonal flooding of the rivers, but regions in the N have only a few inches of rain a year. Niger has some mineral resources, including uranium, tin, and iron. A traditional export is salt from an oasis in the Sahara. Many people, in both N and S, are nomadic herdsmen who travel around to wherever they can find pasturage for their livestock. Even in the more productive areas of the S, most people live by subsistence

Niger. Right: Traders in Niamey display their wares. Niamey, the chief commercial city as well as the country's capital, lies on the road between the historic settlements of Timbuktu, in Mali, and Parakou, in Dahomey. Parakou is connected by rail with Cotonou, which is Niger's nearest outlet to the sea.

farming, growing millet, rice, or maize. The only important cash crop is coffee, although some cotton is grown, too. There are no railways and very few roads. The capital is Niamey, Zinder the main trading town. Area 459,000 sq miles (1,189,000 sq km). For population, see Table on page 149.

Nigeria. Dromedaries —one-humped camels —being given a drink at a well. Camels can travel long distances without eating or drinking, and until recently were the chief means of transport in the desert. But today motor vehicles are taking their place.

Nigeria, in western Africa, has twice as many people as any other country in the continent. It was formerly a British colony. Nigeria has a southern coastline on the Gulf of Guinea. On the N it is bordered by Niger, on the NE by Chad, on the E by Cameroon, and on the W by Dahomey. It takes its name from the River Niger, which enters the country in the NW and flows generally southwards to its large marshy delta on the Gulf of Guinea. A major tributary, the Benue, enters the country from the E and flows westwards to join the Niger in central Nigeria. The coastal lowlands, up to 60 miles (96 km) wide in places, abound in marshes and mangrove swamps. Farther inland is a belt of dense, tropical rain forest. This gives way to the high table-land that forms the greater part of the country. The Niger and its tributaries break the table-land up into distinct parts. To the W and S of the Niger are the Yoruba Highlands, the former kingdom of the Yoruba people. Most of the region lies at between 1,000 and 1,600 ft (300 to 480 m) above sea-level. To the N of the Niger and Benue are the Jos Plateau which rises to some 5,000 ft (1,500 m) and, E of it, the Bauchi Plateau which reaches some 4,000 ft (1,200 m). Both plateaux slope away northwards to about 2,000 ft (600 m). The extreme N of the country is semi-desert. In the NE, Nigeria includes part of the marshy Lake Chad. Some of the rivers of the N flow to the lake; others flow to the Niger. The climate of Nigeria is tropical, but more extreme in the N than in the S. The coastal region has very heavy rainfall, the rainy season lasting from May to October. The country is relatively rich in minerals, the most important being petroleum. Tin and coal are also mined in large amounts, and there is an export trade in columbite, iron, gold, zinc, and wolfram. Industries include the production of steel, cement, and textiles; tin smelting; petrol refining; and food processing. But most people live by agriculture. The main cash crops are palm kernels, cotton, cocoa, and ground-nuts. Other crops include maize, millet, rice, sorghum, cassava, rubber, tobacco, coffee, and gum arabic. The chief towns are Lagos (the cap.), Ibadan, Ogbomosho, Kano, Oshogbo, Abeokuta, Port Harcourt, Zaria, and Ilesha. Area 356,669 sq miles (923,768 sq km). For population, see Table on page 149.

Nigeria. Mountains of ground-nuts awaiting transport from Kano, one of the chief commercial centres of the north. Ground-nuts and cotton are the most valuable cash crops of the northern region.

climate despite being N of the Tropic of Capricorn. The country is rich in minerals, chiefly gold, iron, asbestos, copper, chrome, phosphates, and coal. There are textile, chemical, paper, and other industries. The principal crops are tobacco, maize, cotton, sugar-cane, wheat, and various fruits. The chief towns are Salisbury (the cap.), Bulawayo, Umtali, and Gwelo. Area 150,820 sq miles (390,622 sq km). For population, see Table on page 149.

RHODESIA

Inland country of southern Africa, formerly the British colony of Southern Rhodesia. In 1965, the colonial government declared the country independent in order to preserve the dominant position of the white Rhodesians. Rhodesia is bordered on the NE and E by Mozambique, on the S by South Africa, and on the W by Botswana and Zambia. Most of the country is occupied by a vast plateau, lying between the Zambezi River in the N and the basin of the Limpopo River in the SE. The central part of the plateau, known as the *High Veld*, is some 400 miles (640 km) long and 70 miles (110 km) wide. It rises to more than 4,000 ft (1,200 m) above sea-level. Most of the rest of the country is more than 2,000 ft (600 m) above sea-level. In the NE are the peaks of the Rhodesian Highlands. On the NW border with Zambia is the 175-mile (280-km) long Lake Kariba formed by the building of the Kariba Dam, a project that supplies hydro-electric power to both Zambia and Rhodesia. Because of its altitude, most of Rhodesia has a moderate

RWANDA

Country of E-central Africa, formerly part of the U.N. trust territory of Ruanda-Urundi. It is poor but densely populated. Rwanda lies about 100 miles (160 km) S of the equator. In the N it borders on Uganda, in the E on Tanzania, in the S on Burundi, and in the W on Zaire. Much of its western border runs through Lake Kivu. To the E of the lake is a high ridge of volcanic mountains, marking the edge of the Great African Rift Valley, and separating the basins of the Rivers Congo and Nile. One peak, Mount Karisimbi in the Virunga Range, reaches a height of 14,786 ft (4,507 m) above sea-level. A broken plateau extends eastwards, averaging 3,000 to 6,500 ft (900 to 2,000 m) in height. There are several rivers. Most of the country is hot and wet, but temperatures in the mountains are lower than elsewhere. Many of the people live by subsistence farming, but some cash crops are grown, including coffee, tea, and cotton. Tin and tungsten are mined in small quantities. There are no railways, and roads are poor. The capital is Kigali. Area 10,169 sq miles (26,338 sq km). For population, see Table on page 149.

171

Sierra Leone. The Atlantic port of Freetown, the country's capital. It has one of the finest natural harbours in the world. The city was founded in 1787 by a group of British abolitionists as a settlement for escaped slaves, many of them from the Americas.

SENEGAL

Country on the western coast of Africa, formerly a French colony. The promontory of Cape Vert in Senegal is the continent's most westerly point. The country is bordered on the N by Mauritania, on the E by Mali, and on the S by Guinea and Guinea-Bissau. The Gambia, an independent country, forms an enclave within the territory of Senegal. Most of Senegal consists of grasslands up to 500 ft (150 m) or so above sea-level. But in the SE, where the Fouta Djallon extends into the country from Guinea, there are places at heights of more than 1,600 ft (490 m). The many rivers include the Senegal and the Gambia. Coastal regions have a moderate climate, but inland temperatures are high. There is much rain in the S. There are some mineral deposits and some manufacturing industry, but the country is

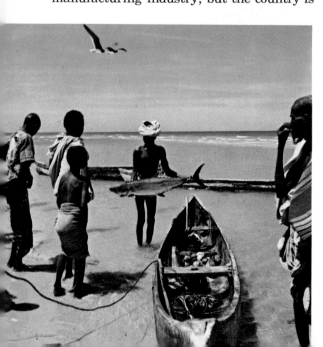

Somalia. Return of the tuna fishermen to a coastal village near Cape Guardafui, in north-eastern Somalia. Fish is an important article of diet for those who live within reach of the sea in this arid corner of Africa.

still almost entirely agricultural. Millet and sorghum are grown for food. The chief cash crop is ground-nuts. Other exports include cotton, hides, palm oil, and gum arabic. Some coastal villages have a trade in fish. The chief towns are Dakar (the cap.), Kaolack, and St. Louis. Area 77,800 sq miles (201,500 sq km). For population, see Table on page 149.

SEYCHELLES

Island republic in the Indian Ocean, until 1976 a British colony. It consists of two groups of islands: the 45 mountainous Mahé islands and the 50 or so flat Coralline islands. Exports include copra and cinnamon bark. The cap., Victoria, is on Mahé, the largest island. Area 270 sq miles (700 sq km). For population, see Table on page 149.

SIERRA LEONE

Country of western Africa, a former British possession. In the W it has coastline on the Atlantic; on the N and NE it borders on Guinea, and in the SE on Liberia. Lagoons fringe the coast. The broad coastal plain is swampy except for the Sierra Leone Peninsula, where Freetown is situated. Inland, the land rises unevenly to the Fouta Djallon plateau. Peaks in the NE reach more than 6,000 ft (1,800 m). The typical vegetation is savanna on the plateau. In the low-lying areas there are still remains of the tropical rain forests that once covered them. Most of the country is hot and wet. Mineral resources include iron and diamonds. Most people live by farming; cash crops include cocoa and coffee. The chief towns are Freetown (the cap.), Bo, and Makeni. Area 27,925 sq miles (72,325 sq km). For population, see Table on page 149.

SOMALIA

The Somali Democratic Republic lies along the coast of the eastern 'horn' of Africa. In the N it has a long coastline on the Gulf of Aden, and in the E a long coastline on the Indian Ocean. On the W it is bordered by Ethiopia and the Territory of the Afars and Issas. It is made up of former British and Italian colonies. The broad, dry coastal plain in the E slopes upwards to the Ogaden plateau of Ethiopia. In the N a narrow coastal plain rises to an escarpment and the hilly northern fringe of the plateau. There are only two permanent rivers, the Juba and the Shibeli. The climate is hot and dry. Many of the people are nomadic herdsmen. Crops include cotton, maize, sorghum, and ground-nuts. The chief towns are Mogadishu (the cap.), Merka, and Hargeisa. Area 246,200 sq miles (637,700 sq km). For population, see Table on page 149.

South Africa is the southernmost country of the African continent, and the most prosperous. The country is ruled by its White inhabitants, who number about one-fifth of the population. South Africa has a coastline on the Atlantic in the W, and on the Indian Ocean in the E. It is bordered on the N by Angola, Botswana, Rhodesia, Mozambique, and Swaziland. It completely encloses the independent country of Lesotho. South-West Africa (Namibia), a territory to the NW, is administered by South Africa. Most of South Africa consists of broad plateaux, called *karroos*. The *Great Karroo* in the S is famous for its merino sheep. In the W is the *Upper Karroo*, dry grasslands suitable chiefly for grazing. The eastern and higher plateau, called the *High Veld*, is the richest and most fertile. It has great mineral resources, thriving industry, and prosperous farms. In the Transvaal, in the N, the High Velt descends through the *Bushvelt* to the hot, damp grasslands of the *Low Velt*. On

SOUTH AFRICA

South Africa. The city of Cape Town, lying below the great rugged wall of Table Mountain. Cape Town, the oldest and second largest city in South Africa, is the country's chief passenger port.

the south-eastern rim of the High Velt is the Drakensberg mountain range. Its highest point is Mont-aux-Sources (10,822 ft; 3,299 m). The range falls away sharply to the narrow coastal strip. The coastal areas of the W are partly in the Namib Desert. The chief rivers are the Orange, the Limpopo, and the Vaal. Most of the country has a temperate climate, though shortage of rain is a problem for farmers in many places. The country's rich natural resources include diamonds, gold, coal, iron, phosphates, uranium, and copper. Important manufactures are steel, engineering products, textiles, and processed foods. Agricultural products include wool, maize, wheat, tobacco, sugar-cane, potatoes, vegetables, fruit, and grapes (for wine). The chief towns are Cape Town (legislative cap.), Pretoria (administrative cap.), Johannesburg, Durban, Port Elizabeth, Germiston, and Bloemfontein. Area 472,360 sq miles (1,223,410 sq km). For population, see Table on page 149.

SUDAN

The Sudan, the largest country in Africa, was governed jointly by Egypt and the United Kingdom until 1956. It is in the NE of the continent, and has a coastline on the Red Sea. On the N it is bordered by Egypt, on the SE by Ethiopia, on the S by Kenya, Uganda, and Zaire, and on the W by the Central African Empire, Chad, and Libya. The land consists largely of mountain-fringed platcaux. The southern, equatorial region contains a vast swamp, called the Sudd, 50,000 sq miles (130,000 sq km) in area. Northwards, a belt of savanna gives way to a wide belt of grassland and scrub. In the centre of this region rise the bare hills of Kordofan. The northern half of the country is desert, the most easterly part of the Sahara: W of the River Nile it is called the Libyan Desert, E of the Nile, the Nubian Desert. Not all the desert is barren; trees, bushes, and crops grow in hilly areas, in the

Sudan. The strange landscape of the Nubian Desert, between the River Nile and the Red Sea. Most of the desert lies on a rocky plateau, broken up by shallow river beds that are dry for the greater part of the year. Very few people live in its 90,000 sq miles (230,000 sq km). A desert railway connects Abu Hamed and Wadi Halfa.

Tanzania. Zebras in one of Tanzania's national parks. Eastern Africa is exceptionally rich in the magnitude and variety of its animal life, and many governments have laws to protect rare animals from hunters and from further devastation of their natural habitats. But despite all efforts at conservation, animals in Africa as elsewhere seem to be fighting a losing battle against the destructive effects of technology.

scattered oases, and in the Nile valley. The Nile and its tributaries extend for thousands of miles through most of the country. The White Nile enters the country in the extreme S, the Blue Nile in the SE. The branches join at Khartoum, and flow north-wards to Egypt. Successful irrigation works have been effected along the course of the river, the earliest being at Gezira, S of Khartoum. All parts of the country are hot, but there are considerable variations in the amount of rainfall, which is greatest in the S. The Sudan has few mineral resources. The economy is almost entirely agricultural. In the desert regions, most poeple are nomadic herdsmen. In more fertile areas, the crops include millet, sorghum, maize, and cassava. Cotton and ground-nuts are grown as cash crops. Other exports are hides and gum arabic. The chief towns are Khartoum (the cap.), Omdurman, El Obeid, Port Sudan, and Wad Medani. Area 967,500 sq miles (2,505,800 sq km). For population, see Table on page 149.

SWAZILAND

Small landlocked kingdom in southern Africa, enclosed by South Africa and Mozambique. The country has four well-defined regions. In the W is the high veld, up to 5,000 ft (1,500 m) above sea-level. Going E, the high veld gives way to the middle veld, and then to the low bushveld. In the E, the Lebombo Mountains rise to more than 9,000 ft (2,700 m). The climate is sub-tropical, with heavy rainfall in the higher areas. Agricultural products include maize, millet, sugar, rice, and fruits. Cattle and sheep are raised. There are valuable mineral deposits; asbestos and iron are among those exploited. The people speak Bantu languages. The chief towns are Mbabane, Manzini, and Stegi. Area 6,704 sq miles (17,363 sq km). For population, see Table on page 149.

TANZANIA

Large country of eastern Africa, formed of the former British trust territory of Tanganyika and a number of offshore islands, of which the chief are Zanzibar and Pemba. It is bordered on the N by Uganda and Kenya, on the S by Mozambique, Malawi, and Zambia, and on the W by Zaire, Burundi, and Rwanda. The mainland part of Tanzania lies in two arms of the Great African Rift Valley. Much of its western border runs through Lake Tanganyika and Lake Malawi. Another great lake, Victoria, lies on the northern boundary. The 500-mile (800-km) long coastal plain is narrow except in the centre, near the mouth of the Rufiji, the largest river. The plain, swampy in many places, rises to the broken plateau that occupies most of the country. The plateau is covered with woods, scrub, and grasslands. Kilimanjaro, Africa's highest mountain, rises to 19,340 ft (5,895 m) in the NE. Near to it is Mount Meru (14,979 ft; 4,566 m). The Livingstone Mountains, at the northern edge of Lake Malawi, rise to some 9,000 ft (2,750 m). The scenery of Tanzania is among the most varied and beautiful in Africa, and the country is famous for its wildlife. It contains Serengeti National Park. Eastern areas have a tropical climate. Inland areas are generally dry, but the mountains tend to have lower temperatures and more rain. Tanzania has good mineral resources, and there are some light manufacturing industries. But the country is still predominantly agricultural. Maize, millet, and cassava are grown for food, and many people in areas free from the tsetse fly keep cattle. The main cash crops are coffee, cotton, tea, sisal, cloves, and cashew nuts. The chief towns are Dar es Salaam (the cap.), Zanzibar, Tanga, Mwanza, Arusha, and Moshi. Area 362,820 sq miles (939,700 sq km). For population, see Table on page 149.

The island of Zanzibar produces cloves, fruit, and copra. It was formerly a British-protected sultanate. Its chief city, also named Zanzibar, was once a rich slave market.

Tanzania. A typical stretch of savanna on the plateau that occupies much of continental Tanzania. The rough grass grows in clumps and patches, and here and there spiky bushes and trees break the monotony. In most regions of savanna, farming consists chiefly of grazing.

TOGO

Small republic in W Africa, long and narrow in shape and with a short coastline on the Gulf of Guinea. The coastal plain is much broken up by lagoons. Along the coast and farther inland in the SE there are large mangrove swamps. The Togo-Atakora range of mountains runs diagonally across the country from SW to NE. In places, its peaks rise to more than 3,000 ft (900 m) above sea level. Many of the lower mountain slopes are covered with tropical rain forest. In the NW is an area of tropical savanna. Togo is

Tunisia. A tempting array of produce for sale. Shoppers consider each item carefully.

extremely hot and humid, with heavy rainfall. It is a poor country. Its chief export crops are cocoa-beans, coffee, and palm nuts. To a large extent, the people live by subsistence farming, growing such crops as maize, yams, cassava, and millet. Phosphates are mined, and the country also has deposits of iron, chromium, and bauxite. The peoples of the S are mainly Bantu, and many in the N are of Hamitic descent. They use a large number of languages. The chief towns are Lomé, Sokodé, and Palimé. Area 21,853 sq miles (56,599 sq km). For population, see Table on page 149.

Tunisia. A street in Tunis, an important trading city as well as the country's capital. Tunis stands near the site of ancient Carthage.

TUNISIA

THE NORTHERN-MOST POINT OF AFRICA
Although Cape Blanc, 5 miles (8 km) NNW of Bizerte, is usually described as the northernmost point of the continent, that distinction really belongs to near-by Ras ben Sekka. The latitude of Cape Blanc is 37°20′N, that of Ras ben Sekka 37°21′.

Tunisia has a Mediterranean coastline on both the N and the E. One of its promontories, Cape Blanc, is the northernmost point of Africa. Another, Cape Bon, is only 90 miles (145 km) from Sicily, in Europe. Tunisia has land borders with Libya in the SE and Algeria in the W. The northern part of the country is in the Atlas Mountains. Just inland from the N coast, the Tell extends into Tunisia from Algeria. S of the Tell are hilly Atlas tablelands, rising in the Jebel Chambi to 5,138 ft (1,566 m). The tablelands fall away in the E to the coastal plain called the Sahel. The country's only important river, the Medjerda, flows across the Atlas region to the Gulf of Tunis. The S part of Tunisia lies in the wastes of the Sahara. There are some oases where cultivation is possible. In the W of this desert region are the great salt flats of Chott Djerid. In coastal and hilly areas, the climate is warm and pleasant, with moderate rainfall. Desert areas are hot and arid. Tunisia has valuable mineral resources, including phosphates, iron, petroleum, and natural gas. Its industries include food processing, sugar refining, and the manufacture of chemicals and paper. Most people live by farming, the productive areas being in the N and NE. The main crops include barley, wheat, grapes (for wine), olives, citrus fruits, vegetables, and dates. The chief towns are Tunis (the cap.), Sfax, Bizerta, Sousse, and Kairouan. Area 45,000 sq miles (116,500 sq km). For population, see Table on page 149.

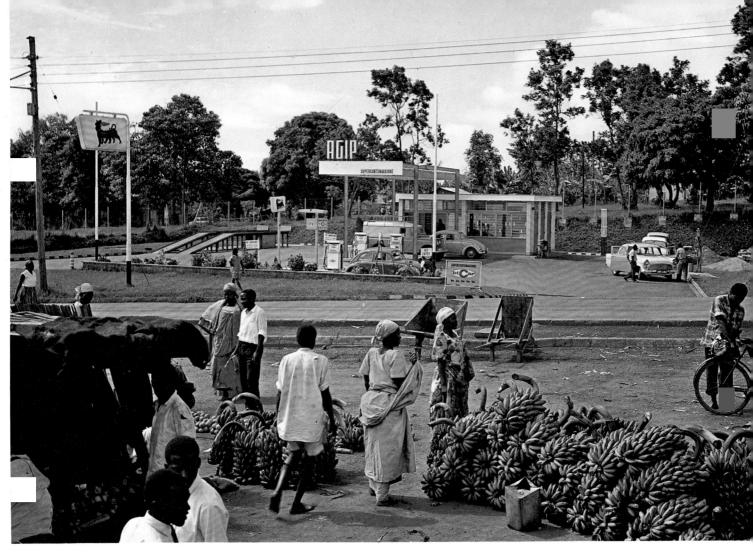

Uganda. The changing face of Africa: a petrol station in Kampala, Uganda's principal commercial centre. Kampala is a well-planned, modern city with fine buildings and good roads. As industrialization advances in Africa, increasing numbers of people move from country areas into the towns.

UGANDA

Country of E-central Africa, lying on the equator. It is surrounded by the territory of other countries, and has no access to the sea. Until 1962 it was a British protectorate. A large part of the country is a high plateau, averaging about 4,000 ft (1,220 m) above sea level. Its characteristic vegetation is lightly-wooded savanna. Branches of the Great Rift Valley enclose the country. There are high mountains on both the W and the E borders. In the W the Ruwenzori Mountains rise to 16,794 ft (5,119 m), and in the E Mount Elgon rises to 14,178 ft (4,321 m). In the Mufumbiro range in the S there are several active volcanoes. In the SE the country's borders enclose a vast expanse of Lake Victoria, Africa's largest lake. Three more great lakes, Albert, George, and Edward, lie in the W, though only Lake George is completely within Uganda's borders. The headwaters of the White Nile drain the country. One of them, the Victoria Nile, flows NW from Lake Victoria—leaving the lake at the Owen Falls, near Jinja—and passes through the vast area of swamp called Lake Kyoga, and from there to Lake Albert. The river then flows N from the lake as the Albert Nile, and joins the White Nile in the Sudan. Because of Uganda's altitude, its climate is not tropical. But it is hot in some areas, and has heavy rainfall in the mountains and in the SSE around Lake Victoria. Most of the people are farmers. They raise cattle, sheep, goats, and pigs, and grow plantains, yams, cassava, and maize. The chief export crops are coffee and cotton, but tea, tobacco, ground-nuts, and sugar are also grown. Uganda has many minerals, but only copper is mined in any quantity. The Owen Falls dam provides hydro-electricity for industry in its part of the country. In the S and centre, most of the people belong to Bantu groups, the largest tribe being the Baganda. In the N there are Nilotic tribes. The chief towns are Kampala (the cap.), Jinja, Mbale, Entebbe, and Kabale. Area 93,981 sq miles (234,410 sq km). For population, see Table on page 149.

Zaire. A boulevard in Kinshasa, formerly Léopoldville. The city, one of the most modern in Africa, lies on the Zaire (Congo) River, opposite Brazzaville.

Below: Zambia. The irresistible, roaring waters of the Victoria Falls on Zambia's southern border.

UPPER VOLTA

Republic of W Africa. It is surrounded by the territory of other countires, and has no access to the sea. Upper Volta consists almost entirely of a wide plateau, which slopes gently from N to S, averaging about 800 ft (244 m) above sea level. Most of it is covered with dry savanna. The country lies across the main headstreams of the River Volta, called the Black, Red, and White Voltas. These headstreams join together in Ghana, to the S. The climate varies considerably. Half of the year is dry, the rest is hot and wet. In the northern savannas the rainfall averages less than 10 in (250 mm) a year, but in the S it averages more than 40 in (1,000 mm). Upper Volta is an almost completely agricultural country, and most of its people live by subsistence farming. They grow sorghum, millet, and guinea corn. Ground-nuts, cotton, and sesame are grown for sale, but the chief source of wealth is livestock. Sheep, goats, and cattle are raised, and meat is exported, chiefly to Ghana. There are some mineral deposits. The largest tribe is the Mossi, whose ancient settlement, Ouagadougou, is the country's cap. Other tribes include the Mande, the Gourma, and the Bobo, who live in the district around Bobo-Dioulasso, the principal market town. Area 105,869 sq miles (274,199 sq km). For population, see Table on page 149.

WESTERN SAHARA
In 1975 Spain abandoned its claim to Western Sahara (then Spanish Sahara) on the NW Atlantic coast of Africa. The territory has two regions: the Seguia al Hamra in the N; and the Rio Rojo in the S. It is mainly desert, but has phosphate deposits. Many of its people are nomads. See Table on page 153.

ZAIRE

One of the largest countries in Africa, Zaire lies in the S-centre of the continent, on the equator. It has only one outlet to the sea: a narrow corridor in the W gives it a 25-mile (40 km) coastline on the Atlantic. The country takes its name from the River Zaire (Congo). It was formerly the *Belgian Congo,* and then the *Democratic Republic of the Congo.* The greater part of Zaire is a vast plateau, up to 3,000 ft (900 m) above sea level. Its high SE part is the Katanga plateau. On the E border is part of the Great Rift Valley, containing Lakes Albert, Edward, Kivu, Tanganyika, and Mweru. Here, too, are great ranges of mountains. The Mitumba Mountains extend for hundreds of miles from the SE border towards Lake Tanganyika. Some of the peaks are more than 10,000 ft (3,000 m) high. The Ruwenzori Mountains, between Lakes Albert and Edward, rise to 16,794 ft (5,119 m). The River Zaire flows through the country for some 2,800 miles (4,500 km). It rises, as the river Lualaba, in the SE, and flows northwards, then westwards, and then south-westwards in a great half-circle to the Atlantic. Towards the end of its course, it forms the boundary with the Republic of Congo (Brazzaville) on the W. Its many tributaries include the Lufira, Lomami, Lulonga, Oubangui, and Kasai Rivers.

178

Dense tropical forests cover much of the low-lying Zaire basin in the W and NW. This region has a hot and humid climate. The higher regions are more temperate. Zaire is extremely rich in minerals, the most important being copper from the Katanga plateau. It also has deposits of cobalt, industrial diamonds, zinc, iron, manganese, tin and gold. Its plantations produce cotton, rubber, coffee, and palm oil. Despite the country's valuable resources, many of its people live by subsistence farming, growing maize, cassava, and millet. Most of the people speak Bantu languages. There are also Hamitic tribes, and small bands of pygmies live in the forests of the Zaire basin. The chief towns are the cap. Kinshasa (formerly Léopoldville), Lubumbashi (formerly Elisabethville), and Kisangani (formerly Stanleyville). Area 905,565 sq miles (2,345,402 sq km). For population, see Table on page 149.

Republic of S-central Africa, on the central African plateau. Before independence it was called Northern Rhodesia. Most of it lies between 3,000 and 5,000 ft (900 and 1,500 m) above sea-level. But in the Muchinga Mountains, which run SW-NE across the eastern half, some peaks rise to over 7,000 ft (2,130 m). The largest rivers are the Zambezi and its tributaries, the Kafue and the

Below:
Zambia. A herd of elephants pursue their orderly way across the savanna. Despite their size, elephants make little noise as they move from place to place. The young elephants are playful, but they are strictly disciplined, and they soon learn the danger of straying away from the herd.

ZAMBIA

Luangwa. The various tributaries of the great Zambezi fan out to almost every part of the country. There are many lakes. In the N, Lake Bangweulu lies completely within the country's boundaries, but Lakes Mweru and Tanganyika are only partly in Zambia. Another large lake, Kariba, lies on the S border. It is artificial, having been formed by the building of the Kariba dam. There are also many huge swamps. Much of the country is covered with lightly-wooded savanna. Although Zambia is in the tropics, its climate is relatively moderate because of its altitude. But there is a hot season from September to November. Zambia is one of the world's leading copper-producing countries. The so-called 'copper belt' is in the N, and much of its electric power comes from the Kariba dam. Cobalt, vanadium, coal, lead, zinc, and manganese are also mined. In general, the country is thinly populated, its areas of heaviest population being in the mining districts. Tobacco, cotton, and fruits are grown in some places, but most people who farm produce only subsistence crops, including millet, maize, and yams. Almost the entire population belongs to Bantu groups. The chief towns include Lusaka (the cap.), Ndola, Kitwe, Mufulira, Luanshya, and Chingola. Area 288,130 sq miles (746,253 sq km). For population, see Table on page 149.

NORTH AMERICA
AND
CENTRAL AMERICA

North America has only about one-tenth of the people of the world, but it is the richest of the continents both in natural resources and in industry. Factories in North America produce around one-third of the world's manufactures. Yet because of extreme cold or desert conditions large parts of the continent are still undeveloped.

North America is the third largest continent. It extends from the Arctic Ocean in the N to the Isthmus of Panama in the S. It is bounded on the E by the Atlantic Ocean, and on the W by the Pacific. Most of its area is occupied by three countries—Canada, the United States, and Mexico. Greenland —which is the world's largest island—the islands of the West Indies, and Central America make up the rest of it. In the NW, at the Bering Strait, North America approaches to within 56 miles (90 km) of Siberia, in Asia. In the E, Greenland is only about 1,000 miles (1,600 km) from Norway and Scotland, in Europe. The total area of the continent is about 9,600,000 sq miles (24,900,000 sq km).

From N to S, North America extends for about 4,500 miles (7,200 km), and the greatest distance W to E is about 4,000 miles (6,400 km). The population in the mid-1970s was more than 330 million.

COUNTRIES OF NORTH AND CENTRAL AMERICA POPULATIONS AND MONETARY UNITS		
Country	Population	Monetary Unit
Bahamas	190,000	Dollar
Barbados	245,000	Dollar
Canada	22,520,000	Dollar
Costa Rica	1,880,000	Colón
Cuba	8,900,000	Peso
Dominican Republic	4,600,000	Peso
El Salvador	3,900,000	Colón
Grenada	105,000	Dollar
Guatemala	5,570,000	Quetzal
Haiti	5,100,000	Gourde
Honduras	2,650,000	Lempira
Jamaica	2,100,000	Dollar
Mexico	56,000,000	Peso
Nicaragua	2,300,000	Córdoba
Panama	1,610,000	Balboa
Trinidad and Tobago	1,140,000	Dollar
United States	211,500,000	Dollar

Physical Features and Climate North America has five major physical regions: the Eastern Highlands; the Western Highlands; the Canadian Shield; the Central Lowlands; and the Highlands of the West Indies.

The Eastern Highlands are formed by the Appalachian Mountains, which run parallel with the Atlantic coast, and the mountains of Newfoundland and Labrador, farther N. They are broken by the valley of the St

The Untamable Mountains. Banff National Park in Alberta, Canada, lies on the eastern slopes of the Rocky Mountains. It is a favourite resort of climbers, fishermen, skiers, and tourists. Glaciers feed its still, forest-fringed lakes; and hot mineral springs form pools at the foot of rugged, icy mountains.

Lawrence River. The Appalachians formed a considerable barrier when the early colonists wanted to move westwards from the Atlantic coastal plain to the rich lands of the interior.

The Western Highlands are a complicated system of mountain chains, high plateaux, and river basins. They run from Alaska in the N to the Isthmus of Panama in the S. The Rocky Mountains, in Canada and the United States, are the easternmost part of the system.

The Canadian Shield is a low area of rock thinly covered with soil. It occupies most of central Canada, forming a great crescent-shaped area around Hudson Bay. Much of the land in the Shield is of poor quality agriculturally.

The Central Lowlands run S from the Canadian Shield to the Gulf of Mexico. They are drained by the Mississippi River and its many tributaries.

The Highlands of the West Indies are the peaks of a mountain chain now submerged. They tail away from the bulk of the continent towards South America.

Most of North America has warm summers and cold winters. But in the N it is cold nearly all year round, while in the S days are mostly hot, except in the mountains.

IMPORTANT DATES IN THE HISTORY OF NORTH AMERICA

18000s B.C. Mongoloid peoples from Asia—the first American Indians—entered Alaska from Siberia.

600-900 A.D. The Maya civilization reached its height.

900s Eric the Red established a colony in Greenland.

1300s-1500s The Aztecs ruled the Valley of Mexico.

1492 Christopher Columbus reached the Americas.

1497 John Cabot landed on Canada's eastern coast.

1518-1521 Hernando Cortes conquered Mexico for Spain.

1607 Settlement established at Jamestown, Virginia.

1608 Samuel de Champlain founded Quebec city.

1620 Pilgrim Fathers landed at Plymouth, Massachusetts.

1642 French colonists founded Montreal.

1689-1763 The British captured the French colonies.

1773 The Boston Tea Party.

1774 The First Continental Congress in Philadelphia resolved to resist the 'Intolerable Acts'.

1775 First shots in the War of Independence, at Lexington, Massachusetts.

1776 The American colonies voted for independence.

1781 Lord Cornwallis surrendered to George Washington at Yorktown—in effect, the end of the war.

1789 George Washington chosen as first President of the United States of America.

1798 Eli Whitney built an arms factory in Connecticut and developed techniques of mass production.

1803 The Louisiana Purchase: President Jefferson's purchase of the Louisiana territory doubled the size of the United States.

1812 The 'War of 1812' between the United States and Britain.

1821-1822 Agustín de Iturbide became Emperor of Mexico after defeating the Spanish garrison.

1803 The Monroe Doctrine proclaimed.

1824 Mexico declared a republic.

1830 The first American railway—the Baltimore and Ohio—began operation.

1845 Texas became the 28th state of the U.S.A.

1846-1848 The Mexican War: the U.S.A. defeated Mexico.

1849 The California gold rush.

1859 Petroleum pumped commercially in Pennsylvania.

1861 The French invaded Mexico.

1861-1865 Civil war in the U.S.A. between the northern Union states and the southern Confederate states. The Union, under President Abraham Lincoln, triumphed.

1864 The French made the Austrian prince, Maximilian, Emperor of Mexico.

1867 Mexican troops executed Emperor Maximilian.

1867 British North America Act: established the Dominion of Canada.

1886 Surrender of the Apache chieftain Geronimo.

1901 Theodore Roosevelt became American President; he offered the working man 'a square deal'.

1910-1920 Revolution in Mexico.

1914-1918 Canada and, later, the United States involved in World War I. President Wilson a chief architect of the peace. But the United States rejected the League of Nations.

1929 Stock market collapsed. Start of the Great Depression.

1931 The Statute of Westminster. The Canadian dominion became independent.

1933 Franklin D. Roosevelt became President of the United States: the 'New Deal'.

1939-1945 Canada and, later, the United States and Mexico involved in World War II.

1954-1959 Building of the St. Lawrence Seaway.

1962 The Trans-Canada Highway opened.

Resources and Industry North America's chief resources are its minerals, its fertile soil, its forests, and its lakes and rivers.

The continent is very rich in minerals, particularly coal, iron, and petroleum. Other minerals of which there are large deposits include asbestos, bauxite, copper, gold, lead, natural gas, nickel, silver, and zinc. Some of these deposits are the largest, or among the largest, in the world. In many cases, the great industrial cities of Canada and the United States owe their development to the fact that they are within easy reach of valuable minerals. The continent's vast natural resources together with the industrial inventiveness of its people and their urge for ever-increasing efficiency have made it pre-eminent in industrial power.

The many lakes and rivers provide water for cities and for irrigation. A series of large lakes stretches for thousands of miles across the northern half of the continent, extending NW to SE. Many of the bigger rivers, such as the Colorado, have been dammed to

Nature's Variety. Despite its general richness, North America has vast barren regions—the hot deserts of the south-west, and the cold, treeless tundra of the north.

THE UNITED STATES AND CANADA: AGRICULTURAL REGIONS

Legend:
- Extensive grazing land
- Dairying with fodder crops
- Tobacco
- Pineapples
- Cash crops—much cotton, tobacco and sub-tropical fruit
- Maize, wheat and livestock
- Sugar cane
- Wheat and livestock
- Cotton
- Forestry products
- Coffee
- Largely unsettled
- Fruit and vegetables
- Irrigated crops

HAWAIIAN ISLANDS: Oahu, Molokai, Maui, Kauai, Hawaii

PUERTO RICO: S.Juan

ALASKA

THE ORGANIZATION OF AMERICAN STATES (OAS)

The OAS is an association of Latin American countries and the United States for mutual defence, co-operation, and the peaceful settlement of disputes. It is organized within the general framework of the United Nations. Its charter states that an act of aggression against any one member is regarded as being directed at all members. The OAS has a permanent council with headquarters in Washington, D.C. A Pan American Conference is held every five years. Urgent matters can be dealt with by specially-convened meetings of Ministers of Foreign Affairs.

provide hydro-electric power. Water from reservoirs formed by damming is used to irrigate dry areas.

Communications Mexico, Central America, and northern Canada have poor communications. Much of northern Canada is, of course, barren and only lightly populated. But southern Canada and the eastern and central parts of the United States have some of the world's best road and rail systems.

The Pan American Highway runs from Alaska in the far NW to Panama in the S. Other highways cross the continent from the Atlantic Ocean to the Pacific. By means of the Great Lakes and the St Lawrence Seaway, cities in the industrial centre of the continent have direct access to the Atlantic. The Mississippi River gives central areas access to the Gulf of Mexico. Airlines connect all the major cities of the continent. In Panama, the Panama Canal—one of man's

UNITED STATES – CANADA: MINERALS AND INDUSTRIAL AREAS

Legend:
- Potential oil producing areas
- Crude oil extraction
- Coal
- Lignite
- Iron
- Gold
- Copper
- Uranium
- Zinc
- Lead
- Bauxite
- Industrial areas

ANIMAL LIFE

The horse, which has played a major part in American folklore, is not indigenous to North America; it was introduced by the Spaniards. But the continent has a wide variety of 'American' animals. They include polar bears, caribou, and Arctic foxes in the north; grizzly bears, mountain lions, and eagles in the western mountains; and black bears, coyotes, and prairie chickens in the east and centre. Some animals that were once common have now disappeared or become rare. Few buffalo are left, and the passenger pigeon is extinct.

greatest feats of engineering—provides a short route between the Atlantic and the Pacific.

Agriculture, Fisheries, and Land Use The United States and Canada have some of the world's largest and most productive farms. On the fertile land wheat is grown in the N, maize (corn) in the midlands, and cotton in the S. The two countries produce much more food than they need, and are able to export vast quantities, particularly of wheat and maize. In the W, the great open ranges, with their sparse, scrubby vegetation, provide good grazing for cattle and sheep.

Mexico and the countries of Central America are less fortunate in their soil than the two northern countries. A large number of their people are poor, and carry on their farming in a primitive, inefficient way. For this reason, the agricultural output is generally only enough for the farmers' own needs—that is, it is subsistence farming.

Forests cover almost one-third of the continent. Their trees provide valuable timber, and also wood for paper-making. Many wild animals are found in the forests, including beavers, porcupines, and various kinds of bears and deer. In the N, where the vegetation is sparse and the weather cold for a large part of the year, many fur-bearing animals live. They are hunted for their skins. Fur seals are found on the islands to the W of Alaska. Other seals live in the waters of the Arctic Ocean, where they provide the main food of the Eskimoes who live there—as well as food for polar bears.

People and Ways of Life Four main groups of peoples live in North America. They are the American Indians; the Eskimoes and Aleuts; the settlers of European origin; and the Negroes.

American Indians were the first people to settle in the continent. Scientists believe that their ancestors may have crossed from Siberia about 20,000 years ago, when there was probably still a land connexion between Asia and North America at what is now the Bering Strait. They are called 'Indians' because the European explorers who reached the Americas in the late 1400s thought they had arrived in India. The American Indians are of Mongoloid stock —that is, they are closely related in type to the peoples of China, Japan, and Korea. Because of their dark, reddish-brown skins they were also known as 'Redskins' or 'Red Indians'. The Indians of South America are closely related to those of the northern continent.

Before the arrival of the white men, there were several hundred tribes of American

A Source of Power. An oil refinery on Delaware Bay, on the Atlantic seaboard of the United States. The industrial might of North America is based on its rich natural resources. They include petroleum, coal, copper, nickel, iron, asbestos, bauxite, and gold.

THE ST. LAWRENCE SEAWAY

THE ST. LAWRENCE SEAWAY

The St. Lawrence Seaway has made it possible for ocean-going ships to sail for 2,300 miles (3,700 km) from the Atlantic to Lake Superior, in the heart of the North American continent. The seaway itself is 182 miles (293 km) long, stretching from Montreal to the entrance to Lake Ontario. Canada and the United States co-operated in the seaway's construction. Work began in 1954, and the opening ceremony took place in 1959. Power installations built at the same time supply hydro-electricity to Ontario in Canada and New York in the United States.

In the Steps of the Conquistadores. An ornate Baroque church typifies the colonial architecture of Latin America.

Below: The Indispensable Assistant. Breaking-in a young horse. Without horses, the great cattle ranches could not operate.

Indians. Generally, each tribe had its own language. But the languages, and the tribes, can be grouped into about 50 families. In each family, the different languages are really dialects (regional variations) of the same basic tongue. Another way of grouping the Indians is by their ways of life. These varied according to the part of the continent in which a tribe lived.

The Indians of the plains, the wide areas of the Central Lowlands, were *nomads* (wanderers) who lived mainly by hunting and fishing. Chiefly, they hunted the bison that roamed the plains in their millions. Before the arrival of the white men, the Indians had no horses, so all their hunting was done on foot. For weapons they used bows and arrows, spears, and tomahawks (axes). At this stage there were comparatively few Plains Indians, but once the Indians acquired horses many more tribes moved into the plains to hunt bison.

Forest Indians such as the Mohicans,

NORTH AMERICA AND CENTRAL AMERICA COLONIES AND OTHER TERRITORIES					
Name	Location, etc	Area sq miles	sq km	Population	Capital
Anguilla (U.K.)	Island in the Lesser Antilles.	35	90	6,000	
Antigua (U.K.)	Island in the Lesser Antilles.	108	280	65,000	St. John's
Belize (U.K.)	Caribbean coast of Central America. Formerly called British Honduras.	8,867	22,965	130,000	Belmopan
Bermuda (U.K.)	Group of 300 small Atlantic islands.	21	54	55,000	Hamilton
Cayman Islands (U.K.)	3 islands; S of Cuba.	100	259	11,000	George Town
Dominica (U.K.)	Island in the Lesser Antilles.	290	750	71,000	Roseau
Greenland (Den.)	World's largest island. *See Denmark.*	840,000	2,176,000	37,000	Godthåb
Guadeloupe (Fr.)	Island & group in the Lesser Antilles.	687	1,779	341,000	Basse-Terre
Martinique (Fr.)	Island in the Lesser Antilles.	435	1,127	350,000	Fort-de-France
Netherlands Antilles (Neth.)	Curaçao & other islands; West Indies.	391	1,013	230,000	Willemstad
Puerto Rico (U.S.)	Island in the West Indies.	3,435	8,897	2,712,000	San Juan
St. Kitts-Nevis (U.K.)	Islands in the Lesser Antilles.	101	261	44,000	Basseterre
St. Lucia (U.K.)	Island in the Lesser Antilles.	238	616	101,000	Castries
St. Pierre & Miquelon (Fr.)	Island group; S of Newfoundland.	93	241	5,000	St. Pierre
St. Vincent (U.K.)	Island group in the Lesser Antilles.	150	388	90,000	Kingstown
Turks & Caicos Islands (U.K.)	Island group in SE Bahamas.	193	500	6,000	Grand Turk
Virgin Islands (U.K.)	Part of West Indies island group.	59	152	12,000	Road Town
Virgin Islands (U.S.)	Part of West Indies island group.	133	344	95,000	Charlotte Amalie

Hurons, and Iroquois also lived mainly by hunting and fishing. Some, however, were farmers.

The Indians of the SW lived in dry areas where game was scarce. They built villages, grew crops, and irrigated the dry land. In Mexico the Aztecs created a great civilization, and built with stone.

The Eskimoes and Aleuts live in the cold lands of the N. They are also of Mongoloid origin. Over hundreds of years they have developed plump bodies whose extra fat helps to keep out the cold. The Eskimoes are found in Alaska, northern Canada, and Greenland. The Aleuts live in the Aleutian Islands, which stretch in a long chain SW of Alaska.

The Europeans first arrived in North America in the late 1400s and early 1500s.

The earliest settlers were Spaniards, who formed colonies in Central America, the West Indies, and Mexico. Settlers from the British Isles occupied the eastern parts of what is now the United States, while French settlers moved into what is now southern Canada. Russians occupied Alaska, but were unable to colonize it successfully. During the 1880s and early 1900s, millions of people from Europe flocked to the United States.

Negroes, enslaved mainly on the W coast of Africa, were carried to America from the 1500s onwards. There they were forced to work on the plantations. The slave trade went on until well into the 1800s. Slavery was abolished in North America in 1865.

As a result of these different strains of settlement, the peoples of North America are today very varied in type. The Spaniards who colonized Mexico, Central America, and the West Indies often intermarried with members of the Indian tribes who lived there. In Mexico, most of the people today are *mestizos*, people of mixed Spanish and Indian blood. There are a few pure-bred Spaniards, some of them forming a wealthy aristocracy, and about 12 people in every hundred are pure-bred Indians.

Most of the countries of Central America also have large numbers of mestizos in their populations. Some of the island countries of the West Indies, such as Jamaica and Haiti, have populations that are mainly Negro.

In the United States, very few settlers intermarried with the American Indians. Today, most of the Indians live in *reservations*, areas set aside for their exclusive use. A few have settled in other parts of the country and integrated with the rest of the population. The Negroes, too, have largely remained a separate community. There are many *mulattos* (half Negro, half white), but they are generally grouped with the Negroes. The white people, who form the largest and most powerful section of the population, have tended to treat the Negroes, and also the American Indians, as second-class citizens. This has led to racial conflict.

In Canada there is cultural conflict of a different kind. There, the tension is between the French-speaking Canadians of Quebec province and the English-speaking Canadians of the rest of the country.

In the United States and Canada, 70 out of every hundred people live in cities or towns. Only about 10 workers out of every hundred are engaged in farming. Most workers are employed in manufacturing, in selling goods, or in government and other services. But in Mexico and Central America about 60 out of every hundred people are farmers or farm workers.

Life on the Ice. Eskimoes in the Arctic north. The Eskimoes of North America are believed to have crossed from Siberia, in Asia, thousands of years ago. For food they rely heavily on walrus, seals, and fish. Their fragile boats, *kayaks* and *umiaks*, are made of walrus hide stretched over a wooden frame.

Right: The First Americans. The American Indians, immigrants from Asia, spread over North and South America and developed many different cultures. Today, Canada and the United States have about 700,000 Indians.

Below: An Unmistakable Skyline. Skyscrapers on Manhattan Island, New York City. In the foreground is the United Nations Secretariat Building. In the background: left, the Empire State Building; right, the Chrysler Building.

THE COUNTRIES OF NORTH AMERICA

AND CENTRAL AMERICA

BAHAMAS

The Bahamas are an archipelago in the north-eastern part of the West Indies, extending NW-SE for about 750 miles (1,200 km). They lie SE of Florida and N of Cuba. The Bahamas include more than 2,700 islands, many of them little more than rocks in the sea. About 30 islands are inhabited. The chief islands include Grand Bahama, Abaco, Acklins, Eleuthera, New Providence, Andros, San Salvador, Exuma, Long Island, Mayaguana, and Inagua. Andros is the largest. Most of the islands are low and flat, with fine beaches. The climate is warm and healthy. Tourism provides most of the country's income. Otherwise the main occupations are farming and fishing. The Bahamas are also a centre of international banking. Industries include cement, food processing, and oil refining. Four-fifths of the people are of African descent, and the rest of British or North American descent. English is spoken and most people belong to the Anglican, Roman Catholic, Baptist, or Methodist churches. The cap. is Nassau on New Providence Island. Area 5,400 sq miles (13,990 sq km). For population, see Table on page 180.

BARBADOS

A small, flattish coral island, Barbados is the most easterly island of the West Indies. It lies about 250 miles (400 km) NE of mainland South America. From the W coast, gently rising terraces culminate in a central ridge where Mount Hillaby reaches 1,104 ft (336 m). The only streams are small. The climate is a warm and tropical but healthy one. Natural resources include fertile soil, limestone, coral, and small amounts of natural gas. Tourism is the most important source of income. The chief crop is sugar cane, and stock raising is important. The principal exports are sugar, molasses, rum, and fish. Almost all the people are of African descent. The others are descended from European, mainly British, settlers. English is the official language. In religion, the population is divided among the various Christian churches. The cap. and largest city is Bridgetown. Area 166 sq miles (430 sq km). For population, see Table on page 180.

Canada. Left: Niagara Falls on the Niagara River, which connects Lake Erie and Lake Ontario. The Falls are on the border between Canada and the United States. Horseshoe Falls, on the Canadian side, are 186 ft (56·7 m) high and 2,100 ft (640 m) wide. Below: Algonquin Provincial Park in Ontario, near the Ottawa River. It covers more than 2,500 sq miles (6,500 sq km), and has about 1,200 lakes.

CANADA

In area, the world's second largest country, covering 3,851,809 sq miles (9,976,139 sq km). For population, see Table on page 180. Canada is in the northern part of the continent of North America, bounded on the S by the United States and on the N by the Arctic Ocean. Its capital is Ottawa. The country is a monarchy, acknowledging Queen Elizabeth II of the United Kingdom as its head of state. She is represented by a governor-general. Canada has 10 provinces, each with its own legislature and government, and two territories administered by councils. A federal parliament and government deals with national and international affairs. Canada is a member of the Commonwealth of Nations.

Land and Climate The northern part of Canada, which includes many islands, lies within the Arctic Circle. This region, about a third of the country, is tundra—cold, treeless plains covered with moss, lichen, and other dwarf plants. The ground is permanently frozen but for the top 3 ft (1 m). Half of the country is forested, many of the trees being commercially valuable. Altogether, some three-quarters of Canada is too cold to be more than sparsely populated. Most people live in the southern 200-mile (320-km) strip along the border with the United States. Even there, cold and snow halt farming for several months in winter.

The southern part of the country has four main regions, marked off by natural barriers. In the W is the high Pacific coast region of British Columbia, bounded on the E by the lofty, rugged Rocky Mountains. Most of Yukon Territory in the NW is also mountainous. The next region includes the prairie provinces, with wide, grassy plains sloping gently from W to E. In the plains there are two large lakes, Winnipeg and Winnipegosis. E of the prairies is the Canadian Shield, a horseshoe of hard rock comprising almost half the land area of the country. The soil is thin in the N.

The third and most important populated region is on the N bank of the St. Lawrence River and bordering the five Great Lakes —Superior, Michigan, Huron, Erie, and Ontario, which form part of the frontier with the United States. The fourth region is around the Gulf of St Lawrence and along the Atlantic coast.

Agriculture and Industry Canada has rich deposits of coal, petroleum, and natural gas, and also copper, gold, iron, lead, nickel, platinum, silver, uranium, and zinc. From the forests come lumber and fur pelts. Cattle graze on the prairies, and Canada is a leading producer of barley, oats, and wheat. The manufacturing industries produce steel, paper, chemicals, textiles, and many other types of products.

Cities and Communications Besides Ottawa, Canada's leading cities are

Canada. Above: The Changing of the Guard in front of the Parliament Buildings in Ottawa. The sandstone buildings stand on three sides of a 35-acre (14-hectare) quadrangle. The cornerstone was laid by the Prince of Wales, later King Edward VII, in 1860. Below: Toronto, the capital of Ontario. The second largest city in Canada, it has a fine harbour on the north-western shore of Lake Ontario. It is an important industrial and cultural centre.

Montreal, Toronto, Vancouver, Winnipeg, Hamilton, and Edmonton. The rail network has more than 43,000 miles (69,000 km) of track, linking the main centres of population, and the Trans-Canada Highway crosses the country from the Atlantic to the Pacific. The St. Lawrence Seaway links the Great Lakes and the Atlantic.

The People The standard of living in Canada is one of the world's highest. About half of the people are descended from settlers from the British Isles. But more than a quarter are of French ancestry and speak

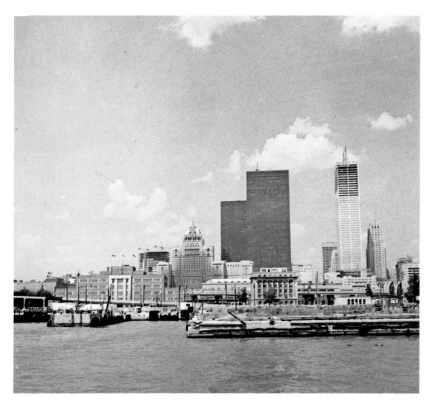

French as their everyday language. Most of them live in Quebec province. One-fifth of all Canadians are from other European countries—Germany, Poland, the U.S.S.R., the Netherlands, and the countries of Scandinavia. The rest are American Indians or Eskimoes, descendants of the country's original inhabitants. English and French are the official languages.

COSTA RICA

Located in Central America, Costa Rica lies between the Caribbean Sea in the E and the Pacific Ocean in the W. On the N it is bordered by Nicaragua, and on the SE by Panama. A wide coastal plain in the E and a narrower one in the W rise to volcanic mountain ranges crossing the country from N to SE. About two-thirds of Costa Rica is forested. The chief rivers are the San Juan and San Carlos. The climate is tropical in the coastlands, but cooler in the mountains. The main agricultural products are coffee, bananas, cocoa, sugar, and rice. There are some manufacturing industries, and some mineral deposits. Most of the people are of European descent, or of mixed Indian-European blood. They speak Spanish, and are mainly Roman Catholics. The cap. is San José. Area 19,650 sq miles (50,890 sq km). For population, see Table on page 180.

CUBA

The largest and most westerly island in the West Indies, Cuba lies between the Caribbean Sea and the Atlantic Ocean. It is about 100 miles (160 km) S of Florida in the United States, at the mouth of the Gulf of Mexico. Gently rolling land, with many wide, fertile valleys and plains, makes up about 60 per cent. of Cuba's surface. The rest of the island is hilly. The highest peak is Turquino (6,560 ft; 2,000 m) in the Sierra Maestra range in the SE. About 760 miles (1,220 km) long, Cuba varies in width from 25 to 125 miles (40 to 200 km), and has a much-indented coastline of about 2,000 miles (3,200 km). The climate is tropical in the lowlands, but cool and moderate in the hills. Cuba's rich mineral deposits include nickel, iron, manganese, cobalt, copper, chromite, silica, gold, and silver. About a third of the people gain their living by farming or working in plantations. The main products are sugar, tobacco, and coffee. Industries include sugar refining, and the manufacture of cigarettes and the world-famous Havana cigars, rum, and textiles. Cuba also has a flourishing film industry. More than two-thirds of the people are of European (mainly Spanish) descent. Of the rest about half are of African ancestry, and half of mixed European-African blood. There is also a small Chinese minority. Spanish is the official language. Many of the people are Roman Catholics, though Cuba has a communist government. The chief cities are Havana (the cap.), Santiago de Cuba, Camagüey, Guantánamo, and Santa Clara. Area 44,218 sq miles (114,524 sq km). For population, see Table on page 180.

Costa Rica. Picking coffee berries on a large plantation. The berries go to a pulping house where the coffee beans are extracted. Each berry contains two beans.

Top:
Canada. The Thompson River in British Columbia. It rises in the Rocky Mountains, and flows 270 miles (434 km) south-west to the Fraser River.

DOMINICAN REPUBLIC

Situated in the West Indies, the Dominican Republic occupies the eastern two-thirds of the island of Hispaniola, which it shares with Haiti. It lies between the Atlantic Ocean on the N, and the Caribbean Sea on the S. On the E the Mona Passage separates it from Puerto Rico. Much of the land is mountainous, with four E to W ranges. The highest peak is Pico Duarte (10,417 ft; 3,175 m) in the Central Cordillera. The rivers include the Yaque del Norte, Yaque del Sur, and Jaina. The climate, tropical in low-lying areas, is cool in the mountains. Mineral deposits include bauxite, nickel, and iron. There may also be oil. Most of the people live by farming. Their main cash crops are sugar, coffee, cocoa, tobacco and bananas. They speak Spanish. The chief cities are Santo Domingo (the cap.) and Santiago. Area 18,800 sq miles (48,700 sq km). For population, see Table on page 180.

EL SALVADOR

The small country of El Salvador in Central America is bordered on the N and E by Honduras and on the NW by Guatemala. On the S it has a coastline on the Pacific Ocean. Mountain ranges running from E to W divide the country into three main natural regions; a narrow Pacific coastal plain; a central region of plateaux and valleys; and mountains in the N. Mineral resources include quartz, china clay, and gypsum. More than half the labour force works in agriculture. The main products are coffee, cotton, sugar, and rice. Livestock are important. Most of the people are of mixed Spanish-Indian blood, and the remainder are either Spanish or Indian. They speak Spanish. The chief cities are San Salvador (the cap.), Santa Ana, San Miguel, and La Union. Area 8,260 sq miles (21,400 sq km). For population, see Table on page 180.

GRENADA

The small country of Grenada in the West Indies consists of the island of Grenada, the most southerly of the Windward Islands, and the islands of the South Grenadines. Grenada is volcanic and mountainous, with fertile soil, and swift streams. Its climate is hot and humid. It is famous for its spices. The people are of African, Carib Indian, and European ancestry. They speak English and a French-African dialect. The cap. is St. George's. Area 133 sq miles (344 sq km). For population, see Table on page 180.

GUATEMALA

The republic of Guatemala in Central America is bordered on the N by Mexico, on the E by Belize, and on the S by Honduras and El Salvador. In the E it has a coastline on the Caribbean Sea, and in the SW on the Pacific. A highland region takes up about a fifth of the country in the SW. There is a narrow Pacific coastal plain, and a wider plain on the Caribbean side with fertile river valleys. The northern part of the country is a forested lowland, some of it thick jungle. The climate ranges from sub-tropical in the lower parts to temperate in mountain regions. The longest rivers are the Motagua, Usumacinta, and Polochic. Guatemala's mineral deposits include zinc, silver, lead, mica, chromite, nickel, and limestone. Among its crops are coffee, cotton, sugar cane, bananas, maize, beans, rice, and wheat. Industries include food processing, and the manufacture of textiles, paper, chemicals, and machinery. About half of the people are Indians, descended from the Mayas, whose civilization is one of the oldest in the Western Hemisphere. The rest are mainly of mixed Indian and European descent. The official language is Spanish, but various Indian dialects are widely spoken. Most of the people are Roman Catholics. The chief cities are Guatemala City (the cap.), Quezaltenango, Mazatenango, Barrios, and Antigua. Area 42,040 sq miles (108,900 sq km). For population, see Table on page 180.

Top:
Cuba. Galleries in a village street offer protection against sun and rain. The island lies in the tropics, and the lowlands are extremely hot. The rainy season lasts from May to October.

191

HAITI

Haiti in the West Indies occupies the western third of the island of Hispaniola, which it shares with the Dominican Republic. On the N is the Atlantic Ocean, and on the S the Caribbean Sea. Much of the land is mountainous and difficult to farm. In lower areas the climate is tropical. Bauxite and copper are mined, but most people live by subsistence farming. The main cash crops are coffee, sugar, sisal, and cotton. Almost all the people are of African descent. Roman Catholicism is the official religion, but most people follow voodoo. They speak French and Creole. The cap. and main city is Port-au-Prince. Area 10,700 sq miles (27,700 sq km). For population, see Table on page 180.

HONDURAS

Honduras is one of the largest and poorest of the countries of Central America. It is bordered on the S by Nicaragua, and on the W by El Salvador and Guatemala. On the N it has a long coastline on the Caribbean Sea, and on the SW a short Pacific coastline. Most of the country is mountainous except for a narrow Pacific coastal plain and a wider Caribbean one. There are many fast rivers. The climate is humid and subtropical in the lower parts. The chief products and crops are bananas, coffee, maize, meat, cotton, and timber. The people speak Spanish and are mostly of mixed Indian-European blood. The cap. is Tegucigalpa. Area 43,300 sq miles (112,000 sq km). For population, see Table on page 180.

JAMAICA

Island country in the Caribbean Sea, the third largest island in the West Indies. It is about 90 miles (145 km) S of Cuba. Together with Cuba, Hispaniola (Haiti and the Dominican Republic), and Puerto Rico, Jamaica is part of the Greater Antilles group. Most of the island is mountainous,

with the main range running from E to W. The highest point is the Blue Mountain Peak (7,388 ft; 2,252 m) in the E. There is a narrow coastal plain in the S, and lowlands stretch across the W end of the island. The island is famous for its golden beaches. There are many short streams; the main ones are the Black and Minho Rivers. The climate is humid and subtropical. Jamaica's mineral resources include bauxite, of which it is the world's largest producer, and gypsum, silica, marble, and clays. The main agricultural products include sugar, bananas, cocoa, coffee, citrus fruits, and copra. Forestry and fishing are also important. Food processing and the production of rum and beer are important industries. Most of the people are of African or mixed descent (Afro-Asian, Afro-European, or all three). English is the official language, but an Afro-English patois is also widely spoken. The chief cities are Kingston (the cap.), Montego Bay, and Spanish Town. Area 4,440 sq miles (11,400 sq km). For population, see Table on page 180.

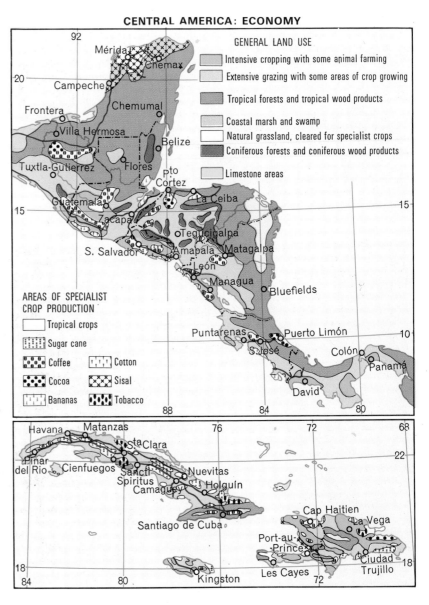

CENTRAL AMERICA: ECONOMY

GENERAL LAND USE
- Intensive cropping with some animal farming
- Extensive grazing with some areas of crop growing
- Tropical forests and tropical wood products
- Coastal marsh and swamp
- Natural grassland, cleared for specialist crops
- Coniferous forests and coniferous wood products
- Limestone areas

AREAS OF SPECIALIST CROP PRODUCTION
- Tropical crops
- Sugar cane
- Coffee
- Cotton
- Cocoa
- Sisal
- Bananas
- Tobacco

The United Mexican States, in the southern part of the North American continent, has a long coastline on the Pacific Ocean in the W, and a coastline on the Gulf of Mexico in the E. In the N it borders on the United States, and in the SE on Belize and Guatemala. Its capital is Mexico City. Mexico's area is 760,000 sq miles (1,968,000 sq km). For population, see Table on page 180.

MEXICO

Mexico. Below: Sightseers and merrymakers take part in a river fiesta. The decoration of the flat-bottomed boats recalls the art of the Aztecs, whose empire had its centre at Tenochtitlán—now Mexico City—until they were conquered by the Spaniards in the 1500s. Bottom: Southern Mexico and Central America were the home of the remarkable Maya civilization that flourished from about the A.D. 300s to the 1400s. Today, the ruins of Maya cities lie deep in the jungles.

Land and Climate   Mexico has more than 6,000 miles (9,600 km) of coastline. The E coast has many sandbanks and lagoons, and few natural harbours. The W coast has better harbours but is cut off from the interior by difficult terrain. Much of Mexico is mountainous, more than half the country being above 3,000 ft (900m) in altitude. The main features are a large, steep-sided central plateau, rising to some 8,000 ft (2,400 m) in the S, and flanked by the Western, Eastern, and Southern Sierra Madre mountain ranges. The plateau has many volcanoes, including the active Popocatepetl (17,887 ft; 5,452 m), Iztaccihuatl (16,883 ft; 5,146 m), and Citlaltepetl (18,700 ft; 5,700 m), Mexico's highest peak. Other features include the rugged peninsula of Lower California, lying to the W of the Gulf of California; the Sonora Desert between the Western Sierra Madre and the coast; and the E coastal plain extending into the low Yucatán peninsula. Mexico's main rivers include the Rio Bravo (Rio Grande), the Lerma, the Santiago, and the Usumacinta. The climate varies with altitude. At low altitudes it is subtropical, ranging from warm to cool temperate and Alpine with increasing altitude.

Agriculture and Industry   Half of the people live by farming. The main cash crops include cotton, coffee, and sugar. Maize is the staple crop. Livestock raising, forestry, and fishing are also important. Mineral resources include silver, of which Mexico is the world's leading producer, and sulphur, lead, zinc, oil, coal, tin, mercury, copper, and iron. The chief industries are the making of textiles and steel.

Cities and Communications   Half of the population lives in towns and cities. Apart from Mexico City, the main centres are Guadalajara, Monterrey, León, Cuidad Juárez, and Mexicali. Road, rail, and air services are well developed.

The People   About two-thirds of Mexicans are *mestizos* of mixed Indian and European descent. Spanish is the official language, but many Indians speak only Indian dialects.

Panama. The Panama Canal cuts through Central America to connect the Atlantic and Pacific Oceans. It saves ships the long journey round the southern tip of South America. By treaty with Panama, the United States governs the Panama Canal Zone, the strip of territory through which the canal runs, except for the cities of Panama and Colón.

NICARAGUA

The largest country of Central America, Nicaragua has coastlines on the Pacific Ocean and the Caribbean Sea. On the N it is bordered by Honduras, and on the S by Costa Rica. The Pacific coastal plain rises gently eastwards towards a volcanic mountain range. Beyond this is the forested interior of plains and hills, cut by rivers. The swampy Caribbean coastal plain—the Mosquito Coast—reaches about 50 miles (80 km) inland. The climate is humid and subtropical. Most people live by working on farms. The main products are cotton, coffee, and sugar. Stock-raising, forestry, and fishing are also important. Two-thirds of the people are of mixed European and Indian ancestry. Spanish is the official language. The chief cities are Managua (the cap.), León, and Granada. Area 57,150 sq miles (148,020 sq km). For population, see Table on page 180.

PANAMA

Panama, in Central America, occupies the southern part of the narrow Isthmus of Panama, which connects North America and South America. On its N is the Caribbean Sea, and on its S the Gulf of Panama and the Pacific Ocean. To the W is Costa Rica, and to the E Colombia. Most of Panama is mountainous, with two main W to E ranges. The highest point is Mount Chiriqui (11,410 ft; 3,478 m) in the W. The country is divided in two by the Panama Canal. The coastal regions are low, the E side being mostly covered by tropical forest. The climate is humid and subtropical. Panama has deposits of several minerals, but most are of low quality. About half of the labour force works in agriculture. The chief crops are rice, sugar, bananas, cocoa, coffee, and tobacco. Industries include food processing, and the manufacture of textiles, cement, and chemicals. Two important sources of income are Canal Zone earnings and payments for the registration of foreign ships under the Panamanian flag. Most people are of mixed Indian-European blood and speak Spanish. The chief cities are Panama City (the cap.) and Colón. Area 29,200 sq miles (75,600 sq km). For population, see Table on page 180.

Trinidad and Tobago. The luxuriant vegetation of Trinidad. The island, which lies 7 miles (11 km) from the coast of Venezuela, is celebrated for its beauty, its sugar, and its asphalt.

TRINIDAD AND TOBAGO

The country of Trinidad and Tobago consists of the two most southerly islands of the Lesser Antilles, just off the coast of Venezuela in South America. On their W is the Caribbean Sea, and on their E the Atlantic Ocean. Most of the people live on the larger island, Trinidad. It is crossed by three mountain ranges, the highest being in the N. There, El Cerro del Aripo rises to 3,085 ft (940 m). Tobago is also mountainous. The islands lie near the equator and have a tropical climate—hot and wet. Their prosperity is based on their oil and asphalt resources, and their sugar and cocoa. The people are mainly of Indian and African ancestry. The cap. is Port of Spain, on Trinidad. Area 1,980 sq miles (5,130 sq km). For population, see Table on page 180.

United States. The Empire State Building in New York City was for long the world's tallest building, and is still probably the most famous of all skyscrapers. It has 102 storeys.

UNITED STATES OF AMERICA

The world's fourth largest country in area and population, the United States is the richest and most powerful of all countries. Its highly developed industries are based on immense natural resources.

It is a country of contrasts. In the NW there are the icy mountains and forests of Alaska. In the S there are burning deserts. Many of its people enjoy the world's highest standard of material prosperity. Others live in the most backward poverty.

The United States consists of the central part of the North American continent, plus Alaska in the NW and the islands of Hawaii in the Pacific Ocean. Its total area is 3,615,211 sq miles (9,363,353 sq km). For population, see Table on page 180.

The mainland of the United States lies between Canada to the N and Mexico to the S, and stretches across the continent from the Atlantic Ocean to the Pacific. Two parts of the country are separated from the rest. Alaska is situated to the NW of Canada and the Hawaiian islands lie 2,400 miles (3,860 km) to the W in the Pacific. The country's capital is Washington, D.C.

The United States is a federation of 50 states, one of which is Alaska and one Hawaii. In addition, there is the federal District of Colombia (D.C.), in which Washington is situated. Each state has its own elected legislative assembly, governor, and state government. And each state has certain powers that the federal (central) government cannot take away from it. Control over the whole country is exercised by the federal government in Washington, D.C.

Land and Climate The mainland of the United States must be considered separately from Alaska and Hawaii.

Along the Pacific coast there is a narrow lowland strip rising sharply to the Western Highlands, a region of mountain ranges and high plateaux. The Pacific Coast Ranges lie very close to the sea, and inland and parallel to them comes a mountain chain made up of the Cascade Mountains and the Sierra Nevada. Still farther inland are the lofty Rocky Mountains, which extend N-S through most of the North American continent. Between the Cascade-Sierra Nevada

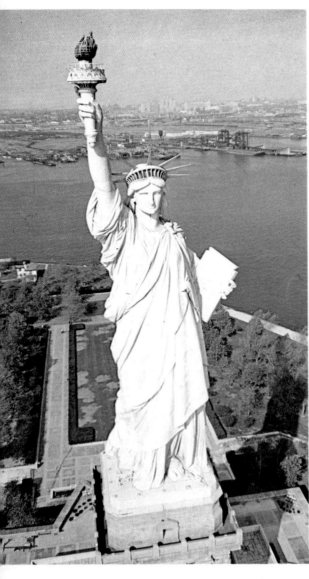

United States. The Statue of Liberty on Liberty Island in New York harbour. The largest statue ever made, it is 151 feet (46 m) high. It is made of beaten copper over an iron framework. The statue was designed by the French sculptor Frédéric Auguste Bartholdi, and was a present from the people of France to the people of the United States. It was unveiled in 1886.

United States. The San Francisco-Oakland Bay Bridge spanning San Francisco Bay between the cities of San Francisco and Oakland. With its approaches, it is some 8 miles (13 km) long, and has a number of sections. Some sections are connected by a tunnel on Yerba Island in the bay. The bridge has two decks, the upper one being a roadway with six traffic lanes. It was opened in 1936, and is one of the longest bridges in the world.

chain and the Rockies lies the Colorado Plateau, parts of which are more than 6,000 feet (1,830 m) above sea-level. There are also basins that once contained salt lakes.

The Great Plains—the prairie country—are the heart of the United States. These vast rolling expanses are watered by

the Mississippi River and its tributaries, the Missouri and the Ohio. To the N of the Great Plains lie the five Great Lakes—Superior, Michigan, Huron, Erie, and Ontario. They measure more than 1,100 miles (1,800 km) from E to W.

East of the Great Plains is another barrier, formed by the Appalachian Mountains, which for a long time limited the westward expansion of the early colonies. From the Appalachians, the Eastern Coastal Plain slopes away to the Atlantic.

The climate of the United States varies considerably, partly because of the wide differences in elevation of various regions of the country, and partly because the United States stretches from cool northern latitudes almost to the Tropic of Cancer. Along the Atlantic coast the climate is hot and humid in summer. But it is cold in winter because of the influence of a cold Atlantic current. On the other hand, Florida, in the SE of the country, is famous for its warm winters. In the Great Plains most areas enjoy hot summers and cool winters. This area has between 20 and 40 inches (500 and 1,000 mm) of rainfall each year, about half the rainfall of the E coast. The Western Highlands have less than 20 inches (500 mm) of rain, and much of the region is desert, or at any rate too dry to be of any use as farmland. Along the NW coast winds blowing in from the Pacific bring the country's highest rainfall.

Alaska, the cold north-western part of North America, has an area almost one-fifth that of the rest of the United States. It is a land of icy mountains and forests. A third of it is N of the Arctic Circle. It has two great mountain ranges. The Alaska Range curves round from the Aleutian Islands in the W to the south-eastern coast. The Brooks Range, in the Arctic N, is part of the Rocky Mountain system. Between the two ranges is the wide Yukon River Valley, a swampy, hilly region, with lichen, shrubs, and a few forests. Reindeer and moose graze in this region.

Hawaii consists of a group of more than 20 volcanic islands, lying in the tropics. The largest of the group is Hawaii Island, but most of the people live on Oahu. Rainfall on the W side of these mountainous and beautiful islands is more than 100 inches (2,500 mm) a year, while in the E it is only about one-third of that.

Agriculture and Industry The United States is one of the world's chief agricultural countries, with millions of acres of rich land. On the Great Plains in the centre, a variety of crops is grown and huge herds of cattle roam. The principal crop in many areas is

United States. Chicago, from the Lake Michigan shore. The city, the second largest in the country, is the commercial and industrial centre of the Midwest. It is also a major port on the Great Lakes.

maize (corn). The United States grows about half of the world's maize. In the N, spring wheat is the chief crop, together with fodder and hay for dairy cattle. In the S, where there are more than 200 frost-free days every year, cotton is the principal crop. Rice is grown on the S coast, along the Gulf of Mexico.

In the eastern states, farmers raise cattle and other livestock, and grow fruit, vegetables, tobacco, and, in the S, cotton. The mountainous regions of the W have little soil suitable for crops, but cattle and sheep graze on the poor pastures. A large area is needed for a few animals. By irrigation schemes from the rivers, such as the Colorado, the amount of arable land has been greatly increased.

The mighty industries of the United States are based on its abundant mineral resources. The chief of these are coal, iron, and petroleum, and the country also has rich supplies of bauxite, copper, gold, lead, phosphorus, silver, uranium, and zinc. But some of these minerals are not plentiful enough to feed the country's developed industries, and the United States imports copper, iron ore, uranium, and petroleum, as well as minerals which it does not possess in great quantities, such as tin.

About one-third of all the world's manufactured goods are produced in the United States. The bedrock of American industry is steel, of which American plants produce as much as the rest of the world put together. It was because of its enormous industrial output and strength that the United States was able to put men on the Moon and lead the world in the 'Space race'. In addition to its actual manufactures, the country operates a continuing programme of research and development, which keeps it in the forefront of the world's scientific discoveries.

Other resources include the great forests that still cover part of the country. Many forests were cleared in the course of settlement, but some are now being replanted.

United States. A steamboat on the Mississippi River. Today, such craft carry mainly tourists, but once they travelled up and down the river transporting cotton, farm produce, and other goods between the cities and small towns along the Mississippi's banks. The author Mark Twain was a pilot on the old Mississippi paddleboats.

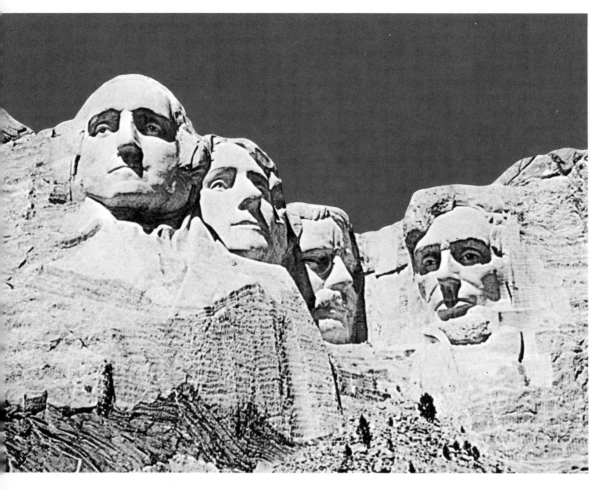

Wild animals such as bison used to be a valuable part of the country's natural resources, but they were slaughtered until only a few hundred were left.

Cities and Communications The United States has more than 50 cities with populations exceeding 250,000. New York City is one of the largest cities in the world, and its metropolitan area may have the greatest population of any city. It is the leading manufacturing city, closely followed by Chicago, Illinois; Los Angeles, California; Philadelphia, Pennsylvania; and Detroit, Michigan. Detroit is the centre of the world's automobile industry, and that industry is one of the country's largest. Very often, American cities are not both administrative centres and manufacturing centres. The fed-

United States. Mount Rushmore National Memorial, a huge carving on a granite cliff in the Black Hills of South Dakota. The carving depicts the faces of four American presidents: George Washington, Thomas Jefferson, Abraham Lincoln, and Theodore Roosevelt. It was designed by the American sculptor Gutzon Borghum, and work on it began in 1927.

United States. The Navajo Indian Tribal Park of Monument Valley in south-eastern Utah.

eral capital, Washington, D.C., has relatively little industry, and many of the state capitals are also basically non-industrial.

Highly-developed networks of roads, railways, waterways, and airlines provide speedy and efficient communication between one part of the country and another. There are more than 3 million miles (5 million km) of roads, including a 41,000-mile (66,000-km) system of inter-state highways. Railways total more than 200,000 miles (320,000 km) but are declining in importance. Airlines link all major cities of the United States, and through international airports provide services to all parts of the world.

America's waterways are based on two great river systems. The St. Lawrence Seaway, a combination of river and canals, links the Great Lakes with the Atlantic Ocean. Other canals link the Great Lakes with the Mississippi River, which with its tributaries provides water transport over a great deal of the central plains, linking the Great Lakes also to the Gulf of Mexico. These waterways provide a cheap and efficient way of moving bulk cargoes.

The People Most people in the United States are immigrants from Europe, or the descendants of immigrants. About one in every 10 is of Negro ancestry, descendants of slaves brought over from Africa. There

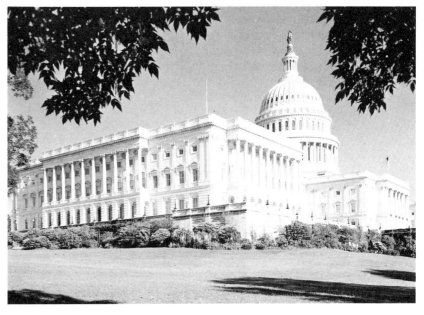

THE STATES OF THE UNITED STATES

The states of the United States (excluding Alaska and Hawaii) are grouped in seven geographical regions. This table shows the states in each region, with the abbreviations of their names that are commonly used and their capitals.

NEW ENGLAND

Connecticut (Conn)	Hartford
Maine (Me)	Augusta
Massachusetts (Mass)	Boston
New Hampshire (N.H.)	Concord
Rhode Island (R.I.)	Providence
Vermont (Vt)	Montpelier

MIDDLE ATLANTIC STATES

New Jersey (N.J.)	Trenton
New York (N.Y.)	Albany
Pennsylvania (Penn)	Harrisburg

MIDWESTERN STATES

Illinois (Ill)	Springfield
Indiana (Ind)	Indianapolis
Iowa (Ia)	Des Moines
Kansas (Kan)	Topeka
Michigan (Mich)	Lansing
Minnesota (Minn)	St Paul
Missouri (Mo)	Jefferson City
Nebraska (Nebr)	Lincoln
North Dakota (N.D.)	Bismarck
Ohio (O.)	Columbus
South Dakota (S.D.)	Pierre
Wisconsin (Wis)	Madison

SOUTHERN STATES

Alabama (Ala)	Montgomery
Arkansas (Ark)	Little Rock
Delaware (Del)	Dover
Florida (Fla)	Tallahassee
Georgia (Ga)	Atlanta
Kentucky (Ky)	Frankfort
Louisiana (La)	Baton Rouge
Maryland (Md)	Annapolis
Mississippi (Miss)	Jackson
North Carolina (N.C.)	Raleigh
South Carolina (S.C.)	Columbia
Tennessee (Tenn)	Nashville
Virginia (Va)	Richmond
West Virginia (W. Va)	Charleston

SOUTH-WESTERN STATES

Oklahoma (Okla)	Oklahoma City
Texas (Tex)	Austin

ROCKY MOUNTAIN STATES

Arizona (Ariz)	Phoenix
Colorado (Colo)	Denver
Idaho (Ida)	Boise
Montana (Mont)	Helena
Nevada (Nev)	Carson City
New Mexico (N. Mex)	Santa Fe
Utah (Ut)	Salt Lake City
Wyoming (Wyo)	Cheyenne

PACIFIC COAST STATES

California (Calif)	Sacramento
Oregon (Ore)	Salem
Washington (Wash)	Olympia

OTHER STATES

Alaska (no abbreviation)	Juneau
Hawaii (no abbreviation)	Honolulu

United States. The United States Capitol in Washington, D.C. The U.S. Senate meets in its northern wing, and the House of Representatives in its southern wing. George Washington laid the cornerstone of the gleaming white building in 1793. In 1814, during the 'War of 1812', the Capitol was burned by British troops, but was later rebuilt. It has been altered and enlarged many times.

United States. Death Valley in east-central California. The parched desert valley lies 282 feet (86 m) below sea-level, and is about 130 miles (210 km) long. Parts of it are spectacularly beautiful.

are small numbers of the early inhabitants of the land—the American Indians and the Eskimoes.

The first settlers in the United States were mostly from Great Britain and Ireland, with some French and Spaniards. Later settlers included a great many Germans, Italians, Poles, Scandinavians, and Russians. All these peoples brought with them their own cultures and traditions, which have helped to make the American culture of today. English is the official language, and all immigrants learn to speak it, even if they still use their own languages in their homes. Certain areas of the country are particularly connected with one ethnic group or another. New Orleans, in the S, is predominantly French, while in parts of Pennsylvania the influence of German settlers remains strong.

SOUTH AMERICA

South America, the fourth largest of the continents, excites the imagination of many people because vast areas of it are still largely unexplored. Potentially, it is very rich, but most of its immense mineral resources have yet to be exploited. It is a turbulent continent. In some of its countries, politics tend to be violent and revolutions are not uncommon.

South America is linked to North America by the narrow Isthmus of Panama. It is roughly triangular in shape, tapering to a point in the S. It reaches farther S than any other continent. Cape Horn, its southern-most tip, is notorious among sailors because of the violence of its storms. At this point, the continent of Antarctica is only about 400 miles (640 km) away. South America is bordered on the W by the Pacific Ocean, and on the E by the Atlantic Ocean. To the N lies the Caribbean Sea.

The coastline of the continent is compara-tively smooth except in the SW, where it is broken by the many inlets and the islands of the Chonos Archipelago. The large island of Tierra del Fuego lies at the extreme S. Other island groups are the Galápagos Islands off the north-western coast, and the Falklands off the south-eastern shore.

The total area of South America is about 6,772,000 sq miles (17,540,000 sq km). The greatest distance N to S is 4,750 miles (7,640 km), and W to E 3,200 miles (5,150 km). The population in the mid-1970s was more than 190 million.

Physical Features and Climate The equator crosses South America just to the N of its widest part. As a result, more than three-quarters of the continent lies between the tropics of Cancer and Capricorn. There are eight well-defined physical regions, but two of them—the Andes Mountains and the basin of the River Amazon—are the domin-ant features of the continent.

The Past. An Indian from a forest tribe makes fire by rubbing two sticks together. South America once had a highly-developed Indian civilization—that of the Incas in present-day Peru, Ecuador, Chile, Bolivia, and Argentina. But most tribal Indians live lives of primitive simplicity.

The Future. The national congress buildings in Brasilia, designed by Oscar Niemeyer. Brasilia, the capital of Brazil since 1960, has some of the most imaginative of contemporary architecture. The city's ground plan resembles a bow and arrow.

The Andes, often referred to as the West-ern Cordillera, are a range of high, jagged mountains close to the Pacific Ocean coast. They stretch along the coast for more than 4,000 miles (6,450 km), and at their widest point are 300 miles (480 km) from W to E. The highest point in the range is Mount Aconcagua in Argentina, which rises to 22,835 ft (6,960 m) above sea-level. More than 50 other mountains rise above the 20,000 ft (6,100 m) level. Many Andean mountains are active volcanoes, and the whole region is shaken by earthquakes from

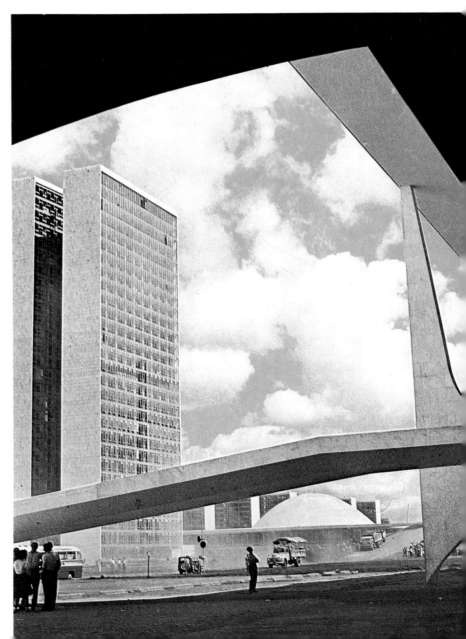

COUNTRIES OF SOUTH AMERICA POPULATIONS AND MONETARY UNITS		
Country	*Population*	*Monetary Unit*
Argentina	25,200,000	Peso
Bolivia	5,100,000	Peso
Brazil	95,000,000	Cruzeiro
Chile	10,500,000	Escudo
Colombia	23,000,000	Peso
Ecuador	7,000,000	Sucre
Guyana	800,000	Dollar
Paraguay	2,600,000	Guaraní
Peru	14,500,000	Sol
Uruguay	2,800,000	Peso
Venezuela	11,600,000	Bolívar

Above:
Close-textured Carpet. The Amazonian forest seen from the air. South America's tropical rain forest covers an area of one million sq miles (2.5 million sq km)—more than 10 times the area of the United Kingdom. Movement through the forest is extremely difficult, and a newly-cut path may disappear in a few weeks.

Leaping Waters. Angel Falls in Venezuela, with a drop of 3,212 ft (979 m), is the highest waterfall in the world. It is located on Mount Auyantepui in the Guiana Highlands. South America also has the world's most powerful waterfall, the Guaíra on the River Paraná.

time to time. The western slopes rise precipitously from the narrow coastal plain to heights of more than 3 miles (4.8 km). On the eastern side, the mountains slope more gradually down to the plains. In the S, the Andes form a single ridge, but farther N they broaden out into two, sometimes three, parallel ridges, with high, level plateaux between them.

The Pacific Coastlands lie between the Andes and the sea. They are very narrow, ranging between 5 miles and 50 miles (8-80 km) wide. They are widest in the N, where they are swampy.

The Orinoco Lowlands lie in the N, around the basin of the Orinoco River. Much of this area is swamp. It is known as the *Llanos*, a word meaning 'plains'.

The Guiana Highlands occupy most of Guyana, Surinam, and Guyane (French Guiana) in the northern part of the continent. Their highest peak, Mount Roraima, reaches 8,620 ft (2,627 m) above sea-level. The Highlands slope gently to the N and drop sharply to the Amazon basin in the S.

The Amazon River Basin covers more than 2 million sq miles (5 million sq km). Most of the region is very low lying, and even at the foothills of the Andes some land is no more than 50 ft (15 m) above sea-level. For almost the whole extent of the Amazon and its tributaries, the region is covered by dense, tropical rain forest, much of it only partly explored.

The Brazilian Highlands, a triangular

region of high ground, lie to the E and S of the Amazon basin. Most of the region is less than 5,000 ft (1,500 m) above sea-level. But the tallest peak, Pico da Bandeira, reaches 9,480 ft (2,890 m).

The Paraná-Paraguay Plains form the basin of the Paraná River and its tributaries, of which the Paraguay is the most important. The plains are low-lying, sloping gently to the S. Part of them is covered by the *Gran Chaco* ('great hunting ground'), a region of forest and swamp which is still mostly a wilderness. Some of it has scrub vegetation.

S of the Gran Chaco lies the eastward-sloping Pampas of Argentina, 250,000 sq miles (650,000 sq km) of rich, fertile land.

The Patagonian Plateau lies in the southernmost part of the continent. It is a flat, irregular rocky tableland, mostly bare and windswept, nowhere more than 3,000 ft (900 m) above sea-level.

Tierra del Fuego is separated from the mainland by the stormy Magellan Strait. The southernmost point of the continent, Cape Horn, is on Horn Island, one of the many smaller islands S and W of Tierra del Fuego. The Falkland Islands, 300 miles (480 km) to the NE of Tierra del Fuego, have deep fiords and wild, grassy moorland.

The Galápagos Islands, 500 miles (800 km) W of Ecuador, consist of 12 large islands and several hundred smaller ones. They are noted for their plant and animal life, which includes many species unique to the islands.

The climate of South America is mostly warm, except in the high Andes plateaux. The Amazon basin has heavy rainfall, but the rest of the continent generally has moderate or light rainfall.

Resources and Industry South America has many valuable deposits of minerals, particularly of iron, manganese, petroleum, copper, tin, and bauxite. There are some active mines producing silver and gold, the lure that drew the early Spanish conquerors to the continent. Colombia is the world's chief source of fine emeralds, and there are diamonds in Brazil. Northern Chile has the world's only natural deposits of sodium nitrate, an important ingredient of fertilizers. Another source of fertilizer is guano, the accumulated droppings of millions of sea-birds, which is found on the Chincha Islands, off Peru. The continent has little coal, which is still one of the mainstays of industrial economies.

One under-developed resource is water power, which can be exploited for generating electricity. But in its rivers South

Snowy Mountains. Mount Condoriri (21,998 ft; 6,705 m), north of La Paz in Bolivia. It is in the Cordillera Real, one of the great Andean ranges.

Past Glories. The ruined Inca city of Machu Picchu, about 50 miles (80 km) from Cuzco, in Peru. It stands between two sharp peaks, about 6,750 ft (2,000 m) up in the Andes Mountains. The ruins of other cities have been discovered near-by.

Below:
The Open Life. Cattle-herding in southern Chile. In the background is the symmetrical cone of Mount Osorno. Most South American farms are small, but the continent also has some of the world's largest plantations and ranches.

Coffee
Cocoa
Cotton
Bananas
Rubber
Alfalfa
Vines and fruits
Sugar cane

Irrigated areas
Plantations
Market gardening
Wheat and maize, with intensive stock raising
Mixed farming
Mainly grazing land
Primitive agriculture—hunting and collecting
Forestry products
Unproductive areas—forests, mountains and deserts

Beef for export. A cattle market in Argentina. Huge amounts of beef are frozen or canned for export to other countries.

America possesses about a quarter of the world's total hydro-electric potential.

South America has tapped some of its other resources, such as timber. It produces many materials for supplying industry, including cotton, wool, and hides.

Industry in South America has developed relatively slowly. Argentina, Brazil, and Chile are the most highly-developed countries industrially. Some other countries are very backward. Until World War I the continent exported most of its mining production, and large amounts of minerals, particularly of petroleum, copper, and iron, are still exported.

Manufacturing industries consist mainly of the processing of raw food materials, such as meat, wheat, and sugar, and textile production. Heavy industries, such as iron and steel production, are still slow to develop, partly because of the coal shortage. Light manufacturing industries producing goods for home consumption, such as clothing, soap, ceramics, furniture, and cigarettes, are flourishing in some countries. And many craftsmen produce fine handmade goods such as leatherwork, silverware, baskets, and wood carvings. The continent has an abundance of labour, but lacks the capital needed to set up new industries on a large scale.

Communications are difficult in many parts of South America. Vast areas, such as the forested parts of the Andes, have few roads and no railways. The chief railway networks are in Argentina and SE Brazil.

Seven different railway gauges are in use—Colombia, for example, with only about 2,000 miles (3,200 km) of track, has three different gauges; while Venezuela, with less than 1,000 miles (1,600 km), has five gauges.

Most of the continent has no motor roads, and even in areas where there are roads many stretches are unpaved or in poor condition. A major road-building programme is going on, but the majority of cities even in Argentina are linked by tracks that are dusty in summer and quagmires in winter.

By contrast, air services are comparatively well developed. Planes can fly over the vast regions of forest and mountain, and air services have done a great deal to open up the continent. Though there are rivers in most areas, many of them have rapids or become shallow in dry weather and consequently are of little use for transport. Some areas with usable rivers have few people or goods to carry.

Only a minority of the people of South

SOUTH AMERICA
COLONIES AND OTHER TERRITORIES

Name	Location, etc.	Area		Population	Capital
		sq miles	sq km		
Guyane or Guiana (Fr.)	Northern coast of South America.	35,000	90,600	51,000	Cayenne
Netherlands Antilles (Neth.)	Two groups of West Indies islands.	395	1,025	225,000	Willemstad
Falkland Islands (U.K.)	Island group in the South Atlantic; called *Islas Malvinas* by Argentina.	4,750	12,300	1,900	Stanley
Surinam (Neth.)	Northern coast of South America.	63,000	163,000	425,000	Paramaribo

SOUTH AMERICA: POPULATION MIGRATIONS AND MODERN SETTLEMENT PATTERNS

Pampas cattle and sheep are raised, while in other parts cash crops such as bananas, coffee, cocoa-beans, cotton, linseed, sugar, sisal, tobacco, wheat, and sunflower seeds are grown. Some countries devote so much attention to growing cash crops that they have to import food, even though there is enough land to grow the food as well. Poor farming methods are largely to blame for this situation.

Another problem is unequal distribution of land. In many countries most of the good land is held by a few rich people, while the rest of the people are poor, landless workers.

People, Ways of Life South America contains the world's most mixed population.

THE EXPANSION OF THE INCA CIVILIZATION

America can afford radio, television, or telephones, even when they live within range of such services. About half the people cannot read. News spreads slowly in many areas.

Agriculture, Fisheries, and Land Use More than half the people of South America live by farming. For most of them, subsistence farming is the rule. They produce just enough food for their own needs. Some primitive Indian tribes in the forests still live by hunting and seed-gathering, as people did in the Stone Age. And huge areas of the continent are useless for agriculture. Some are covered by forests, like most of the Amazon basin; others are mountainous and difficult if not impossible to farm.

Only three countries have well-developed agriculture—Argentina, Uruguay, and Brazil. In those countries and in parts of Chile there are large farms and ranches —some of the biggest in the world. On the

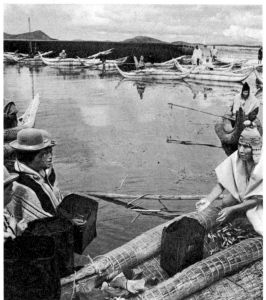

Above and Left: Birthplace of the Incas. Lake Titicaca, 12,500 ft (3,800 m) above sea-level in the Andes, on the boundary between Peru and Bolivia. According to legend, the Incas originated on an island in the lake. The Indian *balsas*—flat-bottomed boats—are made of reeds.

Below and Opposite page: Brazilwood Trees and Skyscrapers. The city of Rio de Janeiro, formerly Brazil's capital. Like other major South American cities, it has wide boulevards and huge modern buildings as well as narrow old streets and stucco colonial houses.

Before Spanish explorers reached the continent in the late 1400s, about 15 million American Indians lived there. These *Amerindians*, as they are often called, were descended from Asian peoples who crossed over what is now the Bering Strait when land still joined Asia and North America. They probably reached South America between 10,000 and 15,000 years ago. Spanish and Portuguese adventurers conquered South America during the 1500s. The earliest of these adventurers arrived without their womenfolk, and set up homes with Amerindian women. So from the very first a strong tradition of mixed marriages began to develop in the continent.

Today the people of South America are of three main stocks: European, Amerindian, and Negro. The Negroes are descended from the millions of slaves imported from Africa. These three stocks have inter-married to produce *mestizos*, people of mixed white and Amerindian blood; *mulattos*, of mixed white and Negro blood; and *zambos*, people of mixed Amerindian and Negro blood. In addition, the mestizos, mulattos, and zambos have also intermarried, producing a racial hotch-potch that will, in time, lead to a completely new racial type.

There is little racialism as such in South America. Any differences are social, and are largely a matter of wealth. As it happens, social divisions are often on racial lines, but there are many exceptions to the rule. The wealthy landowners are mostly white; the professional and business people are white and mestizo; and the bulk of the workers are mestizos, Amerindians, and Negroes, or mixtures of these groups.

The proportions of the different racial

groups vary from country to country. For example, in Argentina about 90 per cent. of the people are white, descendants of Spanish and other European settlers. The rest are mostly mestizos, and there are a few pure-blooded Indians. In Brazil about 20 per cent. of the people are mestizos, and 5 per cent. are Indians or Negroes. In Venezuela, 70 per cent. are mestizos, 20 per cent. are white and 10 per cent. are Negroes.

Most of the countries of South America have much smaller populations than they are capable of supporting, given proper use of the available land. But the population of the continent is growing faster than that of anywhere else in the world. This has several causes, the main one being the continued decline in the death rate due to better hygiene and medical services. Some experts estimate that by the year 2000 the total population of the continent will be more than 300 million.

Most of the people of South America live near the coast, except in the NW. There, the tropical heat and humidity of the coastlands has driven people into the highlands, where the climate is cooler and more healthy. Peru and Bolivia are populated chiefly in the high plateau lands of the Andes. Difficulties of travel have hindered much settlement inland, and where it has occurred it has been mainly along navigable rivers, which provide the only effective means of transport.

Spanish is the principal language spoken in South America, because most of the countries were formerly Spanish colonies. Brazil was a Portuguese colony, and there Portuguese is the main language. In Guyana, a former British colony, the main language is English. Many Indian languages are spoken in the continent, but in only one country, Paraguay, is an Indian language sufficiently common as to be regarded as a second tongue. Most of the people of South America are Christians, and members of the Roman Catholic Church.

About two-thirds of the people live in rural areas, in simple houses made of timber, stone, or brick with thatched roofs. They work on the land. In cities, many people live in cramped, overcrowded conditions, with poor sanitation and water supplies. The better-off people live very comfortably with all modern conveniences.

THE INCAS

The Inca civilization and empire developed in the Andes Mountains, and reached its peak in the late 1400s. Its capital was Cuzco, in present-day Peru. The Incas were skilled builders, weavers, and jewellers. Their language was Quechua. The Spaniards conquered them in the 1500s.

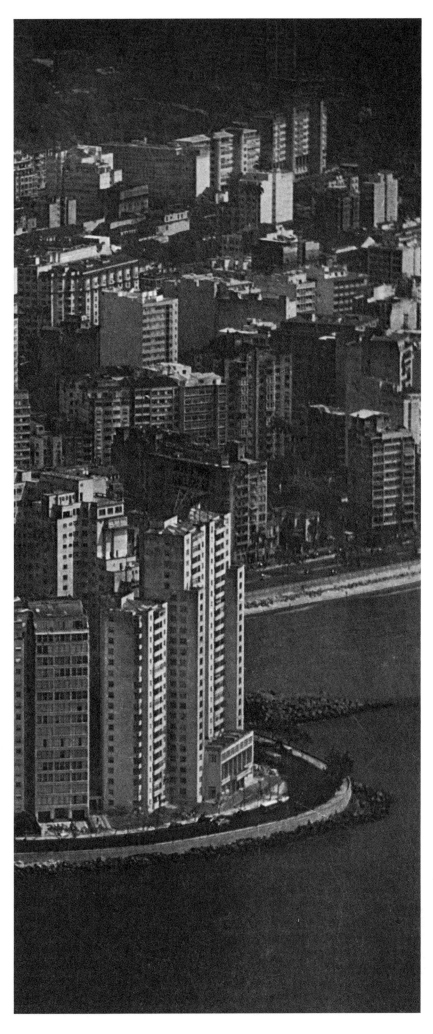

THE COUNTRIES OF SOUTH AMERICA

Argentina. The Iguaçu Falls, on the boundary between Argentina and Brazil, consists of hundreds of separate falls. They extend for more than 2 miles (3 km) across the Iguaçu River. The surrounding countryside is richly wooded, and is celebrated for its exotic plant and animal life.

ARGENTINA

Argentina occupies most of the southern part of South America. It is the second largest and second most populous country in the continent. Only Brazil is larger. Its northern regions swelter in the tropics, while in the S, the coastal areas lie just a few hundred miles from Antarctica. Although it ranks with Brazil as one of the two leading manufacturing nations of South America, Argentina's reputation and wealth are based mainly on livestock and grain.

In addition to its mainland territory, which includes disputed border lands with Chile in the W, Argentina also shares the large island of Tierra del Fuego with Chile in the S. It also claims possession of the offshore British Falkland Islands, which Argentinos call *las Islas Malvinas*. The capital of Argentina is Buenos Aires, the largest city in the Southern Hemisphere. Area 1,073,000 sq miles (2,779,000 sq km). For population, see Table on page 201.

Land and Climate Argentina is bordered on the N by Bolivia and Paraguay, on the NE by Brazil and Uruguay, and on the W by Chile. In the E and S it has a coastline on the South Atlantic.

In the N of the country, heavily wooded plains form the *Gran Chaco*. To the S and SW of Buenos Aires lie the vast stretches of the Pampas, an almost treeless expanse of fertile grasslands. To the W of the Pampas is the Andine region in the shadow of the Andes Mountains. There the people grow citrus fruits and vines, or work in the mines. In the S lies Patagonia, a bleak, windy plateau where the main sources of income are sheep and petroleum.

The towering Andes Mountains, separating Argentina from Chile in the W, include Mount Aconcagua, 22,835 ft (6,960 m) high, the highest peak in the Western Hemisphere. The main rivers are the Paraná and Uruguay. They flow into the Rio de la Plata, a wide freshwater estuary on whose banks stand two capitals—Buenos Aires and Montevideo (Uruguay). In the N, the climate is hot and humid in summer, and mild in winter. In central areas, it is mild all the

year round. Patagonia has hot summers and cold winters. Rainfall is heaviest in the W of the country.

Agriculture and Industry Argentina's agricultural wealth is based mainly on the products of the Pampas. It exports more beef than any other country, and is among the world's leading exporters of sheep and wool. Other important agricultural products include wheat, cotton, barley, rye, oats, alfalfa, linseed, sugar, cotton, and fruit. The principal minerals include petroleum, coal, lead, zinc, iron, copper, silver, and gold. Meat processing and canning is the main industry, although flour-milling, chemicals, textiles, and sugar-refining are growing in importance.

Cities and Communications The chief cities are Buenos Aires, Córdoba, La Plata, Rosario, Tucumán, and Santa Fé. About a third of the population lives in Buenos Aires and its suburbs. There is an excellent, if ageing, rail network, and waterways and airlines provide access to inland areas.

The People There are few Indians, the original inhabitants of Argentina, left. Most of the people are descendants of Spanish and Italian settlers. The official language is Spanish. About 80 per cent. of Argentinos are Roman Catholics. The most highly-developed art forms are painting and sculp-

ture, and music and dance, particularly popular dances.

Argentina. Sheep penned on a ranch in the Pampas. Though Argentina is best known for its cattle, it is also one of the world's principal sheep-rearing countries.

Bolivia is a landlocked country that straddles the Andes Mountains in the heart of South America. It is bordered by Brazil on the N and E, Paraguay and Argentina on the S, and Chile and Peru on the W. Between parallel mountain ranges in the W is the Altiplano, a high, bleak plateau where most of the people live. In the N and E there are forested lowlands, which turn to scrub and desert farther S. Lake Titicaca, the highest navigable stretch of water in the world, lies on the Peruvian border 12,507 ft (3,812 m) above sea-level. The climate is hot in the E and cooler in the W, with heavy rains from December to February. Tin, silver, lead, oil, natural gas, antimony, bismuth, and coffee are the main products. Large numbers of cattle are exported. Bolivia is rich in natural resources, but the difficult nature of the land has hindered their exploitation. Almost all the people are Indians or of mixed Indian and European blood. Most of the Indians are very poor, in spite of the fact that many work in the mines, which are the country's main source of wealth. The official language is Spanish, but Indian tongues are widely spoken. La Paz, the highest capital city in the world, is the seat of government, but Sucre is the official capital. Area 415,000 sq miles (1,075,000 sq km). For population, see Table on page 201.

Bolivia. Llamas are used as beasts of burden on the high, sparsely-vegetated plateau. Their wool—and the wool of their near relatives, alpacas and vicunas—is highly prized. They belong to the same family as the camel.

Brazil. Rio de Janeiro, on Guanabara Bay, has one of the finest sites of any of the world's cities. Beside it rises the unmistakably-shaped Pão de Açúcar —Sugar Loaf Mountain. It is believed that the city's name—meaning 'January River'—derives from the fact that the Portuguese explorer and navigator Gonçalo Coelho first sailed into the bay on January 1, 1502, and decided that it was the mouth of a hitherto-unknown river.

Brazil. A house in the Amazon forest. The great, slow Amazon River—fed by more than 500 smaller streams—flows through many almost-inaccessible regions. Most of the Indians of Amazonia live simple tribal lives. Others are more sophisticated. But many 'centres of population' marked on maps of the Amazon are only the most rudimentary settlements.

BRAZIL

In 1808, the Portuguese Prince Regent declared Rio de Janeiro the capital of Portugal. He had fled to Brazil from Portugal to escape Napoleon. In 1821, as King John VI, he returned to Portugal, leaving his son Pedro as regent of Brazil. But in 1822, Pedro declared Brazil independent of Portugal, and made himself emperor. The empire lasted until 1889, when Brazil became a republic.

Brazil is the fifth largest country in the world, and the largest country in South America both in area and in population. It occupies nearly half of South America, and much of it is still unexplored. Within its borders is the world's greatest river system—the Amazon—and some of the world's largest forests. Brazil grows more coffee than any other country and it supplies the rest of the world with many valuable forest and agricultural products. The capital is Brasília, which superseded Rio de Janeiro in 1960. Area 3,286,488 sq miles (8,511,964 sq km). For population, see Table on page 201.

Land and Climate Brazil occupies the NE and much of the central part of South America. The equator passes through the N of the country. The mouth of the Amazon is on the equator. Brazil is bordered on the N by Colombia, Venezuela, Guyana, Surinam, and French Guiana; on the S by Uruguay; and on the W by Argentina, Paraguay, Bolivia, and Peru. On the E it has a long coastline on the Atlantic.

The land can be divided into three main regions. In the N is the immense, heavily forested Amazon region. It includes highlands near the Venezuelan border. In the NE, some farming is possible on the coastal strip, but most of the area is scrub and farmers suffer severely from drought. The central and southern regions are a series of plateaux crossed by fertile river valleys. Almost half of Brazil's population lives in the S, and 75 per cent. of the country's agriculture and 80 per cent. of its industry are located there.

Some of the world's greatest rivers flow through Brazil. The Amazon discharges more water than any other river on earth and is navigable for its entire length of 1,962 miles (3,158 km) within Brazil. Other important rivers include the Negro, Madeira, Tocantins and São Francisco.

Most of Brazil lies within the tropics and, as a result, the climate is generally hot. It is humid in the Amazon region, hot and dry in the NE, and cooler and dry in the central plateaux and the S.

Agriculture and Industry Brazil is a great storehouse of mineral wealth, still largely untapped. But already it supplies much of the world's quartz crystal, beryl, sheet mica, manganese, columbium, and tantalum. It also has valuable deposits of iron, gold, diamonds, nickel, chrome, tungsten, and oil. The chief agricultural product is coffee, followed by cotton, cocoa, sugar, and soya beans, together with fruit.

Cities and Communications The chief cities are São Paulo (one of the fastest growing cities in the world), Rio de Janeiro, Belo Horizonte, Recife, Fortaleza, Porto Alegre, and Brasília. Communications are poor but improving, being confined mainly to sea and air routes. But there are 26,000 miles (41,800 km) of navigable waterways, and new highways are being carved.

The People Most of the inhabitants are descended from the early Portuguese settlers or from later arrivals such as Spaniards, Germans, Italians, Japanese, and Lebanese. There is also a relatively small number of Indians. A tenth of the people are Negroes. The official language is Portuguese.

CHILE

The Republic of Chile is like a long, narrow ribbon running down the W coast of South America. It stretches about 2,800 miles (4,500 km) from Peru in the N to within 400 miles (640 km) of Antarctica in the S. Hemmed in between the Andes Mountains and the Pacific Ocean, it is nowhere more

TIERRA DEL FUEGO

The archipelago of Tierra del Fuego consists of one large island (also named Tierra del Fuego) and several small islands at the southern tip of South America. Part of it belongs to Chile and part to Argentina. Cape Horn, the most southerly point of the continent, is on Horn Island, which is Chilean. Tierra del Fuego is separated from the mainland by the Strait of Magellan. Drake Passage separates it from the Antarctic Peninsula. Total area 27,500 sq miles (71,000 sq km).

Chile. Bleak, windswept terrain on Tierra del Fuego, the archipelago south of the Strait of Magellan.

than 250 miles (400 km) wide. On the E it borders Bolivia and Argentina. Because Chile is so long, it has many different geographic zones, each with its own climate. In the far N, where most of the copper and nitrate resources are found, there is hot desert. Farther S, this becomes semi-desert with a little vegetation and some crops under irrigation. Then comes a region of fertile valleys washed by rains in winter and warmed by hot sun in summer. This is Chile's heartland, where most of the people live. Still farther S is a region of huge lakes, forests, rivers, and mountains. This part of the country is very popular with tourists. In the far S, stretching all the way to Cape Horn, is a desolate land of wild forests, fiords, islands, and glaciers, which is frequently swept by storms. Chile's main products are copper, nitrates, iron, oil, fishmeal, wool, and fruit. The people are nearly all of European descent, mainly Spanish. There are also about 150,000 pure-blooded Araucanian Indians. Spanish is the official language. The capital and largest city is Santiago. Other large towns include Valparaíso, Antofagasta, and Concepción. Area 292,000 sq miles (756,000 sq km). For population, see Table on page 201.

Ecuador. Indian market in Ambato, a city high in the Andes. Ambato is a holiday resort and is known for its trade in fruit.

COLOMBIA

Colombia is the fourth largest country in South America. It lies in the north-western corner of the continent and has coastlines on the Caribbean Sea and on the Pacific. It is bordered by Venezuela and Brazil on the E, and by Peru and Ecuador on the S. On the NW it has a short border with Panama. Four great mountain ranges, all part of the Andes, run parallel N from the border with Ecuador almost to the Caribbean coast. To the W of them is the Pacific, and to the E the country's capital, Bogotá. Between the ranges are high, narrow valleys where most of the people live. The mountains seriously hamper transport and communication. As a result, air transport is important. To the S and E of the country are grassy plains called *llanos*. They eventually merge into the tropical rain forests of the Amazon basin. The largest river is the Magdalena. It flows N for almost 1,000 miles (1,600 km) to empty into the Caribbean Sea. Its chief tributary, the Cauca, is the second largest river. The fertile river valleys are rich farming regions. Climate varies with altitude. It is tropical in the steaming jungle lowlands, and freezing above the snowline of the Andes. The chief products are oil, gold, emeralds, coal, platinum, coffee, meat, and bananas. Colombia supplies almost all the world's emeralds. Many Colombians are of mixed blood. About a fifth of them have European ancestry, but there are also nearly 400 Indian tribes. The official language is Spanish. Bogotá is the largest city; Barranquilla and Cartagena are the next most important towns. Area 439,737 sq miles (1,138,914 sq km). For population, see Table on page 201.

The port of Santa Marta, on Colombia's N coast, dates from 1525 and claims to be the oldest colonial town in South America.

Ecuador is one of the smallest countries in South America. Its name means 'equator', and it is so named because it straddles the equator on the NW coast of the continent. Ecuador is bordered on the N by Colombia, on the E and S by Peru, and on the W by the Pacific Ocean. It is a land of mountains, valleys, and jungles, with many towns perched high up in the clouds. Ecuador's mountainous 'backbone' is made up of two main ranges of the Andes. These run from NE to SW for about 425 miles (680 km). A great valley, about 30 miles (48 km) wide, lies between them. Most of the people live there. The mountains include more than 30 volcanoes, many of them active. Cotopaxi, 19,344 ft (5,896 m) high, is the highest active volcano in the world. Rivers flow eastwards down the mountain slopes to join the Amazon system or westwards to empty into the Pacific. To the E of the mountains lies the *Oriente,* a sparsely populated, jungle-clad region of lowlands. To the W is the Pacific lowland, about 60 miles (96 km) wide. Climate varies with altitude rather than with the seasons. It ranges from the steaming tropics of the lowlands to the freezing air of the high mountain peaks. The main products are bananas, cocoa, coffee, oil, straw ('Panama') hats, fish, gold, and silver. Ecuador exports more bananas than any other country. But oil, discovered in the eastern lowlands, may soon prove a more important source of income. About 40 per cent. of the people are of pure Indian blood. Slightly more than this have mixed Indian and Spanish blood. Most of the remainder are of Spanish descent. Spanish is the official language. Quito is the country's capital and Guayaquil the main port. Area, 109,500 sq miles (283,600 sq km). For population, see Table on page 201.

ECUADOR

Ecuador. The city of Quito, in the Andes, is Ecuador's capital and one of the world's highest cities. Though it lies almost on the equator, its climate is relatively mild because of its altitude.

city. Area 83,000 sq miles (215,000 sq km). For population, see Table on page 201.

PARAGUAY

Paraguay is a landlocked republic in the centre of the South American continent. It is surrounded by Brazil, Argentina, and Bolivia. The River Paraguay, flowing N and S, cuts the land in two. To the W lies the *Chaco,* a region of jungle, scrub, and marshes. Immediately to the E of the river is a fertile area of gently rolling grasslands that rises to a high, heavily forested plateau. Almost all of the people live in the region to the E of the river. The climate is subtropical, with mild, fairly dry winters. Paraguay has fought two disastrous wars within the last 100 years and, as a result, the country's development has been hindered. The economy depends almost entirely on crops and livestock. Major exports are timber, *quebracho* (a wood extract used for tanning), meat, hides, and *yerba mate* (Paraguayan tea). Spanish is the official language, but most people also speak Guaraní, the language of the conquered Indians. The only large city is Asunción, the capital. Area 157,000 sq miles (406,600 sq km). For population, see Table on page 201.

GUYANA

Paraguay. Traders reach a bargain on the bank of the Paraguay River, near Asunción. Paraguay, which under Jesuit rule in the 1600s and 1700s was one of the most advanced regions of South America, is today backward and impoverished.

Guyana, formerly British Guiana, is located on the NE coast of S America. It is bordered by Surinam, Brazil, and Venezuela. On the N it has a coastline on the Atlantic. The land is hot and humid, with dense forests. The people are a mixture of East Indians, American Indians, Chinese, Europeans, and Africans. The official language is English. Georgetown, the capital, is the only large

PERU

Above:
Peru. The ruins of the Inca stronghold of Sacsahuamán, at Cuzco. It was built in the 1400s, and was captured in the 1530s by the Spaniards besieged in Cuzco by Manco Capac, the last Inca emperor.

Left:
Peru. Life in a sun-baked village. In most rural communities, ponchos (capes) and blankets —woven locally to traditional designs—add colour to the scene.

Peru. Steps from the past. The Incas built in granite and other stones, fitting the unmortared blocks together with extra-ordinary precision.

Peru, the third largest country in South America, lies on the W coast of the continent. It is a land of mountains, deserts, and jungles. Peru is bordered on the N by Ecuador and Colombia, on the E by Brazil and Bolivia, and on the S by Chile. On the W, Peru has a 1,400-mile (2,250-km) coastline on the Pacific Ocean. A narrow desert coastal area, sandwiched between the sea and the mountains inland, contains most of the country's major towns. The mountainous 'backbone' of Peru is made up of several ranges of the Andes Mountains, running N and S, and divided by deep valleys. To the E of the mountain peaks, the lower slopes flatten out into plains and heavily forested, largely unexplored stretches of territory. More than 50 short, swift rivers flow across the coastal plains from the mountains into the ocean. But the two longest rivers, the Marañón and Ucayali, flow on the eastern side of the Andes and are among the headwaters of the Amazon. Lake Titicaca, on the border with Bolivia, lies 12,507 ft (3,812 m) above sea-level. It is the highest navigable body of water in the world. Peru's climate ranges widely, from the icy, snow-covered peaks of the Andes to the almost tropical climate of many of the valleys. The Humboldt Current, flowing northwards from the Antarctic, brings a cold touch to Peruvian shores. Amounts of rainfall vary from 50 inches (1,270 mm) a year to 10 inches (254 mm) in the dry southern part of the country. Peru is rich in antimony, copper, bismuth, gold, lead, silver, zinc, and other minerals. As a result, mining and the processing of minerals are major industries. Minerals, chemicals, and fertilizers (especially fishmeal)—together with some sugar, cotton, wool, and coffee—form the bulk of the exports. About half of the people are Indians, a third are *mestizos* (part Indian and part European), and a small group are *criollos.* Criollos are descendants of the original Spanish settlers, and most of Peru's wealth is still in their hands. The official language is Spanish, but Quechua, the language of the conquered Incas, is also widely spoken. The main cities are Lima, the capital, Callao, Arequipa, Trujillo, Chiclayo, and Cuzco (the old Inca capital). Area

496,000 sq miles (1,285,000 sq km). For population, see Table on page 201.

Uruguay is one of the smallest republics of South America. Its official name, the *Oriental Republic of Uruguay,* refers to its location on the oriental or eastern bank of the estuary of the Rio de la Plata and the River Uruguay. The country lies on the SE coast of the continent. It is bordered by Brazil on the N, and by Argentina on the W. In the E and SE it has a coastline on the Atlantic Ocean. The land is made up of broad, rolling, grassy plains, low hills, and clumps of trees on the banks of the many rivers and streams. Uruguay is unique among South American countries in that virtually the whole of its land area is populated, and there is no great contrast between one part of the country and another. The main rivers are the Rio de la Plata and the Uruguay, which form the boundary with Argentina. Another major river is the Negro, which flows through the centre of the country. The largest of many shallow lakes and lagoons is Lake Mirím, which is about 110 miles (177 km) long and 25 miles (40 km) wide. It lies partly in Uruguay and partly in Brazil. The climate throughout the land is temperate, pleasant, and healthy. Uruguay's greatest natural resources are its fine climate and rich soil. Agriculture is the main industry, and there are more cattle and sheep in Uruguay than there are people. About nine-tenths of the land is devoted to rearing livestock. The remainder is used for growing crops. Cattle are herded by colourful cowboys called *gauchos.* The main exports are wool, beef, and hides. There are also large plants for processing agricultural products. Uruguay has little in the way of mineral resources. Marble and agate are among the few worth mining. The people are almost all of European ancestry, mainly Spanish and Italian. Spanish is the official language. The only large city is Montevideo, the capital. Area 68,536 sq miles (177,507 sq km). For population, see Table on page 201.

Venezuela, the richest country in South America, derives its wealth from its petroleum deposits. Its name means 'Little Venice'. It was given this name by Spanish explorers in 1499. They came across an Indian village built on stilts in the middle of Lake Maracaibo and were reminded of Venice. Venezuela lies on the N coast of South America, facing the Caribbean Sea. It is bordered on the E by Guyana, on the S by Brazil, and on the W by Colombia. The land

URUGUAY

has four main regions. To the S of the river Orinoco lie the Guyana Highlands. This wild, sparsely populated region covers more than half the country. The *llanos* (flat grasslands) of the Orinoco occupy a huge central plain. The third region is made up of the Maracaibo lowlands. This hot, humid area fringes Lake Maracaibo—a vast, shallow, freshwater lake open to the sea. Its bed is the source of Venezuela's petroleum. The Andean or Venezuelan Highlands make up the fourth region. They are an offshoot of the Andes and run across the NW corner of the country. The Orinoco, some 1,600 miles (2,575 km) long, is the eighth longest river in the world. Other major rivers include the

Venezuela. A street in Caracas, named after 'the Liberator', Simón Bolívar, who was born in the city. Although very near the Caribbean coast, Caracas stands some 3,100 ft (945 m) above sea-level.

Caroni, Caura, and Apure. Angel Falls, with a drop of 3,212 ft (979 m), is the highest waterfall in the world. Venezuela lies in the tropics, and the temperature varies more between night and day and according to altitude than between seasons. Low-lying regions are hot and humid. The country's major resource is petroleum, and it exports more oil than any other country. But Venezuelans are also increasing their exploitation of other mineral resources. Among these are iron, asbestos, bauxite, asphalt, nickel, and lead. Farmers produce only enough food for themselves and their families, although cattle are reared on the llanos. Most of the people are of European descent. There are also some blacks and American Indians. The official language is Spanish. The chief town is Caracas, the capital, and the chief port is La Guaira. Area 352,145 sq miles (912,051 sq km). For population, see Table on page 201.

VENEZUELA

AUSTRALASIA

Australasia is the region that includes Australia, New Zealand, and Papua New Guinea. But the term *Australasia* is not a precise one: it can be defined in several different ways. Sometimes it is used in a way that includes a number of the countries of Asia—such as Malaysia and Indonesia—as well as Australia, New Zealand, and Papua New Guinea. Sometimes it includes some of the more important Pacific island groups, or even territories in Antarctica. To zoologists and botanists, Australasia may mean all the land and sea areas east and south of Wallace's Line—an imaginary line drawn N—S through the Celebes Sea and between Borneo and Bali, on one side, and Sulawesi (Celebes island) and Lombok, on the other. The line is named after a British naturalist, Alfred Wallace, and it is said that plant and animal life on one side of it differs significantly from life on the other.

Gateway to a Continent. Sydney, the oldest and biggest city in Australia. The beautiful harbour has an area of more than 20 sq miles (50 sq km).

Woolly Riches. Sheep-shearing in New Zealand. Sheep-farming is the country's main industry.

COUNTRIES OF AUSTRALASIA POPULATIONS AND MONETARY UNITS		
Country	*Population*	*Monetary Unit*
Australia	13,310,000	Dollar
New Zealand	2,980,000	Dollar
Papua New Guinea	2,600,000	Dollar

The only country that is also a continent, Australia has an area of 2,968,000 sq miles (7,687,000 sq km). For population, see Table on page 217.

Australia lies to the SE of Asia. Its W coast is washed by the Indian Ocean and its E coast by the Pacific. The country is a monarchy, acknowledging Queen Elizabeth II of the United Kingdom as queen of Australia. She is represented by a governor-general, who acts as head of state. Australia is a federation of six states, each with its own government and legislature, plus two territories and some overseas territories. The federal government deals with national and international matters. The federal capital is Canberra. The country's official name is the Commonwealth of Australia.

Land and Climate A mountain chain, the Great Dividing Range, lies close to Australia's E coast. This range traps the rain-bearing winds that blow in from the Pacific Ocean. As a result, the E coastal area is fertile and green. Part of the land to the W of the Great Dividing Range is also fertile, and is watered by three great rivers, the

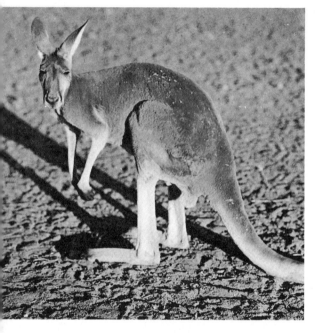

Australia. Above: The famous 'sails' of Sydney Opera House, at Bennelong Point in Sydney Harbour. They cover two main auditoriums. The Opera House, which incorporates several innovations in theatre design, was planned by the Danish architect Joern Utzon. Left: The kangaroo, the largest of the marsupials. It bounds along on its powerful hind legs at speeds of up to 25 m.p.h. (40 k.p.h.). Marsupials are found only in Australasia and the Americas. Below: Ayers Rock, south-west of Alice Springs in central Australia, rises 1,100 ft (335 m) above the plain. It is more than a mile and a half (2.5 km) long.

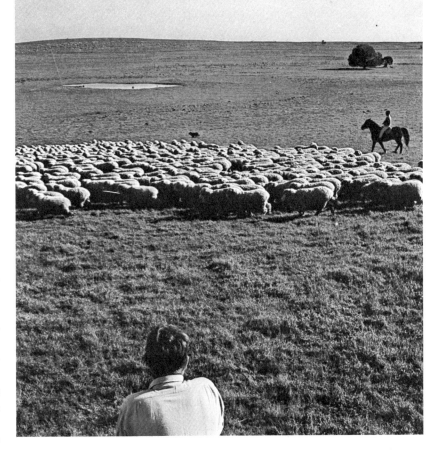

Murrumbidgee, the Darling, and the Murray. The central part of Australia is a vast arid desert, where few plants can grow and no people live. It covers almost two-thirds of the continent. Alice Springs, a town in the centre of the country, is built in one of the few places with enough rainfall to support life.

A large part of Australia to the W of the Great Dividing Range is watered by artesian wells—wells that draw on underground sources of water. The water is full of minerals, but sheep and cattle can drink it.

Along the S coast is another dry desert area, the Nullarbor Plain. The name *Nullarbor* comes from Latin words meaning 'without trees'. The N coastal area is lined with tropical woodlands, parts of the NE coast having dense tropical rain forest. Australia's SW region is also fertile, particularly around Perth.

Australia has a great many lakes, but nearly all of them are dry for part of the year, as are the streams that feed them. When they are full, their water is salty. The largest is Lake Eyre, in South Australia, 39 ft (12 m) below sea-level.

There are several islands off the coast of the Australian mainland. The largest is Tasmania, 124 miles (200 km) to the SE across the Bass Strait. Tasmania is one of the states of the Commonwealth of Australia. It has plenty of rain. In its forested mountains there are many rare animals.

Animals that people particularly associate with Australia are the marsupials. These animals bear their young only partly formed, and rear them in pouches on their mothers' abdomens. Marsupials include the kangaroo, the koala, the wombat, the phalangers, and also the Tasmanian wolf. Two other Australian animals, the duck-billed platypus and the spiny anteater, are mammals that lay eggs. Other animals found only in Australia include black swans, and two flightless birds, the emu and the cassowary.

Agriculture and Industry Farming is the chief source of Australia's wealth. There are more sheep than in any other country, and also great numbers of cattle. Australia exports meat, wool, and dairy products—mainly butter and cheese. It also produces a great amount of wheat, as well as sugar-cane and fruits. Tasmania has long been famous for its apples. The continent is extremely rich in minerals, and new deposits are being found every year. Australia produces and exports gold, lead, silver, and zinc, and has rich reserves of iron, coal, petroleum, natural gas, copper, uranium, and tin.

Australia. Above: Herding sheep. Australia, the world's leading sheep-raising country, produces more than one-quarter of the world's wool. Most of the country's sheep are Merinos. Right: Australian Aborigine. The Aborigines are thought to have arrived in Australia about 16,000 years ago. They moved around from place to place, and lived by hunting, fishing, and plant-gathering. Today, few follow their traditional way of life. Most live on special reserves.

Australia. The Macdonnell Ranges extend for about 100 miles (160 km) across central Australia. Some peaks rise more than 4,000 ft (1,200 m). The ranges are celebrated for their strangely-shaped chasms and bluffs and for their colour changes.

AUSTRALIA: AREAS WITH ARTESIAN WELLS

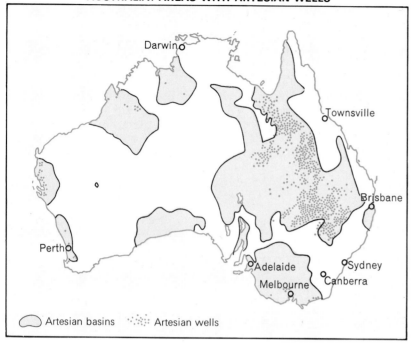

Artesian basins ⠒⠒ Artesian wells

Australasia. Statue of James Cook (1728-1779) in Sydney. Captain Cook, the greatest explorer of the Pacific Ocean, made three long voyages of discovery. He charted the coasts of New Zealand and the eastern coast of Australia, and visited many hitherto 'unknown' islands. He was killed by Hawaiian islanders while trying to recover a stolen boat.

AUSTRALIA, NEW ZEALAND: AGRICULTURE

■ Dairy cattle
▨ Dairy cattle and sugar cane
☐ Wheat growing and sheep raising
☐ Sheep farming
▨ Beef cattle
■ Forestry products
☐ Unproductive areas: desert and mountains

Manufacturing industry has grown rapidly since World War II, and employs more than a quarter of the country's workers. But Australia still imports a high proportion of its manufactured goods.

Cities and Communications Although Australia is renowned for its wide open spaces—the 'outback'—and its great cattle stations, most of its people live in towns and cities. Sydney, in New South Wales, with a population of some 3 million, is the largest city, followed by Melbourne, Brisbane, Adelaide, and Perth.

There are more than 25,000 miles (40,000 km) of rail lines, but the tracks were laid in three different gauges, which makes rail communication difficult on some routes. Many broad and narrow gauge lines have been converted to standard gauge since 1957.

Australians make much use of airlines and private airplanes to link the big cities and the lonely parts of the outback. Medical treatment is provided by the Flying Doctor Service. Remote stations are linked by radio, and children living outside the cities receive much of their education by radio.

The People The original inhabitants of Australia were the Aborigines, dark-skinned people related to some of the people of S India and Malaysia. About 40,000 live in Australia today. Most of the people are descended from settlers from the British Isles, but since World War II more than a million people from continental Europe have emigrated to the country. Under the 'White Australia' policy the government discouraged Asians from settling in the country, but since 1958 this restriction has been considerably relaxed.

The language of Australia is English, but some immigrants from countries outside the English-speaking world still use their own languages among themselves.

NEW ZEALAND

New Zealand, in the South Pacific, lies about 1,200 miles (1,900 km) SE of Australia across the Tasman Sea. It consists of two main islands, the North Island and the South Island, together with Stewart Island, the Chatham Islands, and outlying small islands. The country's capital is Wellington, on the North Island. Its area is 103,736 sq miles (268,675 sq km). For population, see Table on page 217.

Land and Climate The two main islands extend more than 1,000 miles (1,600 km) from N to S. They are separated by the Cook Strait. Their coastlines are irregular and deeply indented in places. Both are mountainous and have fertile coastal plains. The high central plateau of the North Island is of volcanic origin. Its active peaks include Mount Ruapehu (9,175 ft; 2,797 m) and Mount Ngauruhoe (7,515 ft; 2,290 m). The glaciated mountains of the Southern Alps form the backbone of the South Island. Mount Cook (12,349 ft; 3,764 m), at the centre, is New Zealand's highest peak. To the E of the Southern Alps is the Canterbury Plain.

New Zealand has many lakes, the largest being Lake Taupo (238 sq miles; 616 sq km) on the North Island. From it flows the Waikato (270 miles; 435 km), the longest river. The Rotorua area, near the Bay of Plenty on the North Island, is famous for its hot springs and geysers. The country's climate varies from sub-tropical in the far N to temperate in the S.

Agriculture and Industry New Zealand's agriculture is among the world's most efficient and productive. Farm products—mainly meat, dairy produce, and wool—account for most of the country's exports.

New Zealand. Queenstown and Lake Wakatipu, in the South Island. The lake is about 50 miles (80 km) long, and its water level changes considerably. Queenstown is the South Island's most popular tourist resort.

New Zealand. Wellington, the country's capital and second-largest city, on the southern coast of the North Island. Much of it is built on hills overlooking Port Nicholson, usually called Wellington Harbour. The principal business area, however, stands on land reclaimed from the harbour.

Industries include food processing, and the manufacture of textiles, paper, and steel. There are coal, iron, limestone, and oil deposits.

Cities and Communications New Zealand's chief cities include Auckland, Christchurch, Wellington, Dunedin, and Napier-Hastings. Road and rail systems are highly developed. Internal air services link the country's main centres.

The People Nine out of ten New Zealanders are descended from British settlers. About 8 per cent. are Maoris, the country's original Polynesian inhabitants. The country's language is English, but the Maoris also speak Maori. Two-thirds of the people live on the North Island.

PAPUA NEW GUINEA

The country of Papua New Guinea, formerly a trust territory of Australia, consists of the E half of the island of New Guinea and a number of nearby islands. The W half of New Guinea is West Irian, part of Indonesia. Lying just N of Australia across the Torres Strait, Papua New Guinea forms part of Melanesia. It is a wild region of mountains, jungles, and swamps. Mount Wilhelm, in the Central Range of New Guinea, rises to 15,400 ft (4,694 m). Most of the people live by subsistence farming, but there are rich mineral resources. The cap. is Port Moresby. Area 183,000 sq miles (474,000 sq km). For population, see Table on page 217.

PACIFIC ISLANDS

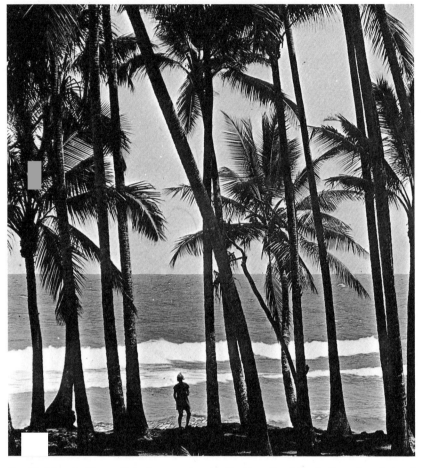

The Pacific Ocean, the largest of the oceans, also has the largest number of islands —more than 30,000 of them. Most are in the South Pacific, the part of the ocean that lies S of the equator. The South Pacific islands together with the islands of the central Pacific are often known as *Oceania*.

Within Oceania there are three main island groups: Melanesia, Micronesia, and Polynesia. These groups are based on the physical characteristics of the original islanders.

Melanesia ('the black islands') is a large island group in the SW Pacific. It includes New Guinea, the Bismarck Archipelago, the Solomon Islands, New Caledonia, and the Fiji Islands. The people of Melanesia are dark-skinned and have frizzy hair. Generally, their islands are hot and wet.

Micronesia ('the little islands') lies NE of Melanesia, and includes the Bonin, Mariana, Caroline, Marshall, and Gilbert Islands. Between the Mariana and Caroline Islands is the Marianas Trench, which has the lowest known point on the earth's surface. The people of Micronesia are copper-coloured and have straight black hair.

Polynesia ('many islands') includes many widely-scattered islands and archipelagos. The chief are the Hawaiian, Line, Marquesas, Ellice, Phoenix, Tokelau, Cook, Society, Tubuai, and Kermadec Islands. New Zealand, too, is usually included in Polynesia. The indigenous peoples of Polynesia are tall and black-haired, and have light skins.

Geologically, the Pacific islands are of two principal kinds. Some are *volcanic islands*, formed by subterranean action, and are mountainous and have fertile soil. Others are *atolls*, formed by coral growth.

Pacific Islands. Above: Palm trees and surf in the Hawaiian Islands. Hawaii, a chain of more than 20 islands, is a state of the United States. Below: Ancient stone carvings on Easter Island. The island has more than 600 of such figures, some more than 40 ft (12 m) high. Their origin is unknown.

Pacific Islands. A village in Fiji, an island group consisting of two large islands and some 300 small islands. Fiji lies just west of the International Date Line. About half of its people are immigrants from India.

Below right:
Pacific Islands. An ancient tomb on the island of Upolu in Western Samoa, reminiscent in style of monuments in South America. There is much evidence of links between Polynesia and Asia and South America.

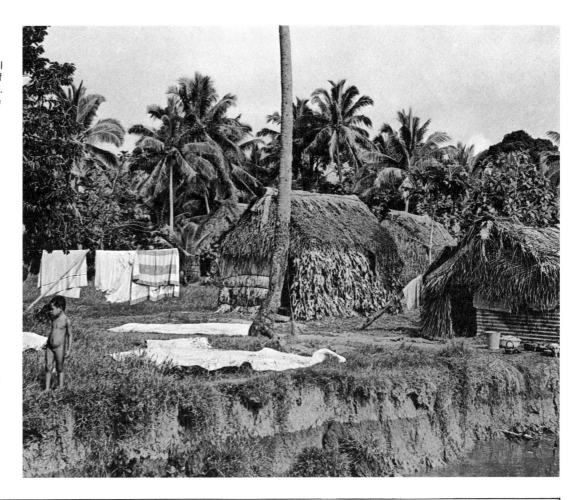

This table lists the principal Pacific islands and island groups, excluding those—such as New Zealand, Indonesia, and Japan—that have individual entries. The islands listed vary in status. Some, such as Nauru and Western Samoa, are independent states controlling their own affairs. Most others are dependencies of large countries, chiefly Australia, France, New Zealand, the United Kingdom, and the United States. Since 1959, the Hawaiian Islands, in Polynesia, have been a state of the United States.

Name	Area		Population
	sq miles	sq km	
ISLANDS OF MELANESIA			
Bismarck Archipelago	22,920	59,363	169,000
D'Entrecasteaux Is.	1,200	3,108	40,000
Fiji	7,040	18,234	518,000
Louisiade Archipelago	600	1,550	10,000
New Caledonia	7,340	19,010	100,000
New Hebrides	5,700	14,763	75,000
Schouten Is.	1,230	3,186	33,000
Solomon Is.	16,120	41,750	173,000
Trobriand Is.	170	440	10,600
ISLANDS OF MICRONESIA			
Bonin Is.	40	104	220
Caroline Is.	500	1,295	59,735
Gilbert Is.	166	430	44,000
Mariana Is.	452	1,170	78,275
Marshall Is.	70	181	18,205
Nauru	8	21	6,800
Wake I.	3	8	1,097
ISLANDS OF POLYNESIA			
Chatham Is.	372	963	300
Cook Is.	93	241	21,000
Easter I.	46	119	1,000
Ellice Is.	14	36	5,100
Gambier Is.	12	31	2,000
Kermadec Is.	13	34	10
Line Is.	240	622	1,000
Marquesas Is.	480	1,243	3,000
Midway Is.	2	5	2,350
Pitcairn I.	2	5	90
Phoenix Is.	19	49	1,200
Samoa, American	76	197	27,800
Samoa, Western	1,097	2,841	141,000
Society Is.	650	1,680	40,000
Tokelau Is.	4	10	1,700
Tonga (Friendly) Is.	270	700	87,000
Tubuai Is.	115	298	3,600
Wallis Is.	40	104	1,000

223

POLAR REGIONS

The Arctic. Igloos, the snow houses in which Eskimoes spend the winter. They are built of blocks of frozen snow, 'mortared' together with loose snow. In summer, Eskimoes live in tents, which they also call igloos.

Opposite page: Antarctica. Penguins, the inquisitive, endearing birds of the icy South. They use their wings to swim in the cold sea, and live by catching fish.

THE ARCTIC

The Arctic is the cold region around the North Pole, extending from the Pole to the Arctic Circle, latitude 66.5°N. At its centre is the Arctic Ocean, which has an area of some 5.5 million sq miles (14 million sq km). Bordering the Ocean are the most northerly parts of Europe, Asia, and North America. Within North America, the greater part of the huge, icy island of Greenland (a territory of Denmark) lies N of the Arctic Circle. Though Greenland is always icy, most of the Arctic lands have no ice or snow in summer. But much of the Arctic Ocean is covered with a permanent sheet of ice, called the *ice pack*. Other parts of the Ocean are frozen only in winter. In spring, the ice starts to melt and forms *floes*—ice islands, which, in turn, are broken up to form *icebergs*. Icebergs are also formed by glaciers entering the Ocean from the Arctic lands.

During the winter there is little sunshine in Arctic regions. But from March to September there is sunshine every day, and on at least one day there is no darkness at all.

Sometimes the term 'Arctic' is used to include regions farther from the Pole than the Arctic Circle—all regions N of the *tree line*, the line N of which forests do not grow. This definition limits the animal life of the Arctic, since many animals (including many birds) are found only where there are trees.

Just S of the Arctic is the *subarctic*, a region that is equally cold in winter, but is warmer in summer. Both the Arctic and the subarctic are within the area of *permafrost*, the regions in which the ground is frozen for some distance down even though the topsoil may thaw during the summer months.

Despite the severe conditions, many people live in the Arctic. They include the Lapps (who live in Lapland—northern Norway, Sweden, and Finland), the Yakuts and other peoples in the U.S.S.R., and the Eskimoes in Greenland, Alaska, and Canada.

MINERAL RESOURCES OF THE ARCTIC REGIONS

- - - - Southern limit of the Arctic Region
———— Southern limit of the sub-Arctic zone

Coal	Lignite	Iron ore
Manganese	Nickel	Vanadium
Molibdenum	Cobalt	Tin
Tungsten	Titanium	Lead
Silver	Platinum	Zinc

Chrome
Copper
Gold
Gold-bearing areas

PETROLEUM
Sedimentary basins
Recently prospected areas
Active oil fields

ANTARCTICA

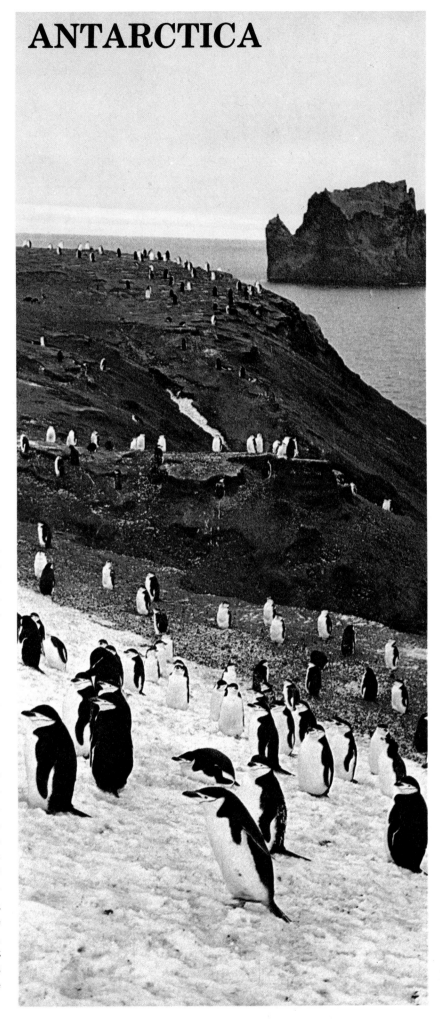

Antarctica, the frozen continent around the South Pole, lies under a great blanket of snow and ice that is a mile deep in many places. The continent is larger than Europe or Australia, but most of it is without life of any kind, animal or plant. Its surface is irregular, with wild hills, enormous glaciers —rivers of ice—and seemingly limitless snow plains buffeted by swirling winds and blizzards. The surrounding seas are sections of the Atlantic, Pacific, and Indian Oceans and are dangerous to shipping, when not actually impassable. In winter they are frozen over with pack ice, and in summer are a heaving mass of icebergs and floes.

Many kinds of animals live in the coastal waters and on the shores. They include various species of fish, whales, and seals, and penguins and other sea birds. A few plants and some insects and spiders also manage to exist in coastal areas. It is believed that the continent is rich in mineral resources, though much of it has yet to be explored. Several countries, including Australia, Argentina, and Norway, claim territory in Antarctica, and many have research stations there. As in the Arctic, there is at least one day a year on which the sun does not shine, and one on which it does not set. The land area of the continent is about 5 million sq miles (13 million sq km).

Antarctica is roughly onion-shaped, and has two huge bays, the Weddell Sea and the Ross Sea. They are so choked with ice on their southern sides that their precise extent is not always clear. The edges of the continent are marked by bare mountains and cliffs, some of which form part of a mountain chain that divides Antarctica into two parts. The smaller part is called *West Antarctica.* From it a long, mountainous tongue of land, the Antarctic Peninsula, extends towards Cape Horn, the tip of South America. It is thought to be part of the Andean system of that continent. At the southern end of the peninsula is the Vinson Massif (16,860 ft; 5,139 m), the highest peak in Antarctica.

Research has shown that under its ice cap much of West Antarctica is below sea-level—that is, that it is not, as it seems to be, a continuous land mass. *East Antarctica,* almost twice the size of West Antarctica, faces Africa, Asia, and Australia. A vast and relatively smooth snow plateau is known to conceal a rugged and mountainous terrain.

The mountains and valleys are hidden under ice and snow several miles thick. The South Pole is in this part of the continent, about 300 miles (480 km) from the great trans-antarctic mountain chain. It was reached for the first time in 1911 by the Norwegian explorer Roald Amundsen.

AVERAGE HEIGHT OF CONTINENTS

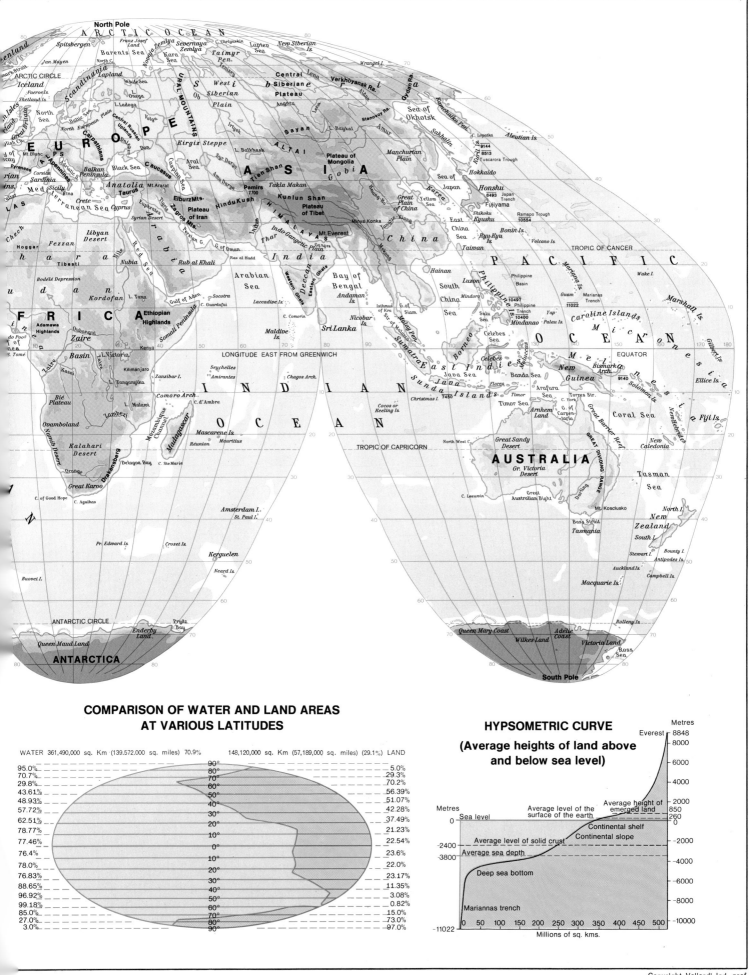

COMPARISON OF WATER AND LAND AREAS
AT VARIOUS LATITUDES

WATER 361,490,000 sq. Km (139.572.000 sq. miles) 70.9%   148,120,000 sq. Km (57,189,000 sq. miles) (29.1%) LAND

Water	Latitude	Land
95.0%	90°	5.0%
70.7%	80°	29.3%
29.8%	70°	70.2%
43.61%	60°	56.39%
48.93%	50°	51.07%
57.72%	40°	42.28%
62.51%	30°	37.49%
78.77%	20°	21.23%
77.46%	10°	22.54%
76.4%	0°	23.6%
78.0%	10°	22.0%
76.83%	20°	23.17%
88.65%	30°	11.35%
96.92%	40°	3.08%
99.18%	50°	0.82%
85.0%	60°	15.0%
27.0%	70°	73.0%
3.0%	80°	97.0%
	90°	

HYPSOMETRIC CURVE
(Average heights of land above and below sea level)

Metres
Everest 8848
8000
6000
4000
Average height of emerged land 850
2000
260
Sea level 0
Average level of the surface of the earth
Continental shelf
Continental slope
-2000
Average level of solid crust -2400
Average sea depth -3800
-4000
Deep sea bottom
-6000
-8000
Mariannas trench
-10000
-11022

0 50 100 150 200 250 300 350 400 450 500
Millions of sq. kms.

MAP 2   227

MAP 3

228

Copyright: Vallardi Ind.Graf.

MAP 4

229

MAP 5

Inset maps:

GANGES DELTA
0   50   100 Kms.
0   50 S.M.

TŌKYŌ
0   10   20   30 Kms.
0   20 S.M.

CANTON, MACAO AND HONG KONG
0   10   20   30 Kms.

SINGAPORE
0   10   20   30 Kms.
0   20 S.M.

MAP 6

Copyright, Vallardi Ind. Graf.

NORTH AMERICA

Scale 1:34 000 000

0   250   500   750   1000   1250   1500 Kms.

0   250   500   750   1000 St.mls.

West from 100 Greenwich

Copyright: Vallardi Ind. Graf.

SOUTH AMERICA

A 90 B 80 C 70 D 60 E 50 F 40 G 30 H

30

UNITED STATES
Tampa
Gulf of Mexico
Gr. Bahama I.
Miami
Gt. Abaco I.
Nassau
S. Salvador or Watling I.
1
ATLANTIC
Tropic of Cancer
HAVANA
Sta. Clara
CUBA
Mérida
I. of Pines
Camagüey
Guantanamo
Yucatan Str.
20
Cayman Is.
(Br.)
Santiago de CHAITI
Port
DOMINICAN
Santo
Virgin Is.
20
MEXICO
JAMAICA
au Prince
Domingo
S. Juan
(U.S.A. - Br.)
Kingston
Puerto Rico (U.S.A.)
Belize
Guadeloupe (Fr.)
BELIZE
Greater
Antilles
Dominica (Br.)
GUATEM
Caribbean Sea
Martinique (Fr.)
HONDURAS
St. Lucia (Br.)
GUATEMALA
Tegucigalpa
Lesser Antilles
BARBADOS
San Salvador
NICARAGUA
GRENADA
2
EL SALVADOR
Bluefields
G. of Venezuela
Curaçao (Neth.)
Margarita
Port of Spain
Managua
L. Nicaragua
Bonaire (Neth.)
TRINIDAD
COSTA RICA
BARRANQUILLA
Maracaibo
CARACAS
Cumaná
AND TOBAGO
San José
Colón
Cartagena
Valencia
Panama Canal
Maracaibo
S. Fernando
Orinoco
Ciudad Bolívar
10
PANAMA
Zone
Barquisimeto
de Apure
Georgetown
10
G. of
S. Cristóbal
VENEZUELA
GUYANA
Paramaribo
Panama
MEDELLIN
Bucaramanga
Cayenne
Manizales
BOGOTÁ
S. Fernando
SURINAM
GUIANA
Cocos I.
de At.
(Neth.)
(Cost.)
Buenaventura
COLOMBIA
Rio Branco
Amapá
St. Peter and
Malpelo
CALI
Neiva
Guaviare
Macapá
Mouths of
St. Paul Rocks
(Col.)
Popayán
S. Carlos
the Amazon
(Braz.)
3
Pasto
Uaupés
Marajó
Equator
Mitú
Negro
Moura
Óbidos
Quito
Japurá
BELÉM
São Luís
ECUADOR
Napo
Manaus
Pará
Parnaíba
GUAYAQUIL
Cúcuta
Iquitos
Leticia
Santarém
Cariús
FORTALEZA
Rocas I.
Fernando
Galapagos Is.
Marañón
Amazon
Teresina
de Noronha
Isabela
Sta. Cruz
Itaituba
Maranhão
Natal
(Braz.)
(Ec.)
Piura
Manicoré
Xingu
Carolina
Piauí
João Pessoa
Chiclayo
Cajamarca
Cruzeiro do Sul
BRAZIL
Conceição do Araguaia
Campina
RECIFE
4
Trujillo
Porto Velho
Humaitá
Porto Nacional
Grande
Maceió
Rio Branco
Barra
Aracajú
Cerro de Pasco
Guajará Mirim
Ilha do
Palma
Bahia
SALVADOR
Callao
Riberalta
Bananal
10
LIMA
Huánuco
Rondônia
Guaporé
Magdalena
Ilhéus
10
Pisco
Ica
Trinidad
Mato
Goiânia
Brasília
Huancavelica
BOLIVIA
Grosso
Arequipa
Puno
La Paz
Cuiabá
Piraporã
Caravelas
Mollendo
L. Titicaca
Cochabamba
Puerto Suarez
Corumbá
Uberaba
Theofilo Otoni
Arica
Oruro
Sucre
Paranaíba
Diamantina
5
Iquique
Potosí
Campo Grande
Uberaba
BELO HORIZONTE
Calama
Uyuni
PARAGUAY
Pto. Casado
Bauru
Vitoria
20
Antofagasta
Jujuy
Paraná
Campos
20
Embarcación
Concepción
CURITIBA
S. PAULO
RIO DE JANEIRO
Trindade
Martin Vaz Is.
Copiapó
Salta
Asunción
Villarrica
Santos
Niteroi
(Braz.)
(Braz.)
Tucumán
Resistencia
Encarnación
Blumenau
La Serena
Santiago
Corrientes
Florianópolis
del Estero
Catamarca
Passo Fundo
La Rioja
Santa Maria
PORTO ALEGRE
CÓRDOBA
Santa Fé
Concordia
Lagoa dos Patos
30
Valparaiso
San Juan
Paraná
Paysandú
Rivera
30
SANTIAGO
Mendoza
ROSARIO
Mercedes
Rio Grande
Mercedes
URUGUAY
L. Mirim
Juan Fernández Is.
Rocha
(Chile)
BUENOS AIRES
MONTEVIDEO
La Plata
River Plate
Talca
San Rafael
(Rio de la Plata)
Concepción
Malargüe
Santa Rosa
Tandil
Chillán
Neuquén
Bahía Blanca
Mar del Plata
Temuco
Zapala
Blanca Bay
Puerto Montt
S. Carlos
Negro
Viedma
de Bariloche
G. of San Matias
Chiloé I.
Esquel
Chubut
Chonos
Arch.
Pto. Aisen
Comodoro Rivadavia
40
Taitao
Desado
Pto. Deseado
40
Pen.
Chico
Santa Cruz
Falkland Is. (Islas Malvinas)
Grande Bay
(Br.)
Rio Gallegos
Port Stanley
Magellan's Str.
Magellan's Str.
Shag Rocks
Punta Arenas
Tierra
(Br.)
del Fuego
South
I. de los Estados
Georgia
50
C. Horn
(Br.)
50

ATLANTIC OCEAN

PACIFIC OCEAN

Drake Passage
South
Sandwich Is.
(Br.)

Scale 1:34 000 000
0 250 500 750 1000 1250 1500 Kms.
0 250 500 750 1000 St. mis.

100 A 90 B 80 C 70 D 60 West from 50 Greenwich 40 G 30 H 20

MAP 8   233

Scale 1:33 000 000

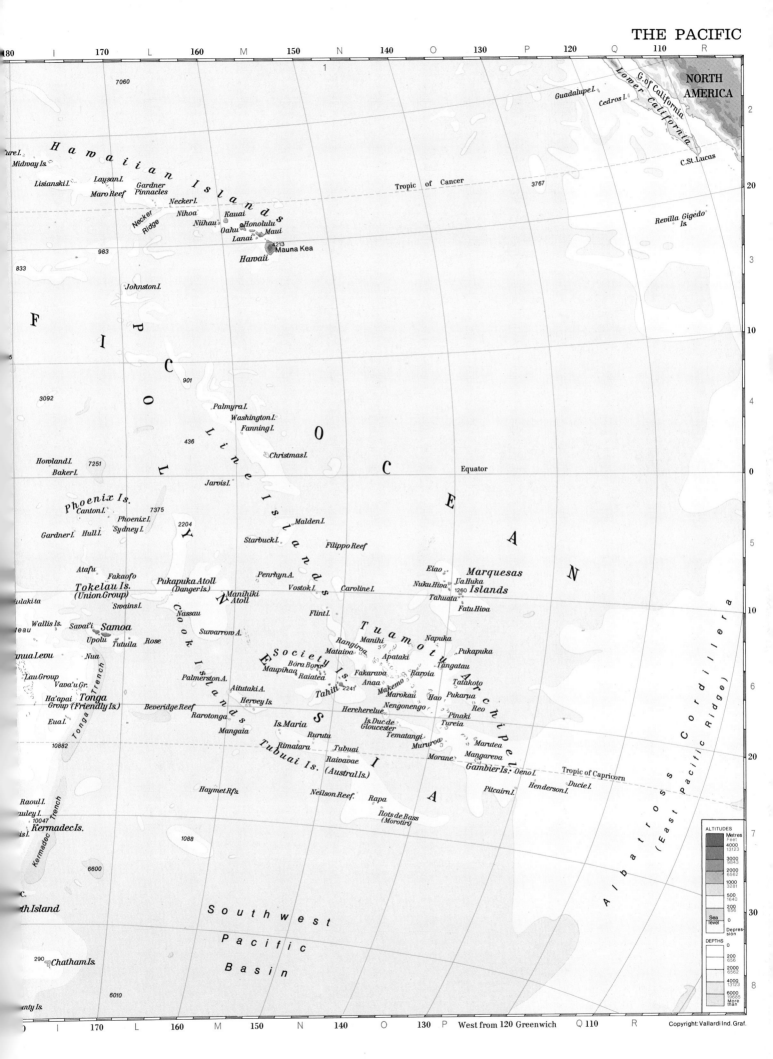

NORTH AMERICA

180   170   160   150   140   130   120   110

7060

G. of California

Lower California

Guadalupe I.

Cedros I.

C. St. Lucas

ure I.

Midway Is.

Lisianski I.   *Laysan I.*

Maro Reef   Gardner Pinnacles

Tropic of Cancer   3767

Revilla Gigedo Is.

H a w a i i a n   I s l a n d s

Necker I.

Nihoa   *Kauai*

Necker Ridge   *Niihau*   Honolulu   *Maui*

Oahu   *Lanai*

983

Hawaii   4213 Mauna Kea

833

Johnston I.

F   I   C   P   O   L   Y

O   C   E   A   N

901

3092

Palmyra I.

Washington I.

Fanning I.

436

Christmas I.

Howland I.   7251

Baker I.

Equator

Jarvis I.

P h o e n i x   I s.

Canton I.   7375

Gardner I.   *Hull I.*   *Phoenix I.*

Sydney I.   2204

Malden I.

Starbuck I.

Filippo Reef

Atafu   *Fakaofo*

Tokelau Is.   *Pukapuka Atoll*

(Union Group)   *(Danger Is.)*

Penrhyn A.

Vostok I.   *Caroline I.*

Eiao   *Marquesas*

Nuku Hiva   *Ua Huka*

1260   *Islands*

Tahuata

ulakita

Swains I.

Manihiki Atoll

Nassau

Flint I.

Fatu Hiva

Wallis Is.   *Savai'i*   Samoa

Upolu   *Tutuila*   *Rose*

Suvarrow A.

Rangiroa   *Manihi*

Mataiva

Napuka

Pukapuka

nua Levu   *Nua*

Society Is.   *Apataki*

Bora Bora

Maupihaa   *Raiatea*

Fakarava   *Raroia*

Tangatau

Lau Group   *Vava'u Gr.*

Aitutaki A.   *Anaa*

Tatakoto

Palmerston A.

Hervey Is.   Tahiti   2241   *Makemo*

Pukarua

Ha'apai   Tonga   *Marokau*   *Hao*

Group (Friendly Is.)   *Beveridge Reef*   *Nengonengo*   *Reo*

Eua I.   *Rarotonga*   *Herehereue*   *Pinaki*

Is. Maria   *Is. Duc de*   *Tureia*

10882   *Mangaia*   *Gloucester*

Rurutu   *Tematangi*

Rimatara   *Mururoa*   *Marutea*

Tubuai   *Morane*   *Mangareva*

Tubuai Is.   *Raivavae*   *Gambier Is.*   *Oeno I.*

(Austral Is.)   Tropic of Capricorn

Raoul I.   *Haymet Rfs.*   *Neilson Reef*   *Rapa*   *Ducie I.*

nuley I.   *Henderson I.*

10047   *Pitcairn I.*

Kermadec Is.   *Ilots de Bass*

(Morotiri)

1088

6600

290 Chatham Is.

S o u t h w e s t

P a c i f i c

B a s i n

th Island

6010

nty Is.

C o o k   I s l a n d s   *T u b u a i   A r c h i p e l*

T u a m o t u   A r c h i p e l

L i n e   I s l a n d s

Tonga Trench   *Kermadec Trench*

P   O   L   Y   N   E   S   I   A

A l b a t r o s s   C o r d i l l e r a

(East Pacific Ridge)

ALTITUDES	Metres	Feet
	4000	13123
	3000	9843
	2000	6562
	1000	3281
	500	1640
	200	656
Sea level	0	
	Depression	

DEPTHS	
0	
200	656
2000	6562
4000	13123
6000	19685
More than	

West from 120 Greenwich

Copyright: Vallardi Ind. Graf.

MAP 10   235

Scale 1:12 000 000

0 50 100   200   300   400   500 Kms.
0   50   100   200   300 St.mls.

236   MAP 11

AUSTRALIA & NEW ZEALAND

Scale 1:12 000 000

Copyright: Vallardi Ind. Graf.

MAP 12

AFRICA

MAP 13

Scale 1:34000000

Copyright: Vallardi Ind. Graf.

ARCTIC REGIONS

PACIFIC OCEAN

Itineraries of the main Arctic expeditions
- – – – Peary-1908-1909
- ·········· Amundsen-Nobile (Norwegian) -1926
- ———— Byrd-1926
- ———— Nobile-1928
- ———— Limit of the inhabited areas

Scale 1:30000000

Copyright: Vallardi Ind.Graf.

SPITZBERGEN
Scale 1:10000000

MAP 14

239

ANTARCTIC REGIONS

MAP 15

GAZETTEER

The first column of figures indicates the map number, and the second (letters and figures) the map reference.

A

Aalborg, Denmark 3 H4
Aarhus, Denmark 3 I4
Abadan, Iran 5 G6
Abadla, Algeria 13 C2
Abakan, U.S.S.R. 5 N4
Abdulino, U.S.S.R. 4 S5
Abeche, Chad 13 F4
Aberdeen, U.K. 3 F4
Abidjan, Ivory Coast 13 C5
Abrolhos I.,
  Australia 11 A5
Abu Hamed, Sudan 13 G4
Acapulco, Mexico 7 L8
Accra, Ghana 13 C5
Aconcagua Mt.,
  Argentina 1
Acre State, Brazil 8 C4
Adamawa Highlands,
  Nigeria/
  Cameroon 2
Adana, Turkey 5 F6
Adare Cape,
  Antarctica 15 A2
Adavale, Australia 12 G5
Ad Dam,
  Saudi Arabia 5 G7
Ad Dawhah, Qatar 5 H7
Addis Ababa,
  Ethiopia 13 G5
Adelaide, Australia 12 F6
Adelaide I.,
  Antarctica 15 O3
Adelaide River,
  Australia 11 E2
Adelie Coast,
  Antarctica 15 bC3
Aden, Gulf of,
  Africa 13 H4
Aden, Yemen 5 G8
Admiralty Gulf,
  Australia 11 D2
Admiralty Is.,
  Pacific Ocean 9 E5
Admiralty Range,
  Antarctica 15 a2
Adriatic Sea, Europe 3 IL7
Adwa, Ethiopia 13 G4
Aegean Sea, Greece 4 N8
Afars and Issas,
  Africa 13 H4
Afghanistan, Asia 5 I6
Agades, Niger 13 D4
Agra, India 5 L7
Agrigento, Sicily 3 I8
Agrihan I.,
  Pacific Ocean 9 E3
Agulhas C.,
  S. Africa 13 F9
Ahmadabad, India 5 L7
Ahvaz, Iran 5 G6
Aileron, Australia 11 E4
Ailinglapalap At.,
  Pacific Ocean 9 G4
Ain Galakko, Chad 13 E4
Ain Sefra, Algeria 13 C2
Aitutaki At.,
  Pacific Ocean 10 M6
Ajaccio, Corsica 3 H7
Ajana, Australia 11 A5
Aketi, Zaire 13 F5
Aklavik, Canada 7 F3
Aksu, China 5 M5
Aktyubinsk,
  U.S.S.R. 4 T5
Akureyri, Iceland 3 C2
Akyab, Burma 5 N7
Alabama State,
  U.S.A. 7 M6
Aland Is., Sweden 3 L3
Alaska, U.S.A. 7 CDE3
Alaska G. of, Alaska 7 E4
Alaska Pen., Alaska 7 CD4
Alaska Range,
  Alaska 14 mM2
Albania, Europe 3 LM7
Albacete, Spain 3 F8
Albany, Australia 11 B6

Albany, U.S.A. 7 N5
Al Basrah, Iraq 5 G6
Albatross Cordillera,
  Pacific Ocean 10 Q67
Al Baida, Libya 13 F2
Alberga R.,
  Australia 11 E5
Albert L.,
  Uganda/Zaire 13 G5
Alberta, Canada 7 H4
Albuquerque,
  U.S.A. 7 I6
Albury, Australia 12 H7
Aldabra Is.,
  Indian Ocean 13 H6
Aldan R., U.S.S.R. 6 R4
Aleksandrovsk,
  U.S.S.R. 6 S4
Alesund, Norway 3 H3
Aleutian Is., Alaska 7 C4
Alexander Arch.,
  Alaska 7 F4
Alexander I.,
  Antarctica 15 o2
Alexandria,
  Australia 11 F3
Alexandria, A.R.E. 13 F2
Algeria, Africa 13 CD23
Algiers, Algeria 13 D2
Al Hufuf,
  Saudi Arabia 5 G7
Alice Springs,
  Australia 11 E4
Al Jawf, Libya 13 F3
Al Jawf,
  Saudi Arabia 5 F7
Allahabad, India 5 M7
Alma Ata, U.S.S.R. 5 L5
Al Manamah,
  Saudi Arabia 5 H7
Al Masira I.,
  Muscat & Oman 5 H7
Almeria, Spain 3 F8
Alps, Europe 2
Al Qunfidha
  Saudi Arabia 5 G8
Amadeus L.,
  Australia 11 E4
Amapa, Brazil 8 E3
Amarillo, U.S.A. 7 I6
Amazon R., Brazil 8 D4
Amazonas region,
  Brazil 8 D4
Amboina, Indonesia 6 Q10
Ambrizete, Angola 13 E6
American Highland,
  Antarctica 15 F2
Amery Ice Shelf,
  Antarctica 15 F3
Amiens, France 3 G6
Amman, Jordan 5 F6
Amoy, China 6 P7
Amritsar, India 5 L6
Amsterdam,
  Netherlands 3 G5
Am Timan, Chad 13 F4
Amu Darya R.,
  U.S.S.R. 5 I5
Amundsen Bay,
  Antarctica 15 g3
Amundsen Gulf,
  Canada 14 nO1
Amundsen Mt.,
  Antarctica 15 d3
Amundsen Sea,
  Antarctica 15 Qq23
Amur R.,
  U.S.S.R./China 6 R4
Anaa I.,
  Pacific Ocean 10 N6
Anadyr Gulf of,
  U.S.S.R. 14 L3
Anadyr R., U.S.S.R. 14 Li3
Anatahan I.,
  Pacific Ocean 9 E3
Anatolia, Turkey 2
Anchorage, Alaska 7 E3
Ancona, Italy 3 I7

Andalsnes, Norway 3 H3
Andaman Is.,
  India 5 N8
Andes Mts.,
  S. America 1
Andizhan, U.S.S.R. 5 L5
Andorra, Europe 3 G7
Andreanof Is., Asia 6 Z4
Andros Is., Bahamas 7 N7
Angara R., U.S.S.R. 5 NO4
Angers, France 3 F6
Angola, Africa 13 EF67
Angouleme, France 3 G6
Ankara, Turkey 5 F6
Ann C., Antarctica 15 G3
Annaba, Algeria 13 D2
Annobon I., Gulf of
  Guinea 13 D6
Annuello, Australia 12 G6
Anshan, China 6 Q5
Ansi, China 5 N5
Anson Bay,
  Australia 11 E2
Antalya, Turkey 5 F6
Antarctic Peninsula,
  Antarctica 15 On3
Anthony Lagoon,
  Australia 11 F3
Anticosti I., Canada 7 O5
Antofagasta, Chile 8 C6
Antwerp, Belgium 3 G5
Anvers I.,
  Antarctica 15 O3
Aomori, Japan 6 S5
Apataki I.,
  Pacific Ocean 10 N6
Apennines Mts.,
  Italy 2
Appalachian Mts.,
  U.S.A. 1
Arabia, Asia 5 G78
Arabian Desert,
  A.R.E. 2
Arabian Sea, Asia 5 HI8
Aracaju, Brazil 8 G5
Arad, Romania 4 M6
Arafura Sea,
  Indonesia 9 D5
Araguaia R., Brazil 8 E5
Araks R., U.S.S.R. 4 R8
Aral Sea, U.S.S.R. 5 HI5
Aralsk, U.S.S.R. 4 U6
Aramac, Australia 12 H4
Araouane, Mali 13 C4
Ararat, Australia 12 G7
Ararat Mt., Turkey 2
Aravalli Range,
  India 36 L7
Archer R., Australia 12 G2
Arel Heiberg I.,
  Canada 14 Pp12
Arequipa, Peru 8 C5
Argentina,
  S. America 8
Argentine I.,
  Antarctica 15 O3
Arica, Chile 8 C5
Arizona State, U.S.A. 7 H6
Arkansas R., U.S.A. 1
Arkansas State,
  U.S.A. 7 L6
Arkhangelsk,
  U.S.S.R. 4 Q3
Armavir, U.S.S.R. 4 Q7
Armenia, U.S.S.R. 4 QR78
Armidale, Australia 12 I6
Arnhem C.,
  Australia 12 F2
Arnhem Land,
  Australia 11 EF2
Arno At.,
  Pacific Ocean 9 H4
Arno R., Italy 3 I7
Arorae I.,
  Pacific Ocean 9 H5
Arshan, China 6 P5
Arta, Greece 4 M8
Aru Is., New Guinea 6 R10

Arusha, Tanzania 13 G6
Aruwimi R., Zaire 13 F5
Arzamas, U.S.S.R. 4 Q4
Ascension Is.,
  Atlantic Ocean 13 B6
Aseb, Ethiopia 13 H4
Ashburton R.,
  Australia 11 B4
Ashley Snow I.,
  Antarctica 15 o2
Ash-Shaab,
  South Yemen 5 G8
Askhabad, U.S.S.R. 5 H6
Asmera, Ethiopia 13 G4
Astrakhan, U.S.S.R. 4 R6
Asuncion I.,
  Pacific Ocean 9 E3
Asuncion, Paraguay 8 E6
Aswan, A.R.E. 13 G3
Asyut, A.R.E. 13 G3
Atafu I.,
  Pacific Ocean 10 I5
Atar, Mauritania 13 B3
Atbara, Sudan 13 G4
Atbara R., Sudan 13 G4
Athabasca L.,
  Canada 7 I4
Athabasca R.,
  Canada 7 H4
Athens, Greece 4 M8
Atlanta, U.S.A. 7 M6
Atlas Mts., Morocco 2
Auckland,
  New Zealand 12 g13
Augsburg, Germany 3 I6
Augusta, Australia 11 B6
Augustus Mt.,
  Australia 11 B4
Aurekun, Australia 12 G2
Aurora I.,
  Pacific Ocean 9 G6
Austin, U.S.A. 7 L6
Austin Lake,
  Australia 11 B5
Austral Downs,
  Australia 12 F4
Australia 11/12
Australian Alps,
  Australia 12 H7
Australian Capital
  Territory,
  Australia 12 H7
Australian
  Dependency,
  Antarctica 15dEeF2
Austria, Europe 3 IL6
Avignon, France 3 G7
Awanui,
  New Zealand 12 g13
Ayan, U.S.S.R. 6 R4
Ayers Rock,
  Australia 11 E5
Ayr, Australia 12 H3
Ayutthaya,
  Thailand 5 O8
Azerbaijan, U.S.S.R. 4 R7
Azov, Sea of,
  U.S.S.R. 4 P6

B

Babuyan Is.,
  Philippines 9 C3
Babylon 13 H2
Badajoz, Spain 3 E8
Baffin Bay, Canada 7 NO2
Baffin I., Canada 7MNO23
Baghdad, Iraq 5 G6
Baguio, Philippines 6 Q8
Bahama Is.,
  Atlantic Ocean 7 N7
Bahia, Brazil 8 F5
Bahia Blanca,
  Argentina 8 D7
Bahrain, Asia 5 H7
Bahr-el-Arib R.,
  Sudan 13 F5
Bahr-el-Jebel R.,
  Sudan 13 G5

Bairnsdale,
  Australia 12 H7
Baker I.,
  Pacific Ocean 10 I4
Baku, U.S.S.R. 4 R7
Balabac Str.,
  Indonesia 9 B4
Balaklava, U.S.S.R. 4 O7
Bald Head,
  Australia 11 B7
Balearic Is.,
  Mediterranean
  Sea 3 G8
Bali I., Indonesia 6 P10
Balikpapan,
  Indonesia 6 P10
Balkan Peninsula 2
Balkhash, U.S.S.R. 5 L5
Balkhash L.,
  U.S.S.R. 5 L5
Balladonia,
  Australia 11 C6
Ballarat, Australia 12 G7
Ballard L.,
  Australia 11 C5
Balleney Is.,
  Antarctica 15 a3
Balonne R.,
  Australia 12 H5
Balovale, Zambia 13 F7
Balranald,
  Australia 12 G6
Baltic Sea, Europe 3 L4
Baltimore, U.S.A. 7 N6
Bamako, Mali 13 C4
Banaras, India 5 M7
Banda Sea, Indonesia 6 Q10
Bandar Abbas, Iran 5 H7
Bandar Atjeh,
  Indonesia 5 N9
Bandirma, Turkey 4 N7
Bandjarmasin,
  Indonesia 6 P10
Bandung, Indonesia 6 O10
Bangalore, India 5 L8
Bangassou,
  Cen. Africa 13 F5
Bangka I., Indonesia 6 O10
Bangkok, Thailand 5 O8
Bangladesh,
  Asia 5 MN7
Bangui, Cen. Africa 13 E5
Bangwelu L.,
  Zambia 13 FG7
Banja Luka,
  Yugoslavia 3 L7
Banjul, Gambia 13 B4
Banks Is.,
  Pacific Ocean 9 G6
Banks I., Australia 12 G12
Banks I., Canada 7 GH2
Banks Pen.,
  New Zealand 12 g14
Banks Str.,
  Australia 12 b9
Banzare Coast,
  Antarctica 15 c3
Baranovichi U.S.S.R. 4 N5
Barbados Is.,
  Caribbean Sea 8 E2
Barcaldine,
  Australia 12 H4
Barcelona, Spain 3 G7
Barcoo R., Australia 12 G4
Bardai, Chad 13 E3
Bareilly, India 5 L7
Barents Sea 3 FG2
Bari, Italy 3 L7
Barlee Lake,
  Australia 11 B5
Barnaul, U.S.S.R. 5 M4
Baroda, India 5 L7
Barquisimeto,
  Venezuela 8 C3
Barra, Brazil 8 F5
Barraba, Australia 12 I6
Barranquilla,
  Colombia 8 C2

Barrow, Alaska 7 D2
Barrow Cape, Alaska 7 D2
Barrow Creek,
  Australia 11 E4
Barrow I., Australia 11 B4
Bartle Frere Mt.,
  Australia 12 H3
Basle, Switzerland 3 H6
Bass Str., Australia 12 H7
Bassas da India,
  Mozambique 13 G8
Bassein, Burma 5 N8
Bastia, Corsica 3 H7
Bata, Equatorial
  Guinea 13 D5
Batan Is., Asia 6 Q7
Bathurst, Australia 12 H6
Bathurst I.,
  Australia 11 E2
Bathurst I., Canada 7 IL2
Baton Rouge, U.S.A. 7 L6
Battle Harbour,
  Canada 7 P4
Batumi, U.S.S.R. 4 Q7
Baykal L., U.S.S.R. 6 O4
Bayonne, France 3 F7
Beach Port,
  Australia 12 F7
Bear I., Norway 5 D2
Beaufort Sea,
  N. America 7 DE2
Bedourie, Australia 12 F4
Bega, Australia 12 H7
Beira, Mozambique 13 G7
Beirut, Lebanon 5 F6
Beja, Portugal 3 E8
Belcher Is., Canada 7 N4
Belem, Brazil 8 F4
Belfast, U.K. 3 E5
Belgium, Europe 3 GH5
Belgorod, U.S.S.R. 4 P5
Belgrade, Yugoslavia 3 M7
Belingshausen Sea,
  Antarctica 15 P3
Belize,
  Central America 7 M8
Belize City, Belize 7 M8
Belle Isle Str.,
  Canada 7 P4
Bellona Reefs,
  Pacific Ocean 9 F7
Belo Horizonte,
  Brazil 8 F5
Bendigo, Australia 12 G7
Bengal, Bay of, Asia 5 MN8
Benghazi, Libya 13 F2
Benguela, Angola 13 E7
Beni R., Bolivia 8 D5
Benin (Dahomey) 13 D45
Benin, Nigeria 13 D5
Bennet I., Australia 11 E4
Bentinck I.,
  Australia 12 F3
Benue R., Nigeria 13 D5
Berbera, Somali 13 H4
Bergen, Norway 3 H3
Bering Sea, Alaska/
  U.S.S.R. 7 BC34
Bering Str., Alaska/
  U.S.S.R. 7 BC3
Berkner I.,
  Antarctica 15 Nn12
Berlin, Germany 3 I5
Bermuda I.,
  Atlantic Ocean 7 O6
Bern, Switzerland 3 H6
Betoota, Australia 12 G5
Beveridge Reef,
  Pacific Ocean 10 L7
Beyla, Guinea 13 C5
Bhopal, India 5 L7
Bhutan, Asia 5 M7
Biak I., Indonesia 6 R10
Bialystok, Poland 4 M5
Bie Plateau, Angola 2
Big Bell, Australia 11 B5
Bikaner, India 5 L7
Bikini At.,
  Pacific Ocean 9 G3
Bilbao, Spain 3 F7
Billiton I., Indonesia 9 A5
Binh Dinh,
  S. Vietnam 6 O8
Birdsville, Australia 12 F5
Birdum, Australia 11 E3
Birimgan, Australia 12 H4
Birmingham, U.K. 3 F5
Biscay, Bay of,
  Europe 3 F67
Biscoe I., Antarctica 15 O3
Biskra, Algeria 13 D2
Bismarck, U.S.A. 7 I5
Bismarck Arch.,
  New Guinea 9 EF5
Bissagos Is.,
  W. Africa 13 B4
Bissau, Guinea 13 B4
Bitola, Yugoslavia 4 M7

Bitterfontein,
  S. Africa 13 E9
Biysk, U.S.S.R. 5 M4
Bizerte, Tunisia 13 D2
Blackall, Australia 12 H4
Black Sea, Europe 4 NOP7
Blackwood R.,
  Australia 11 B6
Blagoveshchensk,
  U.S.S.R. 6 Q4
Blanc C.,
  Western Sahara 13 B3
Blanc Mt., France 2
Blanca Bay,
  Argentina 8 D7
Blanche L.,
  S. Australia 12 F5
Blanche L.,
  W. Australia 11 C4
Blantyre, Malawi 13 G7
Blenheim,
  New Zealand 12 g14
Bloemfontein,
  S. Africa 13 F8
Bluefields,
  Nicaragua 7 M8
Blue Nile R., Sudan 13 G45
Bluff, New Zealand 12 f15
Blumenau, Brazil 8 F6
Bobo Dioulasso,
  Volta 13 C4
Bodele Depression,
  Chad 2
Bodo, Norway 3 I2
Bogota, Colombia 8 C3
Boise, U.S.A. 7 H5
Bolivia,
  S. America 8 D5
Bolivian Plateau,
  S. America 1
Bologna, Italy 3 I7
Bologoye, U.S.S.R. 4 O4
Bolshevik I.,
  U.S.S.R. 5 O2
Bombala, Australia 12 H7
Bombay, India 5 L8
Bonaparte Arch.,
  Australia 11 D2
Bondo, Zaire 13 F5
Bonifacio, Str. of,
  Corsica/Sardinia 3 H7
Bonin Is.,
  Pacific Ocean 9 E2
Bonn, Germany 3 H5
Boothia, Gulf of,
  Canada 7 LM2
Boothia Pen., Canada 7 L2
Bora Bora I.,
  Pacific Ocean 10 M6
Bordeaux, France 3 F7
Borneo, Indonesia 6 P9
Bornholm Is.,
  Denmark 3 L4
Borroloola,
  Australia 11 F3
Boston, U.S.A. 7 N5
Bothnia, Gulf of,
  Europe 3/4 LM23
Botosani, Romania 4 N6
Botswana, Africa 13 F8
Bouake, Ivory Coast 13 C5
Bougainville, I.,
  Pacific Ocean 9 F5
Bougainville Reef,
  Australia 12 H3
Boulder, Australia 11 C6
Boulia, Australia 12 F4
Boulogne, France 3 G5
Bounty I., Pacific
  Ocean 10 H9
Bourem, Mali 13 C4
Bourke, Australia 12 H6
Bouvet I.,
  Antarctica 15 i4
Bowen, Australia 12 H3
Bowman I.,
  Antarctica 15 d3
Braga, Portugal 3 E7
Brahamaputra R.,
  India 5 N7
Bransfield Str.,
  Antarctica 15 On3
Brasilia, Brazil 8 F5
Brasov, Romania 4 N6
Brazil, S. America 8
Brazilian Highlands,
  S. America 1
Brazzaville, Congo 13 E6
Breidhi Fjord,
  Iceland 3 B2
Bremen, Germany 3 H5
Brescia, Italy 3 I6
Brest, France 3 F6
Brest, U.S.S.R. 4 M5
Brewarrina,
  Australia 12 H6
Bright, Australia 12 H7
Brindisi, Italy 3 L7

Brisbane, Australia 12 I5
Bristol, U.K. 3 F5
Bristol Bay, Alaska 14 MI4
Bristol Is.,
  S. Sandwich Is. 15 M4
British Antarctic
  Territory 15 Nm3
British Columbia,
  Canada 7 G4
British Isles, Europe 3 EF45
Brno, Czechoslovakia 3 L6
Broad Arrow,
  Australia 11 C6
Broad Sound,
  Australia 12 HI4
Broken Hill,
  Australia 12 G6
Brooks Range,
  Alaska 14 mM3
Broome, Australia 11 C3
Brunei, Asia 6 P9
Brunswick Bay,
  Australia 11 C3
Bruny I., Australia 12 b9
Brussels, Belgium 3 G5
Bryansk, U.S.S.R. 4 O5
Bucaramanga,
  Colombia 8 C3
Bucharest, Romania 4 N7
Budapest, Hungary 3 L6
Budd Coast,
  Antarctica 15 D3
Buenaventura,
  Colombia 8 C3
Buenos Aires,
  Argentina 8 E7
Buffalo, U.S.A. 7 N5
Bug R.,
  Poland/U.S.S.R. 4 M5
Bujumbura,
  Burundi 13 F6
Bukavu, Zaire 13 F6
Bukhara, U.S.S.R. 5 I6
Bulawayo, Rhodesia 13 F8
Bulgaria, Europe 4 MN7
Buller R.,
  New Zealand 12 g14
Bullfinch, Australia 11 B6
Bulloo, Australia 12 G5
Bulun, U.S.S.R. 5 Q2
Bunbury, Australia 11 B6
Bundaberg,
  Australia 12 I4
Bundooma,
  Australia 11 E4
Bunguran Is.,
  Malaysia 9 A4
Buraida,
  Saudi Arabia 5 G7
Burdekin R.,
  Australia 12 H4
Burgas, Bulgaria 4 N7
Burgos, Spain 3 F7
Burketown,
  Australia 12 F3
Burma, Asia 5 N78
Burnett R.,
  Australia 12 I5
Burnie, Australia 12 b9
Bursa, Turkey 5 F6
Buru I., Indonesia 6 Q10
Burundi, Africa 13 F6
Bushehr, Iran 5 H7
Butte, U.S.A. 7 H5
Bydgoszcz, Poland 3 L5
Byrd Sub-glacial
  Basin, Antarctica 15 q1
Byrock, Australia 12 H6

C
Cabinda, Angola 13 E6
Cabot Str., Canada 7 OP5
Caceres, Spain 3 E8
Cadiz, Spain 3 E8
Cadiz, G. of, Spain 3 E8
Caen, France 3 F6
Cagliari, Sardinia 3 H8
Caird Coast,
  Antarctica 15 M12
Cairns, Australia 12 H3
Cairo, A.R.E. 13 G2
Cajamarca, Peru 8 C4
Calais, France 3 G5
Calama, Chile 8 D6
Calcutta, India 5 M7
Calgary, Canada 7 H4
Cali, Colombia 8 C3
Calicut, India 5 •L8
California, Gulf of,
  Mexico 7 H17
California State,
  U.S.A. 7 GH6
Callabonna I.,
  Australia 12 G5
Callao, Peru 8 C5
Calvert Hills,
  Australia 12 F3
Camaguey, Cuba 7 N7

Cameroon, Africa 13 E5
Camooweal,
  Australia 12 F3
Campania,
  Australia 12 b9
Campbell Town,
  Australia 12 b9
Campeche, Mexico 7 L7
Campeche, Gulf of,
  Mexico 7 L7
Campina Grande,
  Brazil 8 G4
Campinas, Brazil 8 F6
Campo Grande,
  Brazil 8 E6
Campos, Brazil 8 F6
Canada, N. America 7
Canary Is.,
  Atlantic Ocean 13 B3
Canberra, Australia 12 H7
Canterbury Bight,
  New Zealand 12 g14
Canton, China 6 P7
Canton I.,
  Pacific Ocean 10 I5
Cape Barren I.,
  Australia 12 b9
Cape Breton I.,
  Canada 7 O5
Cape Province,
  S. Africa 13 F9
Capetown, S. Africa 13 E9
Cape York Pen.,
  Australia 12 G2
Capricorn Channel,
  Australia 12 I4
Caprivi Strip,
  S.W. Africa 13 F7
Caracas, Venezuela 8 D2
Caravelas, Brazil 8 G5
Cardiff, U.K. 3 F5
Carey L., Australia 11 C5
Caribbean Sea,
  Cen. America 7 MNO8
Carlisle, U.K. 3 F5
Carnarvon,
  Australia 11 A4
Carnegie, Australia 11 C5
Carnegie L.,
  Australia 11 C5
Carnot, Cen. Africa 13 E5
Carolina, Brazil 8 F4
Caroline I.,
  Pacific Ocean 10 M5
Caroline Is.,
  Pacific Ocean 9 DEF4
Carpathian Mts.,
  S. Europe 2
Carpentaria, Gulf of,
  Australia 12 F2
Cartagena, Colombia 8 C2
Cartagena, Spain 3 F8
Casablanca,
  Morocco 13 C2
Caspian Sea,
  U.S.S.R. 4 RS678
Castlereagh, R.,
  Australia 12 H6
Cata I.,
  Pacific Ocean 9 F7
Catamarca,
  Argentina 8 D6
Catania, Sicily 3 L8
Catanzaro, Italy 3 L8
Catastrophe Cape,
  Australia 11 F67
Caucasus Mts.,
  U.S.S.R. 2
Caxias, Brazil 8 F4
Cayenne, Guiana 8 E3
Cayman Is.,
  Caribbean Sea 7 M8
Cebu, Philippines 6 Q8
Cedros I., Mexico 7 H7
Ceduna, Australia 11 E6
Celebes I., Indonesia 6 PQ10
Celebes Sea,
  Indonesia 6 Q9
Central African
  Empire, Africa 13 EF5
Central Range,
  New Guinea 9 DE5
Central Siberian
  Plateau, U.S.S.R. 2
Ceram I., Indonesia 6 Q10
Cerro de Pasco, Peru 8 C5
Ceylon (Sri Lanka),
  Asia 5 M9
Chad, Africa 13 EF4
Chad L., Chad 13 E4
Chalon-sur-Saone,
  France 3 G6
Chamdo, China 5 N6
Changsha, China 6 P7
Channel Is.,
  English Channel 3 F6
Charcot I.,
  Antarctica 15 o23

Charleston, U.S.A. 7 M6
Charleville,
  Australia 12 H5
Charlotte, U.S.A. 7 M6
Charlotte Waters,
  Australia 11 E5
Charlottetown,
  Canada 7 O5
Charters Towers,
  Australia 12 H3
Chatham Is.,
  Pacific Ocean 10 I9
Chatt Djerid L.,
  Tunisia 3 H9
Cheju Do I.,
  S. Korea 6 Q6
Chelkar, U.S.S.R. 4 T6
Chelyabinsk,
  U.S.S.R. 5 I4
Chelyuskin C.,
  U.S.S.R. 5 O2
Chengchow, China 6 P6
Chengtu, China 5 O6
Cherbourg, France 3 F6
Cherchen, China 5 M6
Cheremkhovo,
  U.S.S.R. 5 O4
Cherepovets,
  U.S.S.R. 4 P4
Chernovtsy,
  U.S.S.R. 4 N6
Chersky Range,
  U.S.S.R. 14 hH3
Chesterfield Isle,
  Pacific Ocean 9 F6
Cheyenne, U.S.A. 7 I5
Chiamusze, China 6 R5
Chicago, U.S.A. 7 M5
Chiclayo, Peru 8 C4
Chiengmai, Thailand 5 N8
Chicoutimi, Canada 7 N5
Chihuahua, Mexico 7 I7
Chile, S. America 8 C5-9
Chillagoe, Australia 12 G3
Chillan, Chile 8 C7
Chiloe I., Chile 8 C8
Chimborazo Mt.,
  Ecuador 1
China, Asia 5/6
Chinchou, China 6 Q5
Chirripo Grande,
  Costa Rica 1
Chita, U.S.S.R. 6 P4
Chittagong,
  Bangladesh 5 N7
Choiseul I.,
  Pacific Ocean 9 F5
Chongin, N. Korea 6 Q5
Chonos Arch., Chile 8 C8
Christchurch,
  New Zealand 12 g14
Christensen Mt.,
  Antarctica 15 g3
Christmas I.,
  Pacific Ocean 10 M4
Chubut R.,
  Argentina 8 D8
Chudskoye L.,
  U.S.S.R. 4 N4
Chukchee Pen.,
  U.S.S.R. 14 L3
Chukot Range,
  U.S.S.R. 14 Li3
Chumican, U.S.S.R. 6 R4
Chungking, China 6 O7
Churchill, Canada 7 L4
Cincinnati, U.S.A. 7 M6
Citlaltepetl Vol.,
  Mexico 1
Ciudad Bolivar,
  Venezuela 8 D3
Ciudad Juarez,
  Mexico 7 I6
Ciudad Real, Spain 3 F8
Civita Vecchia, Italy 3 I7
Claire Coast,
  Antarctica 15 C3
Clarence R.,
  Australia 12 I5
Clarence Str.,
  Australia 11 E2
Clear Cape,
  Ireland 3 E5
Clermont-Ferrand,
  France 3 G6
Cleveland, U.S.A. 7 M5
Clipperton Is.,
  Pacific Ocean 7 I8
Cloncurry,
  Australia 12 G4
Cloncurry R.,
  Australia 12 G3
Cluj, Romania 4 M6
Clutha R.,
  New Zealand 12 f15
Coast Range, U.S.A. 1
Coats Land,
  Antarctica 15 M12

Place		
Flensburg, Germany	3	H5
Flinders, Australia	12	G4
Flinders I., Australia	12	H7
Flinders R., Australia	12	G3
Flinders Range, Australia	12	F56
Flinders Reefs, Australia	12	H3
Flint I., Pacific Ocean	10	M6
Florence, Italy	3	I7
Flores I., Indonesia	6	Q10
Flores Sea, Indonesia	6	PQ10
Florianopolis, Brazil	8	F6
Florida State, U.S.A.	7	MN67
Florida Str., N. America	7	MN7
Flying Fish C., Antarctica	15	Q2
Fly R., New Guinea	9	E5
Foggia, Italy	3	L7
Foochow, China	6	P7
Formentera, Balearic Is.	3	G8
Forrest, Australia	11	D6
Forsayth, Australia	12	G3
Fortaleza, Brazil	8	G4
Fort Archambault, Chad	13	E5
Fort Chimo, Canada	7	O4
Fort Crampel, Central Africa	13	E5
Fortescue R., Australia	11	B4
Fort Flatters, Algeria	13	D3
Fort George, Canada	7	N4
Fort Jameson, Zambia	13	G7
Fort McMurray, Canada	7	H4
Fort Resolution, Canada	7	H3
Fort Shevchenko, U.S.S.R.	4	S7
Fort Simpson, Canada	7	G3
Fort Victoria, Rhodesia	13	G8
Fort William, Canada	7	M5
Fort Worth, U.S.A.	7	L6
Fort Yukon, Alaska	7	E5
Fossil Bluff, Antarctica	15	o2
Foulwind C., New Zealand	12	g14
Foumban, Cameroon	13	E5
Foveaux Str., New Zealand	12	f15
Fowlers Bay, Australia	11	E6
Foxe Basin, Canada	7	N3
Foxe Pen., Canada	7	N3
France, Europe	3	FGH67
Frankfurt, Germany	3	H5
Fraser I., Australia	12	I5
Fredericton, Canada	7	O5
Fredrikstad, Norway	3	I4
Freemantle, Australia	11	B6
Freetown, Sierra Leone	13	B5
Freiburg, Germany	3	H6
Fresno, U.S.A.	7	H6
Friendly Is., Pacific Ocean	10	I6
Frisian Islands, Netherlands/Germany	3	H5
Frome L., Australia	12	F6
Fuerteventura, Canary Is.	13	B3
Funafuti At., Pacific Ocean	9	H5
Furneaux Group, Australia	12	b9
Fushun, China	6	Q5
G		
Gaberones, Botswana	13	F8
Gabon, Africa	13	DE56
Gaferut I., Pacific Ocean	9	E4
Gairdner L., Australia	11	F6
Galadi, Ethiopia	13	H5
Galapagos Is., Ecuador	8	AB4
Galati, Romania	4	N6
Galbraith, Australia	12	G3
Galilee L., Australia	12	H4
Galway, Ireland	3	E5
Galveston, U.S.A.	7	L7
Gambia, Africa	13	B4
Gambier Is., Pacific Ocean	10	O7
Ganges R., India/Bangladesh	5	LM67
Gao, Mali	13	C4
Gardner I., Pacific Ocean	10	I5
Gardner Pinnacles, Pacific Ocean	10	L2
Garonne R., France	3	G7
Garoua, Cameroon	13	E5
Gascoyne R., Australia	11	B5
Gaspe, Canada	7	O5
Gaussberg, Antarctica	15	e3
Gavle, Sweden	3	L3
Gdansk, Poland	3	L5
Geelong, Australia	12	G7
Geelvink Channel, Australia	11	A5
Gen. Belgrano, Antarctica	15	m2
Gen. Bernardo O'Higgins, Antarctica	15	n3
Geneva, Switzerland	3	H6
Genoa, Italy	3	H7
Gentil P., Gabon	13	D6
Geographe Bay, Australia	11	B6
Geographe Channel, Australia	11	A4
George V Coast, Antarctica	15	Bb3
George VI Sound, Antarctica	15	O23
Georgetown, Guyana	8	E3
Georgia, U.S.S.R.	4	QR7
Georgia State, U.S.A.	7	M6
Georgina R., Australia	12	F4
Geraldton, Australia	11	A5
Germany, Europe	3	HI56
Ghadanes, Libya	13	D2
Ghana, Africa	13	C5
Ghat, Libya	13	E3
Gibeon Seeheim, S.W. Africa	13	E8
Gibraltar, Europe	3	E8
Gibraltar, Str. of, Europe/Africa	3	E8
Gibson Desert, Australia	11	CD4
Gijon, Spain	3	E7
Gilbert Is., Pacific Ocean	9	H34
Gilbert R., Australia	12	G3
Gilgit, Pakistan	5	L6
Gironde R., France	3	F6
Gisborne, New Zealand	12	h13
Givadar, Pakistan	5	I7
Gizhiga, U.S.S.R.	6	U3
Gladstone, Australia	12	I4
Glasgow, U.K.	3	F4
Glenhope, New Zealand	12	g14
Glen Innes, Australia	12	I5
Glenmorgan, Australia	12	H5
Goa, India	5	L8
Gobabis, S.W. Africa	13	E8
Gobi Desert, China	2	
Godavari R., India	5	LM8
Godhavn, Greenland	7	P3
Godthaab, Greenland	7	P3
Goiania, Brazil	8	F5
Gomel, U.S.S.R.	4	O5
Gonder, Ethiopia	13	G4
Goodenough C., Antarctica	15	c3
Good Hope, Cape of, S. Africa	13	E9
Goondiwindi, Australia	12	I5
Gordon Downs, Australia	11	D3
Gore, New Zealand	12	f15
Gorki, U.S.S.R.	4	Q4
Goteborg, Sweden	3	I4
Gotland I., Sweden	3	L4
Gough I., S. Atlantic	13	C9
Goulburn, Australia	12	H6
Grafton, Australia	12	I5
Graham Bell I., Arctic	14	D1
Graham Land, Antarctica	15	O3
Granada, Spain	3	F8
Gran Chaco Plata Plain, S. America	1	
Grand Bahama I., Atlantic Ocean	7	N7
Grand Canaria I., Canary Is.	13	B3
Grande, Rio, Mexico/U.S.A.	7	IL7
Grande Bay, Argentina	8	D9
Granites, The, Australia	11	E4
Graz, Austria	3	L6
Great Abaco I., N. America	7	N7
Great Australian Bight, Australia	11	DE6
Great Barrier I., New Zealand	12	h13
Great Barrier Reef, Australia	12	GH23
Great Bear L., N.W. Territory	7	G3
Great Britain, Europe	3	EFG45
Great Dividing Range, Australia	12	H54
Greater Antilles, C. America	1	
Greater Sunda Is., Indonesia	6	OP10
Great Inagua I., Bahamas	7	N7
Great Lake, Australia	12	b9
Great Plains, N. America	1	
Great Salt Lake, U.S.A.	7	H5
Great Sandy Desert, Australia	11	CD4
Great Sandy I., Australia	12	I5
Great Slave L., Canada	7	H3
Great Victoria Desert, Australia	11	CD5
Greece, Europe	4	M8
Greenland	7	O23
Greenland Sea, Greenland	14	tA2
Greenough R., Australia	11	B5
Greenville, Liberia	13	C5
Gregory R., Australia	12	F3
Gregory Salt Sea, Australia	11	D4
Grenada, West Indies	8	D2
Greymouth, New Zealand	12	g14
Grey Range, Australia	12	G5
Griffith, Australia	12	H6
Grim C., Australia	12	a9
Groote Eylandt, Australia	12	F2
Grootfontein, S.W. Africa	13	E7
Groznyy, U.S.S.R.	4	R7
Guadalajara, Mexico	7	I7
Guadalcanal, Pacific Ocean	9	FG5
Guadaloupe, Caribbean Sea	8	D2
Guadalquivir R., Spain	3	E8
Guadalupe I., Mexico	7	H7
Guadiana R., Spain	3	E8
Guajara Mirim, Brazil	8	D5
Guam I., Pacific Ocean	9	E3
Guantanamo, Cuba	7	N7
Guardafui C., Somali	13	I4
Guatemala, C. America	7	L8
Guatemala, Guatemala	7	L8
Guaviare R., Colombia	8	C3
Guayaquil, Ecuador	8	C4
Guaymas, Mexico	7	H7
Guiana, S. America	8	E3
Guinea, Africa	13	BC45
Guinea-Bissau, Africa	13	B4
Guryev, U.S.S.R.	4	S6
Guyana, S. America	8	E3
Gympie, Australia	12	I5
H		
Ha'apai Group, Pacific Ocean	10	I6
Hague, The, Netherlands	3	G5
Haig, Australia	11	D6
Hail, Saudi Arabia	5	G7
Hailar, China	6	P5
Hainan, China	6	P8
Haiphong, N. Vietnam	6	O7
Haiti, Caribbean Sea	7	N8
Hakodate, Japan	6	S5
Halab, Syria	5	F6
Halifax, Canada	7	O5
Hall Is., Pacific Ocean	9	F4
Halle, Germany	3	I5
Hallett, Antarctica	15	A2
Halley Base, Antarctica	15	M2
Halmahera I., Indonesia	6	Q9
Hall's Creek, Australia	11	D3
Hamar, Norway	3	I3
Hamadan, Iran	5	G6
Hamburg, Germany	2	
Hamelin Pool, Australia	11	A5
Hamersley Range, Australia	11	B4
Hamilton, Australia	12	G7
Hamilton, New Zealand	12	h13
Hang Chow, China	6	PQ6
Hanoi, N. Vietnam	6	O7
Hanover, Germany	3	H5
Hao I., Pacific Ocean	10	N6
Harbin, China	6	Q5
Hargeisa, Somali	13	H5
Hastings, New Zealand	12	h13
Hatteras, Cape, U.S.A.	7	N6
Haugesund, Norway	3	H4
Hauraki G., New Zealand	12	h13
Havana, Cuba	7	M7
Hawaii, Pacific Ocean	10	M3
Hawaiian I., Pacific Ocean	10	ILM2
Hawke Bay, New Zealand	12	h13
Hay, Australia	12	G6
Hay R., Australia	12	F4
Haymets Reefs, Pacific Ocean	10	L7
Hebrides Is., U.K.	3	E4
Hebron, Canada	7	O4
Hejaz, Saudi Arabia	5	F7
Helmand R., Afghanistan	5	I67
Helsinki, Finland	4	N3
Henderson I., Pacific Ocean	10	P7
Hengyang, China	6	P7
Herat, Afghanistan	5	I6
Herberton, Australia	12	H3
Hereherelue I., Pacific Ocean	10	N6
Hermosillo, Mexico	7	H7
Hervey Bay, Australia	12	I5
Hervey Is., Pacific Ocean	10	M6
Hibernia Reef, Australia	11	C2
Himalayas Mts., Asia	2	
Hindu Kush Range, Pakistan	2	
Hinehinbrook, Australia	12	H3
Hiroshima, Japan	6	R6
Hispaniola I., C. America	1	
Hobart, Australia	12	b9
Hobbs Coast, Antarctica	15	rS2
Hodeida, Yemen	5	G8
Hoggar Mts., Algeria	2	
Hoihow, China	6	P8
Hokitika, New Zealand	12	g14
Hokkaido, Japan	6	S5
Hollick Kenyon Plateau, Antarctica	15	Q1
Holmes Reefs, Australia	12	H3
Homs, Syria	5	F6
Honduras, C. America	7	M8
Honduras, Gulf of, C. America	1	
Hong Kong, Asia	6	P7
Honolulu I., Pacific Ocean	10	M2
Honshu, Japan	6	RS56
Hope Bay, Antarctica	15	n3
Hopetoun, Australia	11	C6
Hopkins L., Australia	11	D4
Horlick Mts., Antarctica	15	R1
Horn C., Iceland	3	B2
Horn C., S. America	8	D9
Hountman Abrolhos Is., Australia	11	A5
Houston, U.S.A.	7	L7
Howe C., Australia	12	H7
Howland I., Pacific Ocean	10	I4
Howrah, India	5	M7
Huancavelica, Peru	8	C5
Hubli, India	5	L8
Hudson Bay, Canada	7	M34
Hudson C., Antarctica	15	B3
Hudson Mts., Antarctica	15	p2
Hudson Str., Canada	7	NO3
Hue, S. Vietnam	6	O8
Huelva, Spain	3	E8
Hughenden, Australia	12	G4
Hull, U.K.	3	F5
Humaita, Brazil	8	D4
Huna Bay, Iceland	3	B2
Hungary, Europe	3	L6
Hungerford, Australia	12	G5
Hurd Point, Australia	11	E2
Huron L., Canada/U.S.A.	7	M5
Hwang Ho, China	6	P6
Hyderabad, India	5	L8
Hyderabad, Pakistan	5	I7
I		
Iasi, Romania	4	N6
Ibadan, Nigeria	13	D5
Iberian Pen., Europe	2	
Ibiza, Balearic Is.	3	G8
Ica, Peru	8	C5
Iceland, Europe	3	BC23
Idaho State, U.S.A.	7	H5
Idjil, Mauritania	13	B3
Iffley, Australia	12	G3
Ifni, Africa	13	B3
Iguatu, Brazil	8	G4
Ilebo, Zaire	13	F6
Ilha do Bananal, Brazil	8	E5
Ilheus, Brazil	8	G5
Illots de Brass, Pacific Ocean	10	N7
Iloilo, Philippines	6	Q8
Imphal, India	5	N7
Inchon, S. Korea	5	Q6
India, Asia	5	
Indianapolis, U.S.A.	7	M6
Indigirka R., U.S.S.R.	6	S23
Indonesia, Asia	6	N-R9,10
Indore, India	5	L7
Indus R., India/Pakistan	5	IL67
Ingham, Australia	12	H3
Ingrid Christensen Land, Antarctica	15	Fe23
Inhambane, Mozambique	13	G8
Injune, Australia	12	H5
Inn R., Austria	3	I6
Innamincka, Australia	12	G5
Inner Mongolia, Asia	5/6	OPQ5
Innisfail, Australia	12	H3
Innsbruck, Austria	3	I6
Invercargill, New Zealand	12	f15
Inverell, Australia	12	I5
Inverness, U.K.	3	F4
Inverway, Australia	11	D3
Ioannina, Greece	4	M8
Ionian Sea, Europe	3	L8
Iowa State, U.S.A.	7	L5
Ipin, China	6	O7
Ipoh, Malaysia	5	O9

Name	Page	Grid
Louisiade Arch., Pacific Ocean	9	F6
Louisiana State, U.S.A.	7	L6
Lourdes, France	3	FG7
Lourenco Marques, Mozambique	13	G8
Lower California, Mexico	7	H67
Lower Hutt, New Zealand	12	g14
Lower Tunguska R., U.S.S.R.	5	N3
Loyalty I., Pacific Ocean	9	G67
Lualaba R., Zaire	13	F6
Luanda, Angola	13	E6
Luang Prabang, Laos	5	O8
Luanshya, Zambia	13	F7
Lubeck, Germany	3	I5
Lublin, Poland	4	M5
Lubumbashi, Zaire	13	F7
Lucknow, India	5	M7
Luderitz, S.W. Africa	13	E8
Lugenda R., Mozambique	13	G7
Lugh Ferrandi, Somali	13	H5
Lulea, Sweden	4	M2
Lusaka, Zambia	13	F7
Lusambo, Zaire	13	F6
Lu-Ta, China	6	Q6
Lutzowholm Bay, Antarctica	15	H3
Luxembourg, Europe	3	H6
Luxembourg City, Luxembourg	3	H6
Luzon Is., Philippines	6	Q8
Lvov, U.S.S.R.	4	M6
Lyndon R., Australia	11	A4
Lyons, France	3	G6
Lyons R., Australia	11	B4

M

Name	Page	Grid
Macao, Asia	6	P7
Macapa, Brazil	8	E3
Macauley I., Pacific Ocean	10	I7
MacDonald L., Australia	11	D4
Macdonnell Ranges, Australia	11	E4
Maceio, Brazil	8	G4
Machattie L., Australia	12	F4
Mackay, Australia	12	H4
Mackay L., Australia	11	D4
MacKenzie Bay, Antarctica	15	F3
MacKenzie Mts., Canada	14	Nn3
MacKenzie R., Canada	14	Nn3
Mac Robertson Land, Antarctica	15	f3
McClintock Mt., Antarctica	15	B2
M'Clure Str., Canada	14	nO2
McKinley Mt., Alaska	7	DE5
McMurdo Scott, Antarctica	15	a2
McMurdo Sound, Antarctica	15	a2
Madagascar Is., Africa	13	H78
Madeira I., Atlantic Ocean	13	B2
Madeira R., Brazil	8	D4
Madras, India	5	M8
Madrid, Spain	3	F7
Madura, India	5	L9
Madura, Indonesia	6	P10
Mafeking, S. Africa	13	F8
Mafia I., Tanzania	13	G6
Magadan, U.S.S.R.	6	T4
Magdalena, Bolivia	8	D5
Magdeburg, Germany	3	I5
Magellan's Str., Chile	8	CD9
Maggie's Springs, Australia	11	E5
Magnitogorsk, U.S.S.R.	4	T5
Mahenge, Tanzania	13	G6
Mahia Pen., New Zealand	12	h13
Maiduguri, Nigeria	13	E4
Maimana, Afghanistan	5	I6
Maindombe Lac, Zaire	13	E6
Maine, U.S.A.	7	O5
Mainland, Orkney Is.	3	F4
Mainland, Shetland Is.	3	F3
Maintirano, Malagasy	13	H7
Maitland, Australia	12	I6
Majorca, Balearic Is.	3	G8
Majunga, Malagasy	13	H7
Makarikari Salt Pan, Botswana	13	F8
Makassar, Indonesia	6	P10
Makassar, Str. of, Indonesia	6	P9/10
Makemo I., Pacific Ocean	10	N6
Makhachkala, U.S.S.R.	4	R7
Makindu, Kenya	13	G6
Makin I., Pacific Ocean	9	H4
Makokou, Gabon	13	E5
Makurdi, Nigeria	13	D5
Malacca, Malaysia	6	O9
Malacca Str., Indonesia	6	h16
Malaga, Spain	3	F8
Malagasy Democratic Republic, Africa	13	H78
Malaita I., Pacific Ocean	9	G5
Malange, Angola	13	E6
Malaren L., Sweden	3	L4
Malargue, Argentina	8	D7
Malawi, Africa	13	G7
Malawi L., Malawi	13	G7
Malaya, Malaysia	6	O9
Malay Pen., Asia	5/6	OP9
Malaysia, Asia	6	OP9
Malden I., Pacific Ocean	10	M5
Maldive Is., Indian Ocean	5	L9
Male, Maldire Is.	5	L9
Malekula I., Pacific Ocean	9	G6
Maleolap At., Pacific Ocean	9	H4
Mali, Africa	13	CD4
Malindi, Kenya	13	G6
Mallowa Well, Australia	9	C4
Malmo, Sweden	3	I4
Malpelo I., Colombia	8	B3
Malta I., Europe	3	I8
Managua, Nicaragua	7	M8
Manakara, Malagasy	13	H8
Manaus, Brazil	8	D4
Manchester, U.K.	3	F5
Manchuria, China	6	QR45
Manchurian Plain, China	2	
Manda, Tanzania	13	G7
Mandalay, Burma	5	N7
Mandonbe L., Zaire	13	E6
Mangaia I., Pacific Ocean	10	M7
Mangalore, India	5	L8
Mangareva I., Pacific Ocean	10	O7
Manicore, Brazil	8	D4
Manifold C., Australia	12	I4
Manihii, I., Pacific Ocean	10	N6
Manila, Philippines	6	Q8
Manitoba, Canada	7	L4
Manizales, Colombia	8	C3
Mannheim, Germany	3	H6
Manono, Zaire	13	F6
Mansfield, Australia	12	H7
Manua Kea, Pacific Ocean	9	M3
Manukau, New Zealand	12	g13
Manych R., U.S.S.R.	4	Q6
Manzanillo, Mexico	7	I8
Mao, Chad	13	E4
Marabo, Brazil	8	F4
Maracaibo, Venezuela	8	C2
Maracaibo L., Venezuela	8	C3
Marajo I., Brazil	8	F4
Maralinga, Australia	11	E5
Maramba, Zambia	13	F7
Maranhao, Brazil	8	F4
Maranoa R., Australia	12	H5
Maranon R., Peru	8	C4
Marble Bar, Australia	11	B4
Mar del Plata, Argentina	11	E7
Mare I., Pacific Ocean	9	G7
Margarita I., Venezuela	8	D2
Marguerite Bay, Antarctica	15	O3
Maria Is., Pacific Ocean	10	M7
Mariana Is., Pacific Ocean	9	E3
Marianas Trench, Pacific Ocean	9	E3
Maria Van Diemen C., New Zealand	12	g12
Marie Byrd Land, Antarctica	15	R2
Marion Is., Antarctica	15	H5
Marion Reef, Australia	12	I3
Markham Mt., Antarctica	15	B1
Marlborough, Australia	12	H4
Marne R., France	3	G6
Maroantsetra, Malagasy	13	H7
Marokau I., Pacific Ocean	10	N6
Maro Reef, Pacific Ocean	10	I2
Marquesas Is., Pacific Ocean	10	O5
Marrakesh, Morocco	13	C2
Marrawah, Australia	12	a9
Marree, Australia	12	F5
Marsabit, Kenya	13	G5
Marsala, Sicily	3	I8
Marseilles, France	3	H7
Marshall Is., Pacific Ocean	9	GH3
Marshall R., Australia	11	F4
Martin Pen., Antarctica	15	q2
Martinique, Caribbean Sea	8	D2
Marvroa I., Pacific Ocean	10	O7
Marutea I., Pacific Ocean	10	O7
Maryborough, Australia	12	I5
Masbate, Philippines	6	Q8
Maseru, Lesotho	13	F8
Mashhad, Iran	5	H6
Masira I., Oman	5	H7
Masson, Antarctica	15	E3
Masterton, New Zealand	12	h14
Matadi, Zaire	13	E6
Matapan C., Greece	4	M8
Mataranka, Australia	11	E2
Matata, New Zealand	12	h13
Matavia I., Pacific Ocean	10	N6
Mato Grosso Plateau, Brazil	8	E5
Matthew I., Pacific Ocean	9	H7
Maui I., Pacific Ocean	10	M2
Maun, Botswana	13	F8
Mauna Kea, Hawaii	10	M3
Maupikao I., Pacific Ocean	10	M6
Maurice L., Australia	11	E5
Mauritania, Africa	13	BC34
Mawsil, Iraq	5	G6
Mawson, Antarctica	15	E3
Maykop, U.S.S.R.	4	Q7
Mazar-i-Sharif, Afghanistan	5	I6
Mazatlan, Mexico	7	I7
Mbabane, S. Africa	13	F8
Mbandaka, Zaire	13	E5
Mecca, Saudi Arabia	5	G7
Medan, Indonesia	5	N9
Medford, U.S.A.	7	G5
Medicine Hat, Canada	7	H4
Medina, Saudi Arabia	5	G7
Mediterranean Sea	3/4	F-O89
Meekatharra, Australia	11	B5
Mekong R., S. Asia	6	O8
Melanesia Is., Pacific Ocean	9	CDEFG 456
Melbourne, Australia	12	G7
Melilla, Morocco	13	C2
Mellis Seamont, Pacific Ocean	9	H1
Mellish Reef, Pacific Ocean	9	F6
Melville Bay, Greenland	7	O2
Melville C., Australia	12	G2
Melville I., Australia	11	E2
Melville I., Canada	7	HI2
Melville Pen., Canada	7	M3
Memphis, U.S.A.	7	M6
Menado, Indonesia	6	Q9
Menam R., Thailand	5	NO8
Mendocino Cape, U.S.A.	7	GH5
Mendoza, Argentina	8	D7
Mengtsz, China	6	O7
Menindee, Australia	12	G6
Menindee L., Australia	12	G6
Mentawai Is., Indonesia	5	NO10
Menzies, Australia	11	C5
Meramangye L., Australia	11	E5
Mercedes, Argentina	8	D7
Mercedes, Uruguay	8	E7
Mercury Is., New Zealand	12	h13
Mergui Arch., Burma	5	N8
Merida, Mexico	7	M7
Merka, Somali	13	H5
Merredin, Australia	11	B6
Mesewa, Ethiopia	13	G4
Messina, S. Africa	13	F8
Messina, Sicily	3	L8
Messina, Str. of, Sicily	3	L8
Methuen, New Zealand	12	g14
Metz, France	3	H6
Mexicali, Mexico	7	H6
Mexico, C. America	7	
Mexico City, Mexico	7	L8
Mexico, Gulf of, North America	7	LM7
Miami, U.S.A.	7	M7
Michigan L., U.S.A.	7	M5
Michigan State, U.S.A.	7	M5
Micronesia, Pacific Ocean	9	EFG4
Middleton, Australia	12	G4
Middleton Reef, Pacific Ocean	9	FG7
Midway I., Pacific Ocean	10	I2
Milan, Italy	3	H6
Mildura, Australia	12	G6
Miles, Australia	12	I5
Mili At., Pacific Ocean	9	H4
Miling, Australia	11	B6
Mill I., Antarctica	15	d3
Millerovo, U.S.S.R.	4	Q6
Milparinka, Australia	12	G5
Milwaukee, U.S.A.	7	M5
Minami Tori Shima, Pacific Ocean	9	F2
Minas Gerais, Brazil	8	F5
Mindanao I., Philippines	6	Q9
Mindoro, Philippines	6	Q8
Mingary, Australia	12	G6
Minigwal L., Australia	11	C5
Minneapolis, U.S.A.	7	L5
Minnesota State, U.S.A.	7	L5
Minnipa, Australia	11	F6
Minorca I., Balearic Is.	3	G8
Minsk, U.S.S.R.	4	N5
Miquelon I., N. America	7	P5
Mirnyy, Antarctica	15	E3
Miskolc, Hungary	4	M6
Mississippi, U.S.A.	7	L6
Mississippi R., U.S.A.	7	L6
Missouri R., U.S.A.	7	L5
Missouri State, U.S.A.	7	L6
Mitchell, Australia	12	H5
Mitchell River, Australia	12	G3
Mitre I., Pacific Ocean	9	H6
Mitu, Colombia	8	C3
Mivanza, Tanzania	13	G6
Mo, Norway	3	I2
Mobara, Japan	6	e13
Mocuba, Mozambique	13	G7
Mogadishu, Somali	13	H5
Mokpo, S. Korea	6	Q6
Moldavia, U.S.S.R.	4	N6
Mollendo, Peru	8	C5
Molucca Sea, Indonesia	9	C45
Molucca Is., Indonesia	6	Q10
Mombasa, Kenya	13	G6
Monaco, Europe	3	H7
Mongalla, Sudan	13	G5
Monger Lake, Australia	11	B5
Mongolia, Asia	5/6	NOP5
Mongolia, Plateau of, Mongolia	2	
Mongu, Zambia	13	F7
Monrovia, Liberia	13	B5
Montagu Is., S. Sandwich Is.	15	M4
Montana State, U.S.A.	7	HI5
Monte Bello I., Australia	11	B4
Monterrey, Mexico	7	I7
Montevideo, Uruguay	8	E7
Montgomery, U.S.A.	7	M6
Montlucon, France	3	G6
Montpellier, France	3	G7
Montreal, Canada	7	N5
Moora, Australia	11	B6
Moore Embaym, Antarctica	15	a2
Moore Lake, Australia	11	B5
Moosonee, Canada	7	M4
Mopti, Mali	13	C4
Morane I., Pacific Ocean	10	O7
Morava R., Czechoslovakia	3	L6
Moray Firth, U.K.	3	F4
Moree, Australia	12	H5
Moreton, Australia	12	G2
Moreton I., Australia	12	I5
Morgan, Australia	12	F6
Morgans, Australia	11	C5
Mornington I., Australia	12	F3
Morocco, Africa	13	C23
Morondava, Malagasy	13	H8
Moroni, Comoro Arch.	13	H7
Mortlock Is., Pacific Ocean	9	F4
Moscow, U.S.S.R.	4	P4
Mossamedes, Angola	13	E7
Mosul, Iraq	5	G6
Mouila, Gabon	13	E6
Moulmein, Burma	5	N8
Mount Douglas, Australia	12	H4
Mount Gambier, Australia	12	G7
Mount Goldsworthy, Australia	11	B4
Mount Isa, Australia	12	F4
Mount Magnet, Australia	11	B5
Mount Morgan, Australia	12	I4
Mt. Panie I., Pacific Ocean	9	G7
Moura, Brazil	8	D4
Moyata, Ethiopia	13	G5
Mozambique, Africa	13	G78
Mozambique Channel	13	GH78
Mtwara, Tanzania	13	H7
Muang Ubon, Thailand	6	O8
Mukalla, South Yemen	5	G8
Mukden, China	6	Q5
Mulhouse, France	3	H6
Mull I., U.K.	3	E4
Mullewa, Australia	11	B5
Mulobezi, Zambia	13	F7
Multan, Pakistan	5	L6
Mundi Windi, Australia	11	C4
Mungbere, Zaire	13	F5
Mungindi, Australia	12	H5
Munich, Germany	3	I6
Munster, Germany	3	H5

Name	Page	Grid
Muonio R., Sweden/Finland	4	MN2
Murchison R., Australia	11	B5
Mures R., Romania	4	M6
Murmansk, U.S.S.R.	4	O2
Murray Bridge, Australia	12	F7
Murray R., Australia	12	H7
Murrumbidgee R., Australia	12	H6
Murtoa, Australia	12	G7
Mururoa, Cameroon	13	E4
Murwillumbah, Australia	12	I5
Muscat, Oman	5	H7
Musgrave, Australia	12	G2
Musgrave Ranges, Australia	11	E5
Muswellbrook, Australia	12	I6
Mutankiang, China	6	Q5
Muttaburra, Australia	12	G4
Mwanza, Tanzania	13	G6
Mweru L., Zambia	13	F6
Myitkyina, Burma	5	N7
Mysore, India	5	L8

N

Name	Page	Grid
Nafud region, Saudi Arabia	5	FG7
Nafud Desert, Saudi Arabia	5	FG7
Nagasaki, Japan	6	Q6
Nagelle, Ethiopia	13	G5
Nagoya, Japan	6	R6
Nagpur, India	5	L7
Naha, Japan	6	Q7
Nairobi, Kenya	13	G6
Nakhichevan, U.S.S.R.	4	R8
Nakumanu Is., Pacific Ocean	9	F5
Nambour, Australia	12	I5
Namib Desert, S.W. Africa	2	
Namoi R., Australia	12	H6
Nampula, Mozambique	13	G7
Namsos, Norway	3	I3
Nanchang, China	6	P7
Nancy, France	3	H6
Nanking, China	6	P6
Nanning, China	6	O7
Nanpo Shafto, Pacific Ocean	9	D1
Nansen Mt., Antarctica	15	t1
Nantes, France	3	F6
Nanumea I., Pacific Ocean	9	H5
Napier, New Zealand	12	h13
Naples, Italy	3	I7
Napo R., Peru	8	C4
Napuka I., Pacific Ocean	10	N6
Naracoorte, Australia	12	G7
Narbonne, France	3	G7
Narrabri, Australia	12	H6
Narran L., Australia	12	H5
Narrogin, Australia	11	B6
Narvik, Norway	3	L2
Naryan-Mar, U.S.S.R.	4	S2
Nashville, U.S.A.	7	M6
Nassau, Bahamas	7	N7
Nassau I., Pacific Ocean	10	L6
Nasser L., A.R.E.	13	G3
Natal, Brazil	8	G4
Natal, S. Africa	13	G8
Naturaliste Cape, Australia	11	AB6
Naturaliste Channel, Australia	11	A5
Nauru I., Pacific Ocean	9	G5
Navarin C., U.S.S.R.	14	Li3
N'djamena, Chad	13	E4
Ndola, Zambia	13	F7
Nebraska State, U.S.A.	7	IL5
Necker I., Pacific Ocean	10	L2
Necker Ridge, Pacific Ocean	10	L2
Negro R., Argentina	8	D78
Negro R., Brazil	8	D4
Negros I., Philippines	6	Q9
Neilsson Reef, Pacific Ocean	10	N7
Neiva, Colombia	8	C3
Nejd region, Saudi Arabia	5	G7
Nelson, New Zealand	12	g14
Nengonengo I., Pacific Ocean	10	N6
Nepal, Asia	5	M7
Netherlands, Europe	3	GH5
Neuquen, Argentina	8	D7
Nevada State, U.S.A.	7	H56
Nevers, France	3	G6
New Britain, New Guinea	9	F5
New Brunswick, Canada	7	O5
New Caledonia, Pacific Ocean	9	G7
Newcastle, Australia	12	I6
Newcastle, U.K.	3	F4
Newcastle Waters, Australia	11	E3
Newdegate, Australia	11	B6
New England Range, Australia	12	I56
Newfoundland I., Canada	7	OP5
New Georgia I., Pacific Ocean	9	F5
New Guinea, Pacific Ocean	9	DE5
New Hanover I., New Guinea	9	F5
New Haven, U.S.A.	7	N5
New Hebrides, Pacific Ocean	9	GH6
New Ireland, New Guinea	9	F5
New Mexico State, U.S.A.	7	I6
New Norfolk, Australia	12	b9
New Orleans, U.S.A.	7	M6
New Plymouth, New Zealand	12	g13
New Schwabenland, Antarctica	15	LiI2
New Siberian Is., U.S.S.R.	5	RS2
New South Wales, Australia	12	GH6
New York State, U.S.A.	7	N5
New York, U.S.A.	7	N5
New Zealand	12	
Ngami L., Botswana	13	F8
N'Guigmi, Niger	13	E4
Nguru, Nigeria	13	E4
Niagara Falls, N. America	7	O N5
Niamey, Niger	13	D4
Nias I., Indonesia	5	N9
Nicaragua, C. America	7	M8
Nicaragua L., Nicaragua	7	M8
Nice, France	3	H7
Nicobar Is., Indonesia	5	N9
Nicosia, Cyprus	5	F6
Nigata, Japan	6	R6
Niger, Africa	13	DE34
Niger R., W. Africa	13	CD45
Nigeria, Africa	13	DE45
Nightcaps, New Zealand	12	f15
Nihoa I., Pacific Ocean	10	L2
Niihau I., Pacific Ocean	10	L2
Nikolayev, U.S.S.R.	4	O6
Nikolayevsk, U.S.S.R.	6	R4
Nile River, Africa	13	G345
Nimes, France	3	G7
Ninigo Group, Pacific Ocean	9	E5
Nipigon L., Canada	7	M5
Nis, Yugoslavia	4	M7
Niulakita, Pacific Ocean	10	H6
Nizhne Kolymsk, U.S.S.R.	6	U3
Nome, Alaska	7	C3
Nonouti I., Pacific Ocean	9	H5
Nordvik, U.S.S.R.	5	P2
Norfolk I., Pacific Ocean	9	G7
Norman R., Australia	12	G3
Normanton, Australia	12	G3
Norman Wells, Canada	7	G3
Nornalup, Australia	11	B6
Norrkoping, Sweden	3	L4
Norseman, Australia	11	C6
Northam, Australia	11	B6
Northampton, Australia	11	A5
North Cape, Antarctica	15	a2
North Cape, New Zealand	12	G12
North Cape, Norway	4	N1
North Carolina State, U.S.A.	7	MN6
North Cliffe, Australia	11	B6
North Dakota State, U.S.A.	7	IL5
North Dvina R., U.S.S.R.	4	Q3
North East Land, Arctic	14	B1
Northern Ireland, U.K.	3	E5
Northern Territory, Australia	11	EF34
North Island, New Zealand	12	g13
North Sea, Europe	3	G4
North Stradbroke I., Australia	12	I5
North Taranaki Bight, New Zealand	12	g13
North West Cape, Australia	11	A4
Northwest Territories, Canada	7	G-N3
Norwegian Sea, N. Europe	3	DEFG3
Norwich, U.K.	3	G5
Nouakchott, Mauritania	13	B4
Nova Scotia, Canada	7	O5
Novaya Zemlya, U.S.S.R.	5	HI2
Novgorod, U.S.S.R.	4	O4
Novi Sad, Yugoslavia	3	L6
Novokuznetsk, U.S.S.R.	5	M4
Novosibirsk, U.S.S.R.	5	M4
Nowra, Australia	12	I6
Nua I., Pacific Ocean	10	I6
Nubia, Africa	2	
Nuevo Laredo, Mexico	7	L7
Nukhd, U.S.S.R.	4	R7
Nuku Hiva I., Pacific Ocean	10	N5
Nukuoro At., Pacific Ocean	9	F4
Nullarbor, Australia	11	E6
Nunkiang, China	6	Q5
Nuoro, Sardinia	3	H7
Nuremberg, Germany	3	I6
Nyala, Sudan	13	F4
Nyngan, Australia	12	H6

O

Name	Page	Grid
Oahu I., Pacific Ocean	10	M2
Oakland, U.S.A.	7	G6
Oamaru, New Zealand	12	g15
Oates Coast, Antarctica	15	aB2
Oaxaca, Mexico	7	L8
Ob R., U.S.S.R.	5	I3
Obbia, Somalia	13	H5
Obidos, Brazil	8	E4
Ocean I., Pacific Ocean	9	G5
October Revolution I., U.S.S.R.	5	N2
Odense, Denmark	3	H4
Odessa, U.S.S.R.	4	O6
Odienne, Ivory Coast	13	C5
Oena, A.R.E.	13	G3
Ogbomosho, Nigeria	13	D5
Ogden, U.S.A.	7	H5
Ohio, U.S.A.	7	M5
Oimyakon, U.S.S.R.	6	S3
Oka R., U.S.S.R.	4	Q45
Okhotsk, U.S.S.R.	6	S4
Okhotsk, Sea of, U.S.S.R.	6	ST4
Okinawa I., Asia	6	Q7
Oklahoma City, U.S.A.	7	L6
Oklahoma State, U.S.A.	7	L6
Oland I., Sweden	3	L4
Olbia, Sardinia	3	H7
Olekminsk, U.S.S.R.	6	Q3
Olenek, U.S.S.R.	5	P3
Oleron, Ile d', France	3	F6
Olsztyn, Poland	4	M5
Omaha, U.S.A.	7	L5
Oman, Asia	5	H78
Omdurman, Sudan	13	G4
Omsk, U.S.S.R.	5	L4
Onega, U.S.S.R.	4	P3
Onega L., U.S.S.R.	4	OP3
Ongerup, Australia	11	B6
Onslow, Australia	11	A4
Ontario, Canada	7	MN45
Ontario L., Canada/U.S.A.	7	N5
Oodnadatta, Australia	11	F5
Ooldea, Australia	11	E6
Oporto, Portugal	3	E7
Opparinna, Australia	11	E5
Opua, New Zealand	12	g13
Oradea, Romania	4	M6
Orange, Australia	12	H6
Orange Free State, S. Africa	13	F8
Orange R., S. Africa	13	EF89
Orbost, Australia	12	H7
Orcadas, S. Orkney Is.	15	N3
Ord R., Australia	11	D3
Ordzhonikidze, U.S.S.R.	4	Q7
Orebro, Sweden	3	L4
Oregon State, U.S.A.	7	G5
Orel, U.S.S.R.	4	P5
Orenburg, U.S.S.R.	4	T5
Orense, Spain	3	E7
Orinoco R., Venezuela	8	D3
Orkney I., U.K.	3	F4
Orleans, France	3	G6
Oroya, Peru	8	C5
Orsk, U.S.S.R.	4	T5
Oruro, Bolivia	8	D5
Osaka, Japan	6	R6
Oslo, Norway	3	I4
Osprey Reef, Australia	12	H2
Ossa Mt., Australia	12	b9
Ostend, Belgium	3	G5
Ostersund, Sweden	3	I3
Ostrava, Czechoslovakia	3	L6
Otranto, Italy	3	L7
Ottawa, Canada	7	N5
Otway C., Australia	12	G7
Ouagadougou, Volta	13	C4
Ouahran, Algeria	13	C2
Ouanda Djale, Cen. Africa	13	F5
Ouesso, Congo	13	E5
Oulu, Finland	4	N3
Outjo, S.W. Africa	13	E7
Ouvea I., Pacific Ocean	9	G67
Ouyen, Australia	12	G7
Ovamboland region, S.W. Africa	2	
Oviedo, Spain	3	E7
Owen Stanley Range, New Guinea	9	EF56
Oxley, Australia	12	G6

P

Name	Page	Grid
Padang, Indonesia	5	O10
Padua, Italy	3	I6
Pagan I., Pacific Ocean	9	E3
Pakistan, Asia	5	ILM67
Palau Is., Caroline Is.	6	R9
Palawan I., Philippines	6	P89
Palembang, Indonesia	6	O10
Palermo, Sicily	3	I8
Palliser C., New Zealand	12	h14
Palma, Brazil	8	F5
Palma, Majorca	3	G8
Palmer Arch., Antarctica	15	O3
Palmer Land, Antarctica	15	n2
Palmerston At., Pacific Ocean	10	L6
Palmerston North, New Zealand	12	h14
Palmyra I., Pacific Ocean	10	L4
Pamirs Mt., Asia	2	
Pampas region, Argentina	1	
Pamplona, Spain	3	F7
Panama, Cen. America	7	MN9
Panama City, Panama	7	N9
Panay, Philippines	6	Q8
Panie Mt., New Caledonia	9	G6
Panjim, India	5	L8
Pantelleria Is., Italy	3	I8
Paotow, China	6	P5
Papua and New Guinea, Australasia	9	DE5
Papua, Gulf of, Papua	9	E5
Para, Brazil	8	E4
Para R., Brazil	8	F4
Parachilna, Australia	12	F6
Paraguana, Brazil	8	F6
Paraguay, S. America	8	DE6
Paraguay R., Paraguay	8	E6
Paramaribo, Surinam	8	E3
Parana, Argentina	8	D7
Parana, Brazil	8	E6
Parana R., S. America	8	E6
Paratoo, Australia	12	F6
Paris, France	3	G6
Parkes, Australia	12	H6
Parma, Italy	3	I7
Parnaiba, Brazil	8	F4
Parnassus, New Zealand	12	g14
Paroo R., Australia	12	H5
Parramatta, Australia	12	I6
Parry Is., Canada	7	H12
Pasley Cape, Australia	11	C6
Passo Fundo, Brazil	8	E6
Pasto, Colombia	8	C3
Patchewollock, Australia	12	G7
Patience Well, Australia	11	D4
Patna, India	5	M7
Patras, Greece	4	M8
Paulding Bay, Antarctica	15	c3
Pavlodar, U.S.S.R.	5	L4
Paysandu, Uruguay	8	E7
Peace R., Canada	7	H4
Peak Hill, Australia	11	B5
Peary Land region, Greenland	7	RS1
Pechenga, U.S.S.R.	4	O2
Pechora R., U.S.S.R.	4	S2
Pecs, Hungary	3	L6
Peebinga, Australia	12	G6
Pegasus Bay, New Zealand	12	g14
Peking, China	6	P6
Pelotas, Brazil	8	E7
Pemba I., Tanzania	13	G6
Penang, Malaysia	5	O9
Penck C., Antarctica	15	e3
Penong, Australia	11	E6
Penrhyn At., Pacific Ocean	10	M5
Pensacola Mts., Antarctica	15	m1
Pentland, Australia	12	H4
Pentland Firth, Channel, U.K.	3	F4
Penza, U.S.S.R.	4	R5
Penzance, U.K.	3	E5
Penzhino, U.S.S.R.	6	U3
Percival Lakes, Australia	11	CD4
Perekop, U.S.S.R.	4	O6
Perigueux, France	3	G6
Perm, U.S.S.R.	4	T4
Persian Gulf, Asia	5	GH7
Perth, Australia	11	B6
Peru, S. America	8	C45
Perugia, Italy	3	I7
Pescara, Italy	3	I7
Peshawar, Pakistan	5	L6
Peterborough, Australia	12	F6
Peter Ist. Is., Antarctica	15	p3
Petropavlovsk, U.S.S.R.	5	I4

Place	No.	Ref.
Petrozavodsk, U.S.S.R.	4	O3
Pevek, U.S.S.R.	6	V3
Phan Rang, S. Vietnam	6	O8
Philadelphia, U.S.A.	7	N6
Philippines, Pacific Ocean	9	C34
Phnom Penh, Khmer Republic	6	O8
Phoenix, U.S.A.	7	H6
Phoenix I., Pacific Ocean	10	I5
Piaui, Brazil	8	F4
Picton, New Zealand	12	g14
Pietermaritzburg, S. Africa	13	G8
Pietersburg, S. Africa	13	F8
Pilsen, Czechoslovakia	3	I6
Pinaki I., Pacific Ocean	10	O6
Pinar del Rio, Cuba	7	M7
Pine Creek, Australia	11	E2
Pine Island Bay, Antarctica	15	Q2
Pines, I. of, Cuba	7	M7
Pins Ile des, Pacific Ocean	9	G7
Piraeus, Greece	4	M8
Pirapora, Brazil	8	F5
Pisco, Peru	8	C5
Pitcairn I., Pacific Ocean	10	O7
Pittsburgh, U.S.A.	7	M5
Piura, Peru	8	B4
Plate R., S. America	8	E7
Plenty, Bay of, New Zealand	12	h13
Pleven, Bulgaria	4	M7
Ploesty, Romania	4	N7
Plovdiv, Bulgaria	4	M7
Plymouth, U.K.	3	F5
Poinsett C., Antarctica	15	D3
Pointe Noire, Congo	13	E6
Pokataroo, Australia	12	H5
Poland, Europe	3/4	LM5
Polar Sub-Glacial Basin, Antarctica	15	Dd1
Poltava, U.S.S.R.	4	O6
Ponape I., Pacific Ocean	9	F4
Pondicherry, India	5	L8
Pontianak, Indonesia	6	O10
Poona, India	5	L8
Poopo L., Bolivia	8	D5
Popayan, Colombia	8	C3
Pori, Finland	4	M3
Porpoise Bay, Antarctica	15	c3
Port Adelaide, Australia	12	F6
Port Albert, Australia	12	H7
Port Augusta, Australia	12	F6
Port au Prince, Haiti	7	N8
Port aux Basques, Canada	7	P5
Port Blair, Indonesia	5	N8
Port Davey, Australia	12	b9
Port Douglas, Australia	12	H3
Port Elizabeth, S. Africa	13	F9
Port Harcourt, Nigeria	13	D5
Port Harrison, Canada	7	N4
Port Hedland, Australia	11	B4
Portland, Australia	12	G7
Portland, Oregon, U.S.A.	7	G5
Portland, Maine, U.S.A.	7	N5
Port Lincoln, Australia	11	F6
Port Moresby, Papua	12	H8
Port Nelson, Canada	7	L4
Port Nolloth, S. Africa	13	E8
Porto Alegre, Brazil	8	E6
Porto Amboin, Angola	13	E7
Porto Amelia, Mozambique	13	H7
Port of Spain, Trinidad	8	D2
Porto Nacional, Brazil	8	F5
Porto Novo, Dahomey	13	D5
Porto Velho, Brazil	8	D4
Port Phillip, Australia	12	G7
Port Pirie, Australia	12	F6
Port Radium, Canada	7	H3
Port Said, A.R.E.	13	G2
Portsmouth, U.K.	3	F5
Port Stanley, Falkland Is.	8	E9
Port Sudan, Sudan	13	G4
Portugal, Europe	3	E78
Portugalia, Angola	13	F6
Potenza, Italy	3	L7
Potosi, Bolivia	8	D5
Powell Creek, Australia	11	E3
Poznan, Poland	3	L5
Prague, Czechoslovakia	3	I5
A. Pratt Cape, S. Shetland Is.	15	n3
Pres Pedro Aguirre Cedra, S. Shetland Is.	15	03
Pres G. Gonzales Videla, Antarctica	15	O3
Pretoria, S. Africa	13	F8
Pribilof I., Alaska	7	BC4
Prince Albert, Canada	7	I4
Prince Albert Mts., Antarctica	15	B2
Prince Charles Mts., Antarctica	15	f2
Prince Edward I., Antarctica	15	H5
Prince Edward I., Canada	7	O5
Prince George, Canada	7	G4
Prince Harald Coast, Antarctica	15	H2
Prince of Wales Cape, Alaska	7	C3
Prince of Wales I., Australia	12	G2
Prince of Wales I., Canada	7	IL2
Prince Olaf Coast, Antarctica	15	g3
Prince Patrick I., Canada	7	GH2
Prince Rupert, Canada	7	G4
Princess Astrid Coast, Antarctica	15	iI2
Princess Charlotte Bay, Australia	12	G2
Princess Martha Coast, Antarctica	15	lLi2
Princess Ragnhild Coast, Antarctica	15	hH23
Pripet Marshes, U.S.S.R.	4	N5
Proclamation I., Antarctica	15	G3
Prome, Burma	5	N8
Proserpine, Australia	12	H4
Proston, Australia	12	I5
Prydz Bay, Antarctica	15	F3
Przemysl, Poland	4	M6
Pskov, U.S.S.R.	4	N4
Pto. Aysen, Chile	8	C8
Pto. Casado, Brazil	8	E6
Pto. Deseado, Argentina	8	D8
Puebla, Mexico	7	L8
Pueblo, U.S.A.	7	I6
Puerto Montt, Chile	8	C8
Puerto Rico, C. America	7	O8
Puerto Suarez, Bolivia	8	E5
Pukapuka At., Pacific Ocean	10	L5
Puka Puka I., Pacific Ocean	10	O6
Pukarua I., Pacific Ocean	10	O6
Pulap At., Pacific Ocean	9	E4
Punakha, Bhutan	5	M7
Puno, Peru	8	C5
Punta Arenas, Chile	8	C9
Pusan, S. Korea	6	Q6
Putumayo R., S. America	8	C4
Pyatigorsk, U.S.S.R.	4	Q7
Pyong Yang, China	6	Q6
Pyrenees Mts., Europe	2	

Q

Place	No.	Ref.
Qatar, Asia	5	H7
Quebec, Canada	7	NO45
Queen Alexandra Range, Antarctica	15	a1
Queen Charlotte Is., Canada	7	F4
Queen Elizabeth Is., Canada	7	H-M2
Queen Mary Coast, Antarctica	15	E3
Queen Maud Land, Antarctica	15	Ii2
Queen Maud Range, Antarctica	15	T1
Queensland, Australia	12	GH4
Queenstown, Australia	12	b9
Quelimane, Mozambique	13	G7
Quetta, Pakistan	5	I6
Quezon City, Philippines	6	Q8
Quilpie, Australia	12	G5
Quito, Ecuador	8	C4

R

Place	No.	Ref.
Rabat, Morocco	13	C2
Raeside L., Australia	11	C5
Raiatea I., Pacific Ocean	10	M6
Raipur, India	5	M7
Raivauar I., Pacific Ocean	10	N7
Ralik Chain, Pacific Ocean	9	G34
Ramapo Trench, Pacific Ocean	9	E2
Ranchi, India	5	M7
Rangiora, New Zealand	12	g14
Rangiroa I., Pacific Ocean	10	N6
Rangitikei R., New Zealand	12	h13/14
Rangoon, Burma	5	N8
Ranken Store, Australia	11	F3
Rantemario Mt., Indonesia	9	B5
Raoul I., Pacific Ocean	10	I7
Rapa I., Pacific Ocean	10	N7
Raroia I., Pacific Ocean	10	N6
Rarotonga I., Pacific Ocean	10	L7
Rasht, Iran	5	G6
Rason L., Australia	11	C5
Ratak Chain, Pacific Ocean	9	H34
Raukumara Range, New Zealand	12	h13
Ravensthorpe, Australia	11	C6
Ravenswood, Australia	12	H4
Rawalpindi, Pakistan	5	L6
Rawlinna, Australia	11	D6
Recherche, Arch. of the, Australia	11	C6
Recife, Brazil	8	G4
Red Sea, Africa/Asia	13	GH34
Regensburg, Germany	3	I6
Reggio, Italy	3	L8
Regina, Canada	7	I4
Renmark, Australia	12	G6
Rennell I., Pacific Ocean	9	G6
Rennes, France	3	F6
Rennick Bay, Antarctica	15	a3
Reo I., Pacific Ocean	10	O6
Resistencia, Argentina	8	E6
Resolution I., Canada	7	O3
Resolution I., New Zealand	12	f15
Reunion, Indian Ocean	2	
Reykjavik, Iceland	3	B3
Rheims, France	3	G6
Rhine R., Europe	3	H56
Rhodes I., Greece	4	N8
Rhodesia, Africa	53	FG78
Rhone R., France	3	G7
Ribeirao Preto, Brazil	8	F6
Riberalta, Bolivia	8	D5
Richmond, Australia	12	G4
Richmond, U.S.A.	7	N6
Riga, U.S.S.R.	4	M4
Rijeka, Yugoslavia	3	I6
Rimatara I., Pacific Ocean	10	M7
Rio Branco, Brazil	8	D3D4
Rio de Janeiro, Brazil	8	F6
Rio Gallegos, Argentina	8	D9
Rio Grande, Brazil	8	E7
Rio Grande R., Mexico/U.S.A.	7	IL7
Ritscher Upland, Antarctica	15	Li2
Rivera, Uruguay	8	E7
Riyadh, Saudi Arabia	5	G7
Robert English Coast, Antarctica	15	oO2
Robertson Bay, Antarctica	15	Aa2
Rocas I., Brazil	8	G4
Rocha, Uruguay	8	E7
Rockefeller Mts., Antarctica	15	S2
Rockefeller Plateau, Antarctica	15	r12
Rockhampton, Australia	12	I4
Rocky Mts., N. America	1	
Roebourne, Australia	11	B4
Roma, Australia	12	H5
Romania, Europe	4	MN67
Rome, Italy	3	I7
Rondonia, Brazil	8	D5
Rongelap At., Pacific Ocean	9	G3
Roosevelt I., Antarctica	15	T2
Roosevelt R., Brazil	84	D45
Roraima, Brazil	8	D3
Roros, Norway	3	I3
Rosario, Argentina	8	D7
Rose I., Pacific Ocean	10	L6
Ross, New Zealand	12	g14
Ross Dependency, Antarctica	15	Tt3
Ross I., Antarctica	15	a2
Ross Ice Shelf, Antarctica	15	TtA1
Ross Sea, Antarctica	15	Tt2
Rostock, Germany	3	I5
Rostov, U.S.S.R.	4	P6
Rotorua, New Zealand	12	h13
Rotterdam, Netherlands	3	G5
Rotuma I., Pacific Ocean	9	H6
Rouen, France	3	G6
Rovaniemi, Finland	4	N2
Rovno, U.S.S.R.	4	N5
Rowley Shoals, Australia	11	B3
Roy Hill, Australia	11	C4
Ruapehu, New Zealand	12	h13
Rubtsovsk, U.S.S.R.	5	M4
Rudolf L., Kenya	13	G5
Rugen I., Germany	3	I5
Rururu I., Pacific Ocean	10	M7
Ruse, Bulgaria	4	N7
Russia (U.S.S.R.), Europe and Asia	5/6	
Rwanda, Africa	13	FG6
Ryazan, U.S.S.R.	4	P5
Ryukyu Is., Pacific Ocean	6	QR7

S

Place	No.	Ref.
Sabah, Malaysia	6	P9
Sabbah, Libya	13	E3
Sabrina Coast, Antarctica	15	cD3
Sacramento, U.S.A.	7	G6
Sahara Desert, N. Africa	1/2	
Saharanpur, India	5	L7
Saigon-Cholon, S. Vietnam	6	O8
St. Etienne, France	3	G6
St. George, Australia	12	H5
St. George's Channel, Ireland/Great Britain	3	EF56
St. Helena I., Atlantic Ocean	13	C7
St. John, Canada	7	O5
St. Johns, Quebec, Canada	7	P5
St. Lawrence, Australia	12	H4
St. Lawrence, Gulf of, Canada	7	O5
St. Lawrence I., Alaska	7	B3
St. Lawrence R., N. America	7	NO5
St. Louis, Senegal	13	B4
St. Louis, U.S.A.	7	L6
St. Mary's, Australia	12	b9
St. Mathieu, Pointe de, France	3	F6
St. Matthias Group, Pacific Ocean	9	E5
St. Nazaire, France	3	F6
St. Paul, U.S.A.	7	L5
St. Pierre and Miquelon Is., N. America	7	P5
St. Quentin, France	3	G6
St. Tome and Principe Is., Africa	13	D5
St. Vincent Cape, Portugal	3	E8
St. Vincent Gulf, Australia	12	F67
Saipan I., Pacific Ocean	9	E3
Sakhalin I., U.S.S.R.	6	S45
Salamanca, Spain	3	E7
Sale, Australia	12	H7
Salekhard, U.S.S.R.	5	I3
Salem, U.S.A.	7	G5
Salerno, Italy	3	I7
Salisbury, Rhodesia	13	G7
Salonika, Greece	4	M7
Salta, Argentina	8	D6
Saltillo, Mexico	7	I7
Salt Lake, Australia	11	A4
Salt Lake City, U.S.A.	7	H5
Salum, A.R.E.	13	F2
Salvador, Brazil	8	G5
Salzburg, Austria	3	I6
Samar I., Philippines	6	Q8
Samarkand, U.S.S.R.	5	I6
Samoa Is., Pacific Ocean	10	I6
Samsun, Turkey	5	F5
San Ambrosio I., Pacific Ocean	8	C6
San Antonio, U.S.A.	7	L7
San Carlos, Venezuela	8	D3
San Carlos de Bariloche, Argentina	8	C8
San Cristobal, Pacific Ocean	9	G6
San Cristobal, Venezuela	8	C3
San Diego, U.S.A.	7	H6
San Felix, Chile	8	B6
San Fernando de Apure, Venezuela	8	D3
San Fernando de Atabapo, Venezuela	8	D3
San Francisco, U.S.A.	7	G6
San Jose, Costa Rica	7	M9
San Juan, Argentina	8	D7
San Juan, Virgin Is.	7	O8
San Luis Potosi, Mexico	7	I7
San Marino, Europe	3	I7
San Matias, G. of, Argentina	8	D8
San Rafael, Argentina	8	D7
San Salvador, El Salvador	7	M8
San Sarvador I., Bahamas	7	N7
Sanae, Antarctica	15	L2
Sandakan, Malaysia	6	P9
Sandover R., Australia	11	F4
Sandy C., Australia	12	I4
Sandy I., Australia	11	C2
Sankuru R., Congo	13	F6
Santa Clara, Cuba	7	N7
Santa Cruz, Argentina	8	D9
Santa Cruz, Bolivia	8	D5
Santa Cruz I., Ecuador	8	B4
Santa Cruz Is., Pacific Ocean	9	G6

Santa Fe, Argentina	8	D7
Santa Fe, U.S.A.	7	I6
Santa Isabel, Fernando Poo I.	13	D5
Santa Isabel I., Pacific Ocean	9	F5
Santa Maria, Brazil	8	E6
Santa Rosa, Argentina	8	D7
Santander, Spain	3	F7
Santarem, Brazil	8	E4
Santiago, Chile	8	C7
Santiago de Cuba, Cuba	7	N7
Santiago del Estero, Argentina	8	D6
Santo Domingo, Dominican Rep.	7	O8
Santos, Brazil	8	F6
Sao Francisco R., Brazil	8	FG45
Sao Luis, Brazil	8	F4
Sao Paulo, Brazil	8	F6
Sao Tome I., Gulf of Guinea	13	D5
Saone R., France	3	G6
Sapporo, Japan	6	S5
Saragossa, Spain	3	F7
Sarajevo, Yugoslavia	3	L7
Saransk, U.S.S.R.	4	R5
Sarapul, U.S.S.R.	4	S4
Saratov, U.S.S.R.	4	R5
Sarawak, Malaysia	6	P9
Sardinia I., Italy	3	H78
Sarema, U.S.S.R.	4	M4
Saskatchewan, Canada	7	I4
Saskatchewan R., Canada	7	I4
Saskatoon, Canada	7	I4
Sasobani, Ethiopia	13	H5
Sassari, Sardinia	3	H7
Saudi Arabia, Asia	5	GH78
Saumarez Reefs, Australia	12	I4
Saunders Is., S. Sandwich Is.	15	M4
Sava R., Yugoslavia	3	L7
Savai'i, I., Pacific Ocean	10	I6
Savannah, U.S.A.	7	M6
Sawknah, Libya	13	E3
Saxby R., Australia	12	G3
Sayn Shanda, Mongolia	6	O5
Scandinavia, N. Europe	2	
Schefferville, Canada	7	O4
Schwangliab, China	6	Q5
Scilly Is., U.K.	3	E6
Scoresby Sound, Greenland	7	S2
Scotia Sea, Antarctica	15	Nm4
Scotland, Europe	3	EF34
Scott I., Antarctica	15	t3
Scott Reef, Australia	11	C2
Seattle, U.S.A.	7	G5
Seeheim, S.W. Africa	13	E8
Seine R., France	3	G6
Sekondi-Takoradi, Ghana	13	C5
Selvas region, S. America	1	
Selwyn, Australia	12	G4
Semarang, Indonesia	6	P10
Semipalatinsk, U.S.S.R.	5	M4
Sendai, Japan	6	S6
Senegal, Africa	13	B4
Senegal R., Senegal	13	B4
Senja, Norway	3	L2
Senyavin Is., Pacific Ocean	9	F4
Seoul, S. Korea	6	Q6
Sept Iles, Canada	7	O4
Serov, U.S.S.R.	5	I4
Serpa Pinto, Angola	13	E7
Serpentine Lakes, Australia	11	D5
Setubal, Portugal	3	E8
Sevastopol, U.S.S.R.	4	O7
Severn R., Canada	7	M4
Severnaya Zemlya Is., U.S.S.R.	5	N12
Seville, Spain	3	E8
Seward, Alaska	7	E3
Seward Pen., Alaska	7	C3
Seychelles I., Indian Ocean	2	
Seydisfjordur, Iceland	3	D2
Sfax, Tunisia	13	E2
Shaba, Zaire	13	F67
Shackleton Ice Shelf, Antarctica	15	E3
Shackleton Inlet, Antarctica	15	A1
Shag Rocks, S. America	8	F9
Shakhty, U.S.S.R.	4	Q6
Shamva, Rhodesia	13	G7
Shanghai, China	6	Q6
Shark Bay, Australia	11	A5
Sheffield, U.K.	3	F5
Shetland Is., U.K.	3	F3
Shibeli R., Somalia	13	H5
Shigatse, China	5	M7
Shikoku, Japan	6	R6
Shillong, India	5	N7
Shiraz, Iran	5	H7
Shkoder, Albania	3	L7
Sholapur, India	5	L8
Siam, Gulf of, Thailand	5/6	O89
Sian, China	6	O6
Siberia, U.S.S.R.	5/6	I-Z23
Sibiu, Romania	4	M6
Sibolga, Indonesia	5	N9
Sicily I., Italy	3	I8
Sidi Ifni, Morocco	13	B3
Sidley Mt., Antarctica	15	R2
Sidney, Canada	7	O5
Sierra da Bandeira, Angola	13	E7
Sierra Leone, Africa	13	B5
Siguiri, Guinea	13	C4
Sikasso, Mali	13	C4
Sikkim, India	5	M7
Simferopol, U.S.S.R.	4	O7
Simpson Desert, Australia	11	F4
Singapore, Asia	6	O9
Sinhailien, China	6	P6
Sining, China	5	O6
Sinkiang-Uigur, China	5	LMN56
Siple Mt., Antarctica	15	R2
Siqui I., S. Orkney Is.	15	N3
Sir Pellew Group, Australia	12	F3
Sirte, Libya	13	E3
Siwa, A.R.E.	13	F3
Skagen, Denmark	3	I4
Skagerrak, Denmark	3	H4
Skagway, Alaska	7	F4
Skoplje, Yugoslavia	4	M7
Skovorodino, U.S.S.R.	6	Q4
Skye I., U.K.	3	E4
Sligo, Ireland	3	E5
Smolensk, U.S.S.R.	4	O5
Snake R., U.S.A.	7	H5
Snowy R., Australia	12	H7
Society Is., Pacific Ocean	10	M6
Socotra I., Yemen	5	H8
Sofia, Bulgaria	4	M7
Sogne Fjord, Norway	3	H3
Sokoto, Nigeria	13	D4
Solomon Is., Pacific Ocean	9	FG5
Somalia (Somali Republic), Africa	13	HI5
Somali Pen., Somalia	2	
Somerset, Australia	12	G2
Somerset I., Canada	7	L2
Sorong, Indonesia	6	R10
Söröy, Norway	3	L1
Sorrento, Australia	12	G7
South Africa, Africa	13	EF89
South Australia, Australia	11	EF56
South Carolina State, U.S.A.	7	M6
South China Sea, Asia	6	P89
South Dakota State, U.S.A.	7	IL5
South East C., Australia	12	b9
South Fiji Basin, Pacific Ocean	9	H7
South Georgia I., S. America	8	G9
South Island, New Zealand	12	g14
South Orkney Is., S. Atlantic Ocean	15	N3
South Sandwich Is., S. Atlantic Ocean	8	H9
South Shetland Is., S. Atlantic Ocean	15	nO3
South Taranaki Bight, New Zealand	12	g13
Southampton I., Canada	7	M3
Southern Alps, New Zealand	12	fg14
Southern Cross, Australia	11	B6
Southport, Australia	12	I5
South-West Africa, Africa	13	E8
Southwest C., New Zealand	12	f15
South Yemen, Asia	5	GH8
Sovetskaya Gavan, U.S.S.R.	6	S5
Soviet Union (U.S.S.R.), Europe and Asia	5/6	
Spain, Europe	3	EF78
Spencer Cape, Australia	11	F7
Spencer Gulf, Australia	11	F6
Spitsbergen	5	DE2
Split, Yugoslavia	3	L7
Spokane, U.S.A.	7	H5
Springsure, Australia	12	H4
Sri Lanka, Asia	5	M9
Srinagar, India	5	L6
Staaten R., Australia	12	G3
Stanley, Australia	12	b9
Stanovoy Range, U.S.S.R.	2	
Starbuck I., Pacific Ocean	10	M5
Stavanger, Norway	3	H4
Stavropol, U.S.S.R.	4	Q6
Ste. Marie C., Malagasy	13	H8
Steep Point, Australia	11	A5
Stelitamak, U.S.S.R.	4	T5
Stepnoy, U.S.S.R.	4	Q6
Stewart I., New Zealand	12	f15
Stockholm, Sweden	3	L4
Stony Tunguska R., U.S.S.R.	5	N3
Storm Bay, Australia	12	b9
Strangeways Springs, Australia	11	F5
Strasbourg, France	3	H6
Streaky Bay, Australia	11	E6
Strzelecki Creek, Australia	12	G5
Sturt Desert, Australia	12	G5
Stuttgart, Germany	3	H6
Subotica, Yugoslavia	3	L6
Suchow, China	6	P6
Sucre, Bolivia	8	D5
Sudan, Africa	13	FG45
Sudan region, Africa	2	
Sudbury, Canada	7	M5
Suez, A.R.E.	13	G2
Suez Canal, A.R.E.	13	G2
Sukarnapura, Indonesia	9	E5
Sukarno Peak, New Guinea	9	D5
Sukhumi, U.S.S.R.	4	Q7
Sukkur, Pakistan	5	I7
Sula Is., Indonesia	6	Q10
Sulawesi I., Indonesia	6	PQ10
Sulu Arch., Philippines	6	Q9
Sulu Sea, Philippines	6	PQ9
Sumatra, Indonesia	5/6	NO9/10
Sumba I., Indonesia	6	PQ10
Sumbawa I., Indonesia	6	P10
Sundsvall, Sweden	3	L3
Sungari R., China	6	QR5
Superior L., Canada/U.S.A.	7	M5
Sur, Oman	5	H7
Surabaja, Indonesia	6	P10
Surgut, U.S.S.R.	5	L3
Surinam, S. America	8	E3
Suwarrow At., Pacific Ocean	10	L6
Svalbard, Norway	5	E2
Sverdlovsk, U.S.S.R.	5	I4
Sverdrup Is., Canada	7	IL2
Swain Reefs, Australia	12	I4
Swains I., Pacific Ocean	10	I6
Swatow, China	6	P7
Swaziland, Africa	13	G8
Sweden, Europe	3	IL34
Switzerland, Europe	3	H6
Sydney, Australia	12	I6
Sydney I., Pacific Ocean	10	I5
Syktyvkar, U.S.S.R.	4	S3
Syr Darya R., U.S.S.R.	5	I5
Syracuse, Sicily	3	L8
Syria, Asia	5	FG6
Syrian Desert, Saudi Arabia	2	
Syrte, Libya	13	E2
Szczecin, Poland	3	I5
Szeged, Hungary	3	M6

T

Tabora, Tanzania	13	G6
Tabriz, Iran	5	G6
Tabuk, Saudi Arabia	5	F7
Tacoma, U.S.A.	7	G5
Tadzhikistan, U.S.S.R.	5	IL6
Tagus R., Portugal/Spain	3	E8
Tahcheng, China	5	M5
Tahiti I., Pacific Ocean	10	N6
Tahoua, Niger	13	D4
Tahuata, Pacific Ocean	10	O6
Taimyr Pen., U.S.S.R.	5	NO2
Tainan, Taiwan	6	Q7
Taipei, Taiwan	6	Q7
Taitao Pen., Chile	8	C8
Taiwan, Asia	6	Q7
Taiyuan, China	6	P6
Ta'iz, Yemen	5	G8
Takaka, New Zealand	12	g14
Takla Makan Desert, China	2	
Talaud Is., Philippines	6	Q9
Talbot Cape, Australia	11	D2
Talca, Chile	8	C7
Tallin, U.S.S.R.	4	M4
Tamale, Ghana	13	C5
Tamanrasset, Algeria	13	D3
Tamatave, Malagasy	13	H7
Tambo, Australia	12	H4
Tambov, U.S.S.R.	4	Q5
Tambura, Sudan	13	F5
Tampa, U.S.A.	7	M7
Tampere, Finland	4	M3
Tampico, Mexico	7	L7
Tamsag Bulag, Mongolia	6	P5
Tamworth, Australia	12	I6
Tana L., Ethiopia	13	G4
Tana R., Kenya	13	G6
Tanami, Australia	11	D3
Tananarive, Malagasy	13	H7
Tandil, Argentina	8	E7
Tanga, Tanzania	13	G6
Tanganyika L., Tanzania/Zaire/ Burundi/Zambia	13	FG6
Tangarog, U.S.S.R.	4	P6
Tangatau I., Pacific Ocean	10	O6
Tangier, Morocco	13	C2
Tang-Shan, China	6	P6
Tanimbar Is., New Guinea	6	R10
Tanzania, Africa	13	G6
Taongi At., Pacific Ocean	9	G3
Taoudenni, Mali	13	C3
Tapajos R., Brazil	8	E4
Tarakan, Indonesia	6	P9
Taranto, Italy	3	L7
Tarawa At., Pacific Ocean	9	H4
Tarbes, France	3	G7
Tarcoola, Australia	11	E6
Taree, Australia	12	I6
Tarim R., China	5	LM56
Tarnow, Poland	4	M5
Taroom, Australia	12	H5
Tarragona, Spain	3	G7
Tartary Str., U.S.S.R.	14	H4
Tartu, U.S.S.R.	4	N4
Tashkent, U.S.S.R.	5	I5
Tasman Bay, New Zealand	12	g14
Tasman Mts., New Zealand	12	g14
Tasman Pen., Australia	12	b9
Tasman Sea, New Zealand	12	fg13/14
Tasmania, Australia	12	e9
Tatakoto I., Pacific Ocean	10	O6
Tatung, China	6	P6
Taumarunui, New Zealand	12	h13
Taupo L., New Zealand	12	h13
Tazovskoye, U.S.S.R.	5	L3
Tbilisi, U.S.S.R.	4	Q7
Te Anau, New Zealand	12	f15
Tegucigalpa, Honduras	7	M8
Tehran, Iran	5	H6
Tehuantepec, Mexico	7	L8
Tehuantepec, Gulf of Mexico	7	L8
Tekapo L., New Zealand	12	g14
Tel Aviv, Israel	5	F6
Teles Pires R., Brazil	8	E4
Teluk Betung, Indonesia	6	O10
Tematangi I., Pacific Ocean	10	N7
Temora, Australia	12	H6
Tenerife, Canary Is.	13	B3
Tengiz L., U.S.S.R.	5	I4
Ten. Matienzo, Antarctica	15	n3
Tennant Creek, Australia	11	E3
Teofilo Otoni, Brazil	8	F5
Teresina, Brazil	8	F4
Terewah L., Australia	12	H5
Ternate, Indonesia	6	Q9
Terra Nova Bay, Antarctica	15	a2
Teruel, Spain	3	F7
Tete, Mozambique	13	G7
Texas State, U.S.A.	7	L6
Thailand, Asia	5/6	NO8
Thames, New Zealand	12	h13
Thames R., U.K.	3	F5
Thar Desert, India/Pakistan	2	
Thargomindah, Australia	12	G5
Theodore, Australia	12	I4
Theron Mts., Antarctica	15	M1/12
Thessaloniki, Greece	4	M7
Thiel Mts., Antarctica	15	P1
Thomson R., Australia	12	G4
Three Kings Is., New Zealand	12	g12
Thule, Greenland	7	N2
Thule Is., S. Sandwich Is.	15	M4
Thurston I., Antarctica	15	p2
Tibati, Cameroon	13	E5
Tiber R., Italy	3	I7
Tibesti Mt., Libya/Chad	2	
Tibet, Asia	5	MN6
Tibet, Plateau of, Tibet	2	
Tien Shan, China	2	
Tientsin, China	6	P6
Tierra del Fuego I., S. America	8	D9
Tigris R., Asia Minor	13	H2
Timaru, New Zealand	12	g14
Timboon, Australia	12	G7
Timbuktu, Mali	13	C4
Timisoara, Romania	4	M6
Timor, Indonesia	6	Q10
Timor Sea, Australia	11	D2
Tindouf, Algeria	13	C3
Tinian I., Pacific Ocean	9	E3

Place	Map	Ref
Tirane, Albania	3	L7
Tiruchirapalli, India	5	L8
Titicaca L., Peru	8	D5
Titograd, Yugoslavia	3	L7
Tlemcen, Algeria	13	C2
Tobi I., Pacific Ocean	9	D4
Tobolsk, U.S.S.R.	5	I4
Tobruk, Libya	13	F2
Tocantins R., Brazil	8	F45
Togo, Africa	13	D5
Tokara Is., Asia	6	R7
Tokelau I., Pacific Ocean	10	I5
Tokyo, Japan	6	R6
Toledo, Spain	3	F8
Toledo, U.S.A.	7	M5
Tomini, Gulf of, Indonesia	9	C45
Tomsk, U.S.S.R.	5	M4
Tonga, Pacific Ocean	10	I6
Tonga Trench, Pacific Ocean	10	I67
Toowoomba, Australia	12	I5
Torne R., Finland/Sweden	4	M2
Tornio, Finland	4	N2
Toronto, Canada	7	N5
Torrens L., Australia	12	F6
Torreon, Mexico	7	I7
Torres Str., Australia	12	G2
Torrowangee, Australia	12	G6
Tottenham, Australia	12	H6
Touggourt, Algeria	13	D2
Toulon, France	3	H7
Toulouse, France	3	G7
Tours, France	3	G6
Townsville, Australia	12	H3
Transvaal, S. Africa	13	FG8
Trapani, Sicily	3	I8
Traversay Is., S. Sandwich Is.	15	M4
Trento, Italy	3	I6
Trieste, Yugoslavia	3	I6
Trikkala, Greece	4	M8
Trinidad, Bolivia	8	D5
Trinidad, U.S.A.	7	I6
Trinidad I., S. Atlantic Ocean	8	GH6
Trinidad and Tobago, Caribbean Sea	8	D2
Tripoli, Libya	13	E2
Tripolitania, Libya	13	E2
Tristan da Cunha Is., S. Atlantic Ocean	13	B9
Trivandrum, India	5	L9
Trobriand Is., Pacific Ocean	9	F5
Tromso, Norway	3	L2
Trondheim, Norway	3	I3
Trondheim Fjord, Norway	3	H3
Troyes, France	3	G6
Trujillo, Peru	8	C4
Truk Is., Pacific Ocean	9	F4
Tsamkong, China	6	OP7
Tselinograd, U.S.S.R.	5	L4
Tsinan, China	6	P6
Tsingtao, China	6	Q6
Tsitsihar, China	6	Q5
Tsumeb, S.W. Africa	13	E7
Tuamotu Arch., Pacific Ocean	10	NO67
Tubuai Is., Pacific Ocean	10	MN67
Tucson, U.S.A.	7	H6
Tucuman, Argentina	8	D6
Tula, U.S.S.R.	4	P5
Tulear, Malagasy	13	H8
Tulsa, U.S.A.	7	L6
Tummo, Libya	13	E3
Tunisia, Africa	13	DE2
Tunis, Tunisia	13	E2
Turakh, U.S.S.R.	5	Q2
Tureia I., Pacific Ocean	10	O7
Turfan, China	5	M5
Turin, Italy	3	H6
Turkey, Asia	5	FGH56
Turkmenistan, U.S.S.R.	5	HI56
Turku, Finland	4	M3
Turnu Severin, Romania	4	M7
Turukhansk, U.S.S.R.	5	M3
Tutuila I., Pacific Ocean	10	I6
Tuva, U.S.S.R.	5	N4
Tyrrell L., Australia	12	G7
Tyrrhenian Sea, Italy	3	I8
Tyumen, U.S.S.R.	5	I4
U		
Ua Huka, Pacific Ocean	10	O5
Uaupes, Brazil	8	D3
Uberaba, Brazil	8	F5
Ucayali R., Peru	8	C4
Udaipur, India	5	L7
Udine, Italy	3	I6
Uelen, U.S.S.R.	6	Z3
Ufa, U.S.S.R.	4	T5
Uganda, Africa	13	G5
Ukhta, U.S.S.R.	4	S3
Ukraine region, U.S.S.R.	4	M-P6
Ulan Bator, Mongolia	5	O5
Ulan Kom, Mongolia	5	N5
Ulan Ude, U.S.S.R.	5	O4
Ulithia At., Pacific Ocean	9	E4
Ulyanovsk, U.S.S.R.	4	R5
Ume R., Sweden	3	L3
Umea, Sweden	3	M3
Ungava Bay, Canada	7	O34
Ungava Pen., Canada	7	N3
United Arab Emirates, Asia	5	H7
United Kingdom, Europe		3EFG45
United States of America, N. America	7	
Upernavik, Greenland	7	P2
Upington, S. Africa	13	F8
Upolu I., Pacific Ocean	10	I6
Upper Volta, Africa	13	CD4
Uppsala, Sweden	3	L4
Ural Mts., U.S.S.R.	2	
Ural R., U.S.S.R.	4	S6
Uralsk, U.S.S.R.	4	S5
Urandangi, Australia	12	F4
Uranium City, Canada	7	I4
Uruguay, S. America	8	E7
Uruguay R., S. America	8	E67
Urumchi, China	5	M5
Urup I., U.S.S.R.	6	ST5
Ushuahia, Argentina	8	D9
U.S.S.R., Europe and Asia	5/6	
Ussuriysk, U.S.S.R.	6	R5
Ust Kut, U.S.S.R.	5	O4
Ust Maya, U.S.S.R.	6	R3
Utah State, U.S.A.	7	H6
Uyuni, Bolivia	8	D6
Uzbekistan, U.S.S.R.	5	HI56
V		
Vaasa, Finland	4	M3
Vadso, Norway	4	N1
Vaitupu I., Pacific Ocean	9	H5
Valencia, Spain	3	F8
Valencia, Venezuela	8	D3
Valladolid, Spain	3	F7
Valletta, Malta	3	I8
Valparaiso, Chile	8	C7
Van, Turkey	5	G6
Vancouver, Canada	7	G5
Vancouver I., Canada	7	G5
Van Diemen Gulf, Australia	11	E2
Vanern L., Sweden	3	I4
Vanua Levu I., Pacific Ocean	10	H6
Varanasi, India	5	M7
Varna, Bulgaria	4	N7
Vatican City, Europe	3	I7
Vattern L., Sweden	3	L4
Vava'u Group, Pacific Ocean	10	I6
Velikie Luki, U.S.S.R.	4	O4
Vel Ustyug, U.S.S.R.	4	R3
Venezuela, S. America	8	D3
Venezuela, G. of, Venezuela	8	C2
Venice, Italy	3	I6
Veracruz, Mexico	7	L8
Verkhoyansk, U.S.S.R.	6	R3
Verkhoyansk Range, U.S.S.R.	2	
Vert C., Senegal	13	B4
Vesteralen, Norway	3	IL2
Vestfjord, Norway	3	I2
Vestmannaeyjar, Iceland	3	B3
Victor Harbour, Australia	12	F7
Victoria, Australia	12	GH7
Victoria, Canada	7	G5
Victoria I., Canada	7	HI23
Victoria Land, Antarctica	15	B2
Victoria L., Tanzania	13	G6
Victoria R., Australia	11	E3
Victoria River Downs, Australia	11	E3
Viedma, Argentina	8	D8
Vienna, Austria	3	L6
Vienne R., France	3	G6
Vientiane, Laos	6	O8
Vietnam, Asia	6	O78
Vigo, Spain	3	E7
Vijayawada, India	5	M8
Vila Cabral, Mozambique	13	G7
Villa Cisneros, Spanish Sahara	13	B3
Villach, Austria	3	I6
Villarica, Argentina	8	E6
Vilnius, U.S.S.R.	4	N5
Vilyuisk, U.S.S.R.	5	P3
Vincennes Bay, Antarctica	15	d3
Vinson Massif, Antarctica	15	Pp2
Virgin Is., Caribbean Sea	7	O8
Virginia State, U.S.A.	7	N6
Viscount Melville Sound, Canada	7	H12
Vistula R., Poland	3	L5
Vitebsk, U.S.S.R.	4	O4
Viti Levu I., Pacific Ocean	9	H6
Vitoria, Brazil	8	F6
Vizhakhapatnam, India	5	M8
Vladivostok, U.S.S.R.	6	R5
Vlone, Albania	3	L7
Volcano Is., Pacific Ocean	9	E2
Volga R., U.S.S.R.	4	R456
Volgograd, U.S.S.R.	4	Q6
Vologda, U.S.S.R.	4	P4
Volta L., Ghana	13	C5
Vorkuta, U.S.S.R.	5	I3
Voronezh, U.S.S.R.	4	P5
Vostok I., Pacific Ocean	10	M5
Vyazma, U.S.S.R.	4	O4
Vyborg, U.S.S.R.	4	N3
W		
Wadi Halfa, Sudan	13	G3
Wad Medani, Sudan	13	G4
Wagga Wagga, Australia	12	H7
Wagin, Australia	11	B6
Waiau, New Zealand	12	g14
Waiau R., New Zealand	12	f15
Waigeo I., Indonesia	9	D5
Waikerie, Australia	12	G6
Waitaki R., New Zealand	12	g14
Wakatipu L., New Zealand	12	f14
Wake Is., Pacific Ocean	9	G3
Wakkanai, Japan	6	S5
Wales, Europe	3	EF5
Walgett, Australia	12	H6
Walgreen Coast, Antarctica	15	Qq2
Wallaroo, Australia	12	F6
Wallis Is., Pacific Ocean	10	I6
Walsh, Australia	12	G3
Walvis Bay, S.W. Africa	13	E8
Wanaaring, Australia	12	G5
Wanaka L., New Zealand	12	f14
Wandunya, Australia	11	D5
Wanganui, New Zealand	12	g13
Wanganui R., New Zealand	12	g13
Wangaratta, Australia	12	H7
Wanhsien, China	6	O6
Warburton R., Australia	12	F5
Warrego R., Australia	12	H5
Warrina, Australia	11	F5
Warrnambool, Australia	12	G7
Warsaw, Poland	4	M5
Warta R., Poland	3	L5
Warwick, Australia	12	I5
Washington, U.S.A.	7	N6
Washington I., Pacific Ocean	10	M4
Waterford, Ireland	3	E5
Watlings I., Bahamas	7	N7
Wau, Sudan	13	F5
Wave Hill, Australia	11	E3
Wdirau, New Zealand	12	g14
Weddell Sea, Antarctica	15	Nm2
Weld Springs, Australia	11	C4
Wellesley I., Australia	12	F3
Wellington, New Zealand	12	g14
Wells L., Australia	11	C5
Wenchow, China	6	Q7
Wenlock R., Australia	12	G2
Wentworth, Australia	12	G6
Wessel Is., Australia	11	F2
Western Australia, Australia	11	BCD45
Western Ghats Mts., India	2	
Western Sahara, Africa	13	B3
Western Samoa, Pacific Ocean	10	I6
West Irian, Indonesia	6	R10
Westport, New Zealand	12	g14
West Siberian Plain, U.S.S.R.	2	
West Spitsbergen I., Norway	5	D2
West Wyalong, Australia	12	H6
Wetar I., Indonesia	6	Q10
Whangerei, New Zealand	12	g13
Wharanui, New Zealand	12	g14
Whitehorse, Canada	7	F3
White L., Australia	11	D4
White Nile R., Sudan	13	G45
White Russia, U.S.S.R.	4	N5
White Sea, U.S.S.R.	4	P2
Whitney Mt., U.S.A.	7	H6
Whyalla, Australia	12	F6
Wichita, U.S.A.	7	L6
Wick, U.K.	3	F4
Wight, I. of, U.K.	3	F5
Wilcannia, Australia	12	G6
Wilhelm II Coast, Antarctica	15	E3
Wilkes Land, Antarctica	15	CcDd2
Wilkes Sub-Glacial Basin, Antarctica	15	bC2
Wilkins Coast, Antarctica	15	O23
Willis Group, Australia	12	I3
Wilson's Prom., Australia	12	H7
Wilton R., Australia	11	E2
Wiluna, Australia	11	C5
Windhoek, S.W. Africa	13	E8
Windorah, Australia	12	G5
Windward Is., Caribbean Sea	8	D2
Winnipeg, Canada	7	L4
Winnipeg L., Canada	7	L4
Winton, Australia	12	G4
Wisconsin State, U.S.A.	7	LM5
Wohlthat Mts., Antarctica	15	I2
Woleai At., Pacific Ocean	9	E4
Wollongong, Australia	12	I6
Wolseley, Australia	12	G7
Wonthaggi, Australia	12	H7
Wontha I., Australia	12	a8
Woods Lake, Australia	11	E3
Woolgar, Australia	12	G3
Woomera, Australia	11	F6
Wooramel, Australia	11	A5
Worcester, S. Africa	13	E9
Wrangel I., U.S.S.R.	6	VZ2
Wreck Reef, Australia	12	L4
Wright L., Australia	11	D5
Wrigle Gulf, Antarctica	15	R2
Wroclaw, Poland	3	L5
Wuhan, China	6	P6
Wyndham, Australia	11	D3
Wyoming State, U.S.A.	7	I5
X		
Xingu R., Brazil	8	E4
Y		
Yaapeet, Australia	12	G7
Yakutsk, U.S.S.R.	6	Q3
Yalgoo, Australia	11	B5
Yalta, U.S.S.R.	4	O7
Yalu R., China	6	Q56
Yamal Pen., U.S.S.R.	14	dD23
Yamma Yamma L., Australia	12	G5
Yana R., U.S.S.R.	5	R2
Yangtse-Kiang, China	6	P67
Yanrey, Australia	11	A4
Yaounde, Cameroon	13	E5
Yap Is., Pacific Ocean	9	D4
Yaraka, Australia	12	G4
Yaroslavl, U.S.S.R.	4	P4
Yelets, U.S.S.R.	4	P5
Yellow R., China	6	P6
Yellow Sea, Asia	6	Q6
Yellowstone R., U.S.A.	7	I5
Yemen, Asia	5	G8
Yenisey R., U.S.S.R.	5	MN3
Yeniseysk, U.S.S.R.	5	N4
Yeo L., Australia	11	C5
Yeppoon, Australia	12	I4
Yerevan, U.S.S.R.	4	Q7
Yeysk, U.S.S.R.	4	P6
Yinchwan, China	6	O6
Yokohama, Japan	6	R6
Yola, Nigeria	13	E5
York C., Australia	12	G2
Yorke Pen., Australia	12	F6
Yucatan Str., Cen. America	7	M8
Yugoslavia, Europe	3/4	LM67
Yukon R., Alaska	7	CD3
Yukon Territory, Canada	7	F3
Yuzhno Sakhalinsk, U.S.S.R.	6	S5
Z		
Zadar, Yugoslavia	3	L7
Zagreb, Yugoslavia	3	L6
Zahedan, Iran	5	I7
Zaire, Africa	13	F56
Zambezi R., Africa	13	F7
Zamboanga, Philippines	6	Q9
Zanthus, Australia	11	C6
Zanzibar, Tanzania	13	G6
Zapala, Argentina	8	C7
Zaporozhye, U.S.S.R.	4	P6
Zawi, Rhodesia	13	F7
Zealand I., Denmark	3	I4
Zeehan, Australia	12	b9
Zhdanov, U.S.S.R.	4	P6
Zhitomir, U.S.S.R.	4	N5
Zinder, Niger	13	D4
Zomba, Malawi	13	G7
Zonguldak, Turkey	5	F5
Zungeru, Nigeria	13	D5

INDEX

252